Challenges, Strategies, and Resiliency in Disaster and Risk Management

Zamokuhle Mbandlwa
Durban University of Technology, South Africa

A volume in the Advances in
Human Services and Public Health
(AHSPH) Book Series

IGI Global
PUBLISHER of TIMELY KNOWLEDGE

Published in the United States of America by
IGI Global
Information Science Reference (an imprint of IGI Global)
701 E. Chocolate Avenue
Hershey PA, USA 17033
Tel: 717-533-8845
Fax: 717-533-8661
E-mail: cust@igi-global.com
Web site: http://www.igi-global.com

Library of Congress Cataloging-in-Publication Data

Names: Mbandlwa, Zamokuhle, 1984- editor.
Title: Challenges, strategies, and resiliency in disaster and risk management / edited by Zamokuhle Mbandlwa.
Description: Hershey, PA : Information Science Reference, [2024] | Includes bibliographical references and index. | Summary: "The objectives of the book typically revolve around providing in-depth insights and knowledge in the field of disaster and risk management"-- Provided by publisher.
Identifiers: LCCN 2024002262 (print) | LCCN 2024002263 (ebook) | ISBN 9798369327210 (hardcover) | ISBN 9798369327227 (ebook)
Subjects: LCSH: Emergency management. | Crisis management.
Classification: LCC HD49 .C428 2024 (print) | LCC HD49 (ebook) | DDC 658.4/056--dc23/eng/20240228
LC record available at https://lccn.loc.gov/2024002262
LC ebook record available at https://lccn.loc.gov/2024002263

This book is published in the IGI Global book series Advances in Human Services and Public Health (AHSPH) (ISSN: 2475-6571; eISSN: 2475-658X)

British Cataloguing in Publication Data
A Cataloguing in Publication record for this book is available from the British Library.

All work contributed to this book is new, previously-unpublished material.
The views expressed in this book are those of the authors, but not necessarily of the publisher.

For electronic access to this publication, please contact: eresources@igi-global.com.

Advances in Human Services and Public Health (AHSPH) Book Series

ISSN:2475-6571
EISSN:2475-658X

Editor-in-Chief: Jennifer Martin, RMIT University, Australia

MISSION

The well-being of the general public should be a primary concern for any modern civilization. Ongoing research in the field of human services and public healthcare is necessary to evaluate, manage, and respond to the health and social needs of the global population.

The **Advances in Human Services and Public Health (AHSPH)** book series aims to publish high-quality reference publications focused on the latest methodologies, tools, issues, and strategies for managing the health and social welfare of the public. The AHSPH book series will be especially relevant for healthcare professionals, policy makers, government officials, and students seeking the latest research in this field.

COVERAGE

- Youth Development
- Assistance Programs
- Social Welfare Policy
- Poverty
- Domestic Violence
- Access to Healthcare Services
- Health Policy
- Public Funding
- Healthcare Reform
- Medicare and Medicaid

IGI Global is currently accepting manuscripts for publication within this series. To submit a proposal for a volume in this series, please contact our Acquisition Editors at Acquisitions@igi-global.com or visit: http://www.igi-global.com/publish/.

The Advances in Human Services and Public Health (AHSPH) Book Series (ISSN 2475-6571) is published by IGI Global, 701 E. Chocolate Avenue, Hershey, PA 17033-1240, USA, www.igi-global.com. This series is composed of titles available for purchase individually; each title is edited to be contextually exclusive from any other title within the series. For pricing and ordering information please visit http://www.igi-global.com/book-series/advances-human-services-public-health/102256. Postmaster: Send all address changes to above address.

Titles in this Series

For a list of additional titles in this series, please visit:
http://www.igi-global.com/book-series/advances-human-services-public-health/102256

Examining Corruption and the Sustainable Development Goals
Ibrahim Nandom Yakubu (University for Development Studies, Ghana)
Business Science Reference • copyright 2024 • 334pp • H/C (ISBN: 9798369321010) • US $255.00 (our price)

Analyzing Global Responses to Contemporary Regional Conflicts
Piotr Pietrzak (Sofia University, Bulgaria)
Information Science Reference • copyright 2024 • 395pp • H/C (ISBN: 9798369328378) • US $245.00 (our price)

Dealing With Regional Conflicts of Global Importance
Piotr Pietrzak (Sofia University, Bulgaria)
Information Science Reference • copyright 2024 • 411pp • H/C (ISBN: 9781668494677) • US $215.00 (our price)

The Role of Health Literacy in Major Healthcare Crises
Vassilios Papalois (Royal Berkshire Hospital NHS Foundation Trust, UK) and Kyriaki Papalois (Royal Berkshire Hospital NHS Foundation Trust, UK)
Information Science Reference • copyright 2024 • 326pp • H/C (ISBN: 9781799896524) • US $235.00 (our price)

Ecological and Evolutionary Perspectives on Infections and Morbidity
P.A. Azeez (Salim Ali Centre for Ornithology and Natural History, Coimbatore, India) P.P. Nikhil Raj (Amrita School of Engineering, Amrita Vishwa Vidyapeetham, India) and R. Mohanraj (Bharathidasan University, India)
Medical Information Science Reference • copyright 2023 • 331pp • H/C (ISBN: 9781799894148) • US $325.00 (our price)

Transformation and Efficiency Enhancement of Public Utilities Systems Multidimensional Aspects and Perspectives

For an entire list of titles in this series, please visit:
http://www.igi-global.com/book-series/advances-human-services-public-health/102256

IGI Global
PUBLISHER of TIMELY KNOWLEDGE

701 East Chocolate Avenue, Hershey, PA 17033, USA
Tel: 717-533-8845 x100 • Fax: 717-533-8661
E-Mail: cust@igi-global.com • www.igi-global.com

Table of Contents

Detailed Table of Contents

South Africa constantly experiences disaster incidents linked to extreme weather, fires, structural collapse, and drowning. While the prevalence of these incidents varies in each region, KwaZulu-Natal, which is one of South Africa's coastal provinces, usually records several cases. For example, between 2018 and 2022 the province reported 5428 disaster incidents that affected 66026 households, damaged 73433 houses, and caused 805 fatalities. The total estimated costs of damages from these incidents over the period were more than R12.21 billion (about US$666.7 million). This chapter examines the effects of selected socio-economic variables on disaster incidents in the KwaZulu-Natal province of South Africa. Panel data for the province's district and metropolitan municipalities over the period 2018 to 2022 are estimated using fixed and random effects regression models. Results from the chapter shed light on the policy direction necessary to mitigate the prevalence and impact of disaster incidents both in the province and other similar environments.

This chapter explores community resilience in urban environments focusing on communities in the city of Harare in Zimbabwe. Despite its increased use in research and practical application for disaster risk management, resilience is a complex and contested issue. In this work, resilience is examined using a concurrent nested mixed method approach for data collection and analysis through the use of random and purposive sampling and questionnaire surveys. Some of the key findings include the diverse understanding of resilience by communities and urban local authorities leading to interventions mostly by urban planners that do not holistically promote community resilience. Urban informality is not always a hazard, but can promote community resilience if well managed, and social capital is one of the most important pillars for supporting community resilience building. The chapter concludes by emphasizing the importance of risk localization and a holistic understanding of resilience at community levels if city authorities are to reduce disasters and strengthen community resilience.

Zimbabwe is prone to a variety of natural and climate-related hazards, which disproportionately affect communities. While community-based disaster risk reduction (CBDRR) initiatives exist in the country, their effectiveness and sustainability in

Landslides can be considered one of the most severe natural hazards globally, and their management has a key role to inman safety. Landslide susceptibility maps can help the sustainable management of peri-urban areas such as determining the target areas of projects to develop landslide resistant areas, forest planning, infrastructure planning, and land use zoning. This study aims to analyse the relationship between landslide occurrence and its determinants in a peri-urban region, namely Taşlıdere Basin located in Güneysu District of Rize, Turkey, through providing both global and local regression models. To consider spatial non-stationarity, geographically weighted regression was used as a local model. It was found that the local model outperformed the global model in the estimation of landslide susceptibility determinants. The spatial analysis results indicated that almost all variables had heterogeneous variation over the study area. Therefore, this study provides a methodology for understanding the local dynamics of landslide occurrence.

The tourism sector in South Africa plays a crucial role in contributing significantly to economic growth and employment. However, the COVID-19 pandemic has had an unparalleled impact on the global tourism industry, exposing vulnerabilities and uncertainties. This study addresses a gap in the literature by examining the effects of the pandemic on employment within the accommodation sector in South Africa. Online surveys were conducted with managers using a quantitative research methodology. The findings reveal a significant decline in employment, salary reductions, and widespread redundancies. Notably, the study finds that more experienced establishments demonstrated resilience against job losses, and chain-owned establishments exhibited greater resilience in mitigating salary reductions. The study's implications extend to informing strategic measures for recovery, resilience, and long-term viability within the accommodation sector.

Appropriate and fast response to the emergency is the most important aspect of disaster management. Gharda Institute of Technology (GIT) is situated near Chiplun, Maharashtra, India. The predominant natural disasters which are frequent in the nearby vicinity like floods and water logging are not seen in the institute area due to its elevation above sea level. The possible disasters are chemical spillage, fire, and snake related issues (snake bite). Periodic mock drills and demo sessions are conducted for preparedness related to these disasters. The snake bite injections are made available at the nearby hospital. The disaster management cell of GIT has handled the COVID-19 situation with appropriate precaution. The cell conducts frequent checks on the preparedness of the concern taskforce to combat any untoward incident. The display of contact numbers of hospital, police, district level authorities brings about the safe feeling among the students and staff. The institute has chemical department and hence handling of chemicals is an important aspect.

The chapter addresses resilience engineering's role in sustainable and disaster-resistant transportation infrastructure. Integrating autonomous cars, intelligent transportation systems (ITS), and network reconfiguration methodologies improves transportation network flexibility and recovery. The technology may decrease traffic, increase safety, and improve inhabitants' commuting experiences. Integration requires stakeholder involvement, infrastructure development, and strategic planning, according to the research. Contraflow, crossing removal at junctions, optimisation models, traffic assignment problem (TAP), and gradient projection increase traffic flow, congestion, and evacuation efficiency in resilient engineering. It strengthens transportation systems' ability to adapt, respond, and recover from disturbances to protect communities and ensure infrastructure sustainability. Finally, resilience engineering reduces transportation system disruptions and makes infrastructure more adaptive, durable, and sustainable.

The COVID-19 pandemic has impacted the hospitality industry, prompting the implementation of health and safety measures. This study examines COVID-19 health and safety measures used in South Africa's accommodation sector. Data was obtained from 477 establishments through online surveys. The findings reveal a robust commitment to health and safety practices driven by adherence to regulatory standards. Disparate levels of social distancing measures highlight several challenges in implementation. A strong association with star grading indicates the use of measures tailored to each type of establishment. High-end establishments are more inclined to implement comprehensive measures. Larger facilities use floor marks more frequently, whereas smaller ones rely on personal communication. The number of employees influences the adoption of measures, highlighting the importance of ongoing training and communication on health issues. The study advocates for proactive health practices, customised strategies, financial support mechanisms, and a commitment to long-term preparedness.

Although there has been significant global recognition of the South African disaster management legislation and the policy framework, its implementation approaches are flawed across spheres of government, with municipalities being the worst. Municipalities face various obstacles to effectively integrating disaster risk management into their development planning, leaving marginalized populations such as undocumented immigrants vulnerable during disasters. This study examines the challenges faced by municipalities in integrating disaster risk management into their development planning. A qualitative case-study approach following desktop research was used in this study. The secondary data was collected from peer-reviewed articles and government documents such as the city of eThekwini integrated development plan 2023-2024 to 2027-2028 and the disaster management sector plan. Using two case studies, notable challenges were revealed, such as financial and capacity constraints (technical and non-technical), incomplete or not approved plans inertia, public awareness deficits, absence of a disaster management advisory forum, communication breakdowns, and the often-overlooked impacts of sand mining and undocumented

Preface

As Editor of *Challenges, Strategies, and Resiliency in Disaster and Risk Management*, it is my privilege to present this comprehensive resource to scholars, professionals, and stakeholders invested in the critical field of disaster and risk management.

Disasters and risks are inherent facets of our complex world, presenting multifaceted challenges that demand innovative strategies and unwavering resilience. From natural calamities to technological hazards, from pandemics to climate change, our ability to anticipate, mitigate, respond to, and recover from these events profoundly impacts lives, communities, and economies.

This edited volume embarks on a journey through the intricate landscape of disaster and risk management, offering insights gleaned from diverse perspectives, experiences, and expertise. Through meticulous scholarship and real-world case studies, our contributors illuminate the historical evolution, conceptual frameworks, and practical applications underpinning effective disaster management practices.

Each chapter serves as a beacon of knowledge, guiding readers through topics ranging from risk assessment and emergency preparedness to community resilience building and international cooperation. With a steadfast commitment to fostering understanding and driving positive change, this book navigates the intricacies of disaster management with clarity and depth.

Our target audience encompasses a spectrum of stakeholders, including emergency managers, policymakers, scholars, and students, as well as practitioners and organizations engaged in disaster response and risk mitigation efforts. By fostering interdisciplinary dialogue and sharing best practices, we aspire to empower individuals and institutions to confront the challenges of tomorrow with confidence and foresight.

As we embark on this collective endeavor, let us heed the lessons of the past, embrace the opportunities of the present, and chart a course toward a safer, more resilient future for all.

Chapter 1, titled "Comparative Study of Disaster Management and Skills Development Between Selected Developing and Developed Countries," authored by

Pauline Edwige Ngo-Henha and Zamokuhle Mbandlwa, delves into the importance of skills development in disaster management frameworks through a comparative analysis of empirical data from Mozambique and Japan. The chapter sheds light on the pivotal role of skills development in mitigating the impacts of natural disasters and fostering resilience.

Chapter 2, "Socio-Economic Variables and Disaster Incidents in KwaZulu-Natal, South Africa," by Maliga Reddy, Zamokuhle Mbandlwa, Strinivasan Pillay, Genius Murwirapachena, and Sibongiseni Ngema, examines the influence of socio-economic factors on disaster incidents in KwaZulu-Natal, offering insights essential for informed policy-making and effective mitigation strategies.

In Chapter 3, "Building Resilient Communities Through Social Capital in Cities: Reflections From City of Harare, Zimbabwe," authored by Elias Mabaso, the dynamics of community resilience in urban environments are explored through the lens of social capital. The chapter unveils the nuances of resilience-building efforts, emphasizing the pivotal role of social cohesion and community engagement in disaster preparedness and response.

Chapter 4, "Re-Imagining Community-Based Disaster Risk Reduction Towards Building Resilient Rural Communities," authored by Francis Ziwanai, investigates innovative approaches to enhance the efficacy of resilience-building initiatives in rural areas prone to natural hazards. The chapter offers practical recommendations for strengthening community-based disaster risk reduction efforts and bolstering the resilience of vulnerable communities.

Drawing inspiration from the works of Kenji Miyazawa, Chapter 5, titled "People, Culture, Knowledge, and Skill: A Philosophical Approach to Disaster and Environment (With Special Reference to the Works of Kenji Miyazawa)," authored by Anjali Sharma and Merry Ray, embarks on a philosophical exploration of disaster management, highlighting the interplay between culture, knowledge, and resilience.

In Chapter 6, "Examining Impact of Landslide Hazards in The PeriUrban Areas byGlobal and Local Modelling Techniques," authored by Arif Aydinoglu, Suleyman Sisman, Eda Ustaoglu, and Gehver Aydogdu, the focus shifts to the peril of landslides in peri-urban areas, offering insights into landslide susceptibility and determinants essential for informed land use planning and disaster risk management.

Chapter 7, "Empirical Insights Into the Impact of the COVID-19 on Employment Patterns in South Africa's Accommodation Sector," authored by Reshma Sucheran, examines the profound socio-economic impacts of the COVID-19 pandemic on employment within South Africa's accommodation sector, offering insights into strategic interventions to safeguard livelihoods and promote resilience.

Chapter 8, "Disaster Management Activities in an Indian Engineering Institute: A Case Study of Gharda Institute of Technology," authored by Sunil Kulkarni and

Vivek Talnikar, offers a case study of proactive disaster management activities at Gharda Institute of Technology in Maharashtra, India, highlighting strategic preparedness efforts and measures taken to ensure the safety of students and staff.

In Chapter 9, "Fostering a Sustainable and Disaster Resilient Transportation Infrastructure," authored by Syed Arif and Shalom Akhai, resilience engineering's role in enhancing the adaptive capacity of transportation infrastructure is explored, offering insights into innovative approaches to ensure the sustainability and resilience of critical transportation systems.

Chapter 10, "Pandemic Resilience Crisis Management and COVID-19 Health and Safety Responses in the Hospitality Industry," authored by Reshma Sucheran, examines COVID-19 health and safety responses within South Africa's accommodation sector, offering a nuanced understanding of the challenges and opportunities for resilience-building in the face of a global health crisis.

Chapter 11, titled "Disaster Risk Management Challenges Towards Integrated Disaster Risk Management at the South African Municipalities-What Left Untouched?" authored by Mosekama Mokhele, Alice Ncube, and Olivia Kunguma, explores the complex challenges facing South African municipalities in implementing integrated disaster risk management, offering insights from case studies and advocating for collaborative interventions to strengthen disaster resilience at the local level.

Finally, in Chapter 12, "Resilience in Compound Disasters: A Case Study of KwaZulu-Natal Province in South Africa," authored by Obianuju Okeke-Uzodike, Tyolome Kinga-Upaa, and Chika Eze, resilience-building efforts in the face of compound disasters in the KwaZulu-Natal province of South Africa are examined, offering insights into robust disaster management frameworks and stakeholder engagement essential for building resilience in disruptive times.

As Editor of *Challenges, Strategies, and Resiliency in Disaster and Risk Management*, I am honored to present this comprehensive volume to our esteemed readership. In traversing the intricate terrain of disaster and risk management, our contributors have illuminated essential facets of this critical field, offering insights gleaned from diverse perspectives, experiences, and expertise.

From the comparative analysis of disaster management frameworks in developing and developed countries to the exploration of socio-economic variables influencing disaster incidents, each chapter serves as a beacon of knowledge, guiding readers through topics ranging from community resilience-building to pandemic crisis management.

Throughout this journey, we have underscored the pivotal role of skills development, social capital, and innovative approaches in enhancing disaster resilience. By examining real-world case studies and advocating for collaborative interventions, we aim to empower individuals and institutions to confront the challenges of tomorrow with confidence and foresight.

As we conclude this collective endeavor, let us heed the lessons of the past, embrace the opportunities of the present, and chart a course toward a safer, more resilient future for all.

Zamokuhle Mbandlwa
Durban University of Technology, South Africa

Chapter 1
Comparative Study of Disaster Management and Skills Development Between Selected Developing and Developed Countries

Pauline Edwige Ngo-Henha
Durban University of Technology, South Africa

Zamokuhle Mbandlwa
Durban University of Technology, South Africa

ABSTRACT

The occurrence of natural disasters has become increasingly prevalent worldwide, prompting a heightened focus on readiness and response strategies for future events. This study aimed to explore the importance of skills development and training in disaster management from both theoretical and empirical perspectives. The research included a thorough literature review of disaster management models, as well as an analysis of empirical studies conducted in Mozambique and Japan. Additionally, the literature underscores the repercussions of natural disasters on the susceptibilities of both developed and developing nations. The findings of this study indicate that skills development plays a crucial role in current disaster management frameworks, such as the Sendai Framework for Disaster Risk Reduction (SFDRR) and the Capability Approach model. Additionally, the study highlights various skills development and training initiatives present in Japan's disaster management practices and policies, as well as in Mozambique's disaster management projects.

DOI: 10.4018/979-8-3693-2721-0.ch001

INTRODUCTION

The paramount significance of Disaster Management (DM) in protecting communities and individuals during natural disasters cannot be overemphasized. According to the World Risk Report 2023 (WRR, 2023), certain countries such as Japan, the Philippines, Indonesia, India, Mexico, Columbia, Myanmar, Mozambique, Russia, Bangladesh, and China are identified as having the highest disaster risk values. These risk factors may be attributed to shifts in population and national wealth density, global warming, rapid industrialization, environmental degradation, climate change, and geographical positioning (Wood, 2018). Furthermore, the same report underlines the exacerbation of existing disparities during crises. In line with the perspectives of management as proposed by Drucker (1974), disasters and risks management involve the facilitation of individuals in engaging in collective performance through the establishment of shared objectives, principles, appropriate organizational structures, and ongoing training and development opportunities that are vital for an effective adaptation and response to environmental dynamics. Therefore, the domain of disasters and risks management encompasses a range of strategies, skillsets, and proactive and reactive approaches undertaken collectively to mitigate risks within communities, as well as to prevent and alleviate the detrimental impact of natural disasters. This chapter, therefore, focuses on comparative studies of disaster management and skills development between selected developing and developed countries. This chapter provides an analysis of empirical studies conducted in Mozambique and Japan.

Economic Disparities and Vulnerability to Natural Disasters

Vulnerability to disasters is a critical determinant of the risks facing communities. Noy and Yonson (2018) define vulnerability as a pre-disaster concern that encompasses the conditions leading to a hazard impacting a system, thereby resulting in a disaster. Vulnerability indicates the susceptibility or fragility of communities, systems or elements at risk, as well as their ability to cope with hazardous conditions (Birkmann et al., 2013). The concept of vulnerability elucidates why similar hazards such as earthquakes, storms or floods can yield varied negative impacts on communities, societies and infrastructures exposed and affected by them (Birkmann, Sorg & Welle, 2017).

The vulnerability of a community to disasters is influenced by historical, cultural, socio-economic and political factors. Scholars such as Tseliosa & Tompkins (2018) and Ramani et al. (2022) argue that a higher level of economic development correlates with reduced vulnerability to disasters.

Socio-economic disparities in disaster and risk management pertain to the unequal distribution of resources, opportunities, and capabilities amongst distinct socioeconomic groups or regions in terms of preparation for, response to, and recovery from disasters and risks (Celse & Kensen, 2023). Individuals and communities with varying levels of economic well-being encounter distinct challenges and vulnerabilities when confronted with disasters. Empirical studies have demonstrated that the severity of a disaster's impact on a particular area is influenced by the area's socio-economic conditions, with developing countries being more adversely affected compared to developed nations (Tseliosa & Tompkins, 2018; Tasri, Karimi & Muslim, 2022; Rahmani et al., 2022).

Consequently, economic disparities manifest in the susceptibility of communities; their access to financial resources, infrastructure and technologies; as well as their access to education and the relevant skillsets necessary to effectively prepare for and mitigate the effects of disasters. Noy and Yonson (2018) posit that the ability of individuals and economies to avoid adverse impacts and recover from disasters is influenced by socio-economic factors.

Economically disadvantaged communities, primarily in developing countries, are more vulnerable to and severely impacted by disasters (Felbermayr & Gröschl, 2014; Ward & Shively, 2017; Kim & Bostwick, 2020; Thompson, 2010; Adams et al., 2022). Data from the World Bank Report in 2018 reveals that 89% of disaster-related fatalities occurred in developing countries from 1995 to 2014 (Thompson, 2010).

In the context of disaster and risk management, socioeconomic vulnerability evaluates the well-being of individuals, communities, or nations in terms of literacy levels, governance, economic resources, infrastructure and technology (Mwale et al., 2015).

Studies assert that developing nations contain diverse socioeconomically disadvantaged communities living in precarious environments that are susceptible to natural disasters such as floods, landslides, or coastal storms (Kim & Bostwick, 2020; Adams et al., 2022). Limited financial resources in these impoverished communities often impede their capacity to secure safer housing or to relocate to less vulnerable areas, consequently heightening their susceptibility to disasters. For instance, as depicted in pictures 1 and 2, the recent catastrophe of November 2023 in Yaoundé, the capital of Cameroon, resulting from prolonged heavy rainfall causing mudslides and floods, impacted over 700 individuals from 110 households in impoverished hillside communities (Internet 1). A similar incident occurred in March 2023 in Buea City, the capital of the Southwest Region of Cameroon, where heavy rainfall triggered floods and landslides, devastating numerous buildings and leaving underprivileged families homeless (Internet 2).

Likewise, the substantial rainfall experienced in Mozambique between 31 January and 12 March 2023 had a profound impact on a significant portion of the population, affecting 255 786 individuals, resulting in 186 fatalities, and displacing

Figure 1. Images showing the heavy rains and flooding leading to landslides
Source: Davies, R. (2023).

200 870 people as reported by the World International Organization on 18 April 2023 (Internet 3).

Situated on the South-eastern coast of Africa, Mozambique emerges as one of the most susceptible nations on the continent to frequent natural calamities, having endured 75 such events in the past 35 years. These disasters include 13 instances of drought, 25 floods, 14 tropical cyclones, and 23 epidemics (UN Global Assessment Report on Disaster Risk Reduction, 2019). Notably, cyclones have proven to be the most devastating form of disaster in Mozambique, impacting around 2 million individuals annually. Following closely behind are droughts affecting 600 000 people and floods affecting 200 000 individuals each year (Global Facility for Disaster Reduction and Recovery (GFDRR), 2019). As depicted in pictures 3 and 4 below, these recurrent disasters have a significant and deleterious effect on the lives of the most vulnerable populations (International Bank for Reconstruction and Development Report, 2019). This impact is exacerbated as the country is classified amongst the poorest nations globally, with a Human Development Index of 0.432 and 46% of the populace residing below the poverty threshold (UN Global Assessment Report on Disaster Risk Reduction, 2019; Salvucci & Santos, 2020).

In addition to widespread poverty, the frequent occurrence of natural disasters in Mozambique can be attributed to a variety of factors, including the country's expansive 2 515 km coastline; its nine international river systems; deficient infrastructure; a community lacking in disaster management preparedness and skills; and a heavy dependence on agriculture (Mugabe et al., 2021). It is important to note that these challenges are not unique to Mozambique but are prevalent in many developing nations.

Figure 2. Images showing natural disaster events
https://www.aljazeera.com/gallery/2021/1/25/in-pictures-cyclone-eloise-displaced-thousands-in-mozambique

The vulnerability of developing nations to natural disasters is largely attributable to a lack of infrastructure aimed at the prevention, mitigation, and response to such events; inadequate governance characterized by a lack of effective disaster management policies; and unsustainable agricultural practices that contribute to soil degradation and flooding (Zorn, 2018; Glaco, 2021). The phenomenon of rapid urbanization driven by poverty compels rural residents to migrate to urban areas and settle in high-risk areas prone to flooding and landslides, which is exacerbated by deficiencies in drainage infrastructure, poor water management practices, inconsistencies in urban planning, and the absence of natural drainage systems (Glaco, 2021).

Both developing and developed countries are vulnerable to natural disasters. For instance, Japan, located in a seismic and volcanic zone, has faced numerous natural calamities such as earthquakes, tsunamis, volcanic eruptions, and typhoons. The most devastating event depicted in Pictures 5 and 6 was the Tohoku Earthquake and Tsunami in March 2011, which struck the north-eastern part of Japan and resulted in the loss of over 20 000 lives (Banerji & Singh, 2013; Goto et al., 2021), leaving more than 250 000 people homeless (Banerji & Singh, 2013). This catastrophic event severely impacted the country's economy, with damages amounting to 16–25 trillion yen in terms of lives lost and property destruction (Koshimura, Hayashi & Gokon, 2014).

Although both developed and developing nations are vulnerable to natural disasters, developed countries possess not only resources and technologies, but also a range of skills and expertise that enable them to effectively prepare for, respond to, and recover from such disasters. Suindramedhi (2015) asserts that Japan, for example, is a well-prepared country with advanced physical infrastructure and specialized skills for crisis management and disaster recovery, allowing for the development

Figure 3. Natural disasters
https://www.britannica.com/event/

of more efficient basic needs following each disaster. The technology that is used by Japan to respond to disasters is also more advance.

Perera et al. (2013) recognize that the mitigation of catastrophes can be achieved through the diligent development of skills. Kong et al. (2020) support this assertion by highlighting capacity development as a prevalent strategy in disaster risk reduction efforts. Furthermore, Cvetković et al. (2021) assert that the enhancement of capacities through skills development is essential for effective disaster mitigation to minimize the adverse effects on individuals, organizations and society as a whole.

Theoretical Review of Disaster and Risk Management Models

Disaster management encompasses the policies, administrative decisions, operational activities, stakeholders, and technologies involved in addressing all stages of a disaster across different levels (Surianto et al., 2019). Disaster management models are based on the principle that disasters have a significant impact on development. Therefore, it is the duty of disaster planners, managers, and practitioners to implement appropriate measures to facilitate recovery and the continuity of development (Hussain, 2013). Thus, this section presents the Disaster Management models that serve as a theoretical foundation for disaster management, with a focus on models relating to skills development or training.

Based on an analysis of existing research on disaster and risk management, Asghar et al. (2006) as cited by Nojavan et al. (2018) categorized disaster management models into four main groups, namely logical models, integrated models, causal models, and combinatorial models. In addition, Alrehaili et al. (2022) introduced a fifth group known as the uncategorized model. These groups of models will be briefly discussed below, with a subsequent presentation of Table 1, providing a summary of these model classifications and the frameworks or models associated therewith.

Logical models, according to Asghar et al. (2006), provide a framework for defining the various stages of a disaster, highlighting the fundamental events and actions that contribute to the occurrence of a disaster, with one prominent logical model being the Traditional Model of disaster management that divides disaster management into three distinct phases: pre-disaster, during a disaster, and post-disaster (Asghar et al., 2006; Nojavan et al., 2018; Alrehaili et al., 2022). According to ADPC (2000), the pre-disaster phase involves activities centred on prevention, mitigation, and preparedness. During the disaster phase, the focus shifts to reaction and response activities. Finally, the post-disaster phase encompasses activities related to recovery, reconstruction, and overall development (Alrehaili et al., 2022).

The *Causal models* assume that there are underlying factors contributing to disasters. The Crunch Cause Model (2000) appeared to be the most popular and assumes that vulnerability to disasters can be caused by a multitude of factors, whereby a catastrophic event occurs because of the convergence of a hazard with vulnerable elements, including populations, property, infrastructure and the environment (Nojavan et al., 2018; Alrehaili et al., 2022). Subsequently, the model delineates community vulnerability, identifies its root causes, and assesses potential hazardous circumstances.

Integrated models provide a rational and systematic approach to achieving comprehensive disaster management. As such, these models suggest that various activities should be organized and coordinated to ensure the effective implementation of risk assessment, hazard management, mitigation, and preparedness (Nojavan et al., 2018). An example of an Integrated Emergency Model is The Monitoring and Evaluating Disaster Risk Management Model developed by Scott et al. (2016). The model presents a framework designed for utilization by disaster risk management programs to effectively monitor the outcomes of their interventions and ultimately enhance the quality of practices. Another popular integrated model is the Manitoba Model (2002), which outlines the various stages of disaster management to facilitate the effective management of disasters and establish flexible linkages between the processes and activities involved in disaster management. Furthermore, the Manitoba Model (2002) sought to establish a balance between readiness and adaptability to effectively address the unique requirements brought about by the disaster situation (Nojavan et al., 2018).

The *Combinatorial models* encompass a blend of logical, causal and/or integrated models to develop a more comprehensive disaster management model. One such example is the Wheel-Shape Disaster Management Model by Rowshandel et al. (2008), which combined logical and integrated frameworks drawing upon the life cycle of disasters and crises, along with their respective stages (Nojavan et al., 2018; Alrehaili et al., 2022).

In contrast to the aforementioned models, *uncategorized models,* according to scholars, do not adhere to the criteria established by the previously mentioned groups. The Octopus Model developed by Shi et al. (2011) is an example of an uncategorized model. It assumes that disasters involve complex systems. Therefore, effective risk management strategies should be rooted in a multidimensional approach in policy-making to achieve success (Shi et al., 2011; Nojavan et al., 2018; Alrehaili et al., 2022).

Skills Development and Disaster Management Models

Skills development and education are crucial factors to consider in the management of disasters and mitigation of risks. Unfortunately, many existing disaster

Table 1. Classification of disaster management models

Logical Models		
Name of Model	**Author and Year of Publication**	**Description**
Iceberg model	Heinrich (1941)	The most critical aspect of this model is its focus on structure and the appearance of a model template.
Lechat model	Lechat (1990)	This model begins with disaster preparedness and finishes with disaster recovery.
The five-stage model	Mitroff and Pearson (1993)	The detection and learning processes are emphasised in this model.
Traditional model	DPLG-2 (1998)	The various phases of disaster management run in a sequential order in this model but with differing degrees of focus.
Expand and contract model	DPLG-2 (1998)	The distinction of this traditional model is that the action sequences occur simultaneously.
Circular model	Kelly (1999)	The ability of this model to learn from actual disasters is its most significant aspect.
Mitroff model	Mitroff (2000)	This is a proactive model that focuses mainly on the learning stage.
The four-phase disaster management model	Kimberly (2003)	The response phase is the most significant part of this model. It focuses on disasters in healthcare facilities and requires appropriately trained staff.
The four-stage model of Tuscaloosa	Tuscaloosa (2003)	This model revolves around the response phase.
Two-part model	Hosseini and Jedi (2006)	A variety of operational and logistical elements are contained in this model. As a result, it is referred to as a two-part model.
Gupta stair model	Gupta (2010)	Pre-disaster phases are not given significant consideration in this model.
Contreras model	Contreras (2016)	Several indicators have been established in this model to measure disaster vulnerability. However, the most critical aspect of this model is its focus on disaster recovery.

Table 1. (continued)

Causal Models		
Name of Model	**Author and Year of Publication**	**Description**
Littlejohn six-stage model	Littlejohn (1983)	This model is a framework for disaster management that includes fundamental guidelines.
Fink's comprehensive audit model	Fink (1986)	This model predicts what occurrences might lead to a disaster, so action plans can be created and scenarios developed.
Crunch cause model	Asian Disaster Preparedness Centre (2000)	This model presents a foundation for understanding the causes of a disaster. Its structure is explained by the equation: Hazard + Vulnerability = Disaster Risk.
Pressure and release (PAR) model	Blaikie et al. (2005)	This model employs preventative actions in an attempt to reduce the risk of disaster.

Table 1. (continued)

Integrated Models		
Name of Model	**Author and Year of Publication**	**Description**
Onion model	Mitroff et al. (1987)	This model lays out a framework for crisis preparedness in organisations.
McConkey linear model	McConkey (1987)	In four stages, the McConkey model gives considerable emphasis to pre-disaster management.
The Deming cycle model (PDCA model)	Aguayo (1991)	The PDCA model was created as a continuous improvement cycle of planning, implementing, and assessing.
Manitoba Health model	Manitoba Health Disaster Management (2002)	This model provides a balance between preparedness and resilience in order to adapt to the disaster's specific requirements.
Integrated model	Moe and Pathranarakul (2006)	This model demonstrates the necessity of both proactive and reactive measures in the management of natural disasters.
McEntire et al. integrated model	McEntire et al. (2010)	This integrated model for modelling vulnerability takes into account social science research, engineering, and physics.
Integrated system-oriented Model	Meshkati and Tabibzadeh (2016)	This model's most distinguishing characteristic is its focus on emergency response.
Monitoring and evaluating framework for disaster risk management	Scott et al. (2016)	This model is a one-of-a-kind framework for monitoring and assessing disaster risk management plans that such programmes may use to track the effects of their interventions and, in turn, increase standards in their field.

management models have overlooked these fundamental aspects. Nonetheless, a small number of scholars have proposed models or frameworks that prioritize skills development as a key element in disaster management and risk reduction.

Table 1. (continued)

Combinatorial Models		
Name of Model	**Author and Year of Publication**	**Description**
Cuny comprehensive model	Cuny (1998)	Cuny presented one of the most comprehensive disaster management cycles. Using a combination of logical, integrated, and causal models, this model addresses the managerial and administrative procedures of disaster management.
Risk management proactive model	Australian Development Gateway (2008)	This model aims to blend logical and integrated models.
Disaster risk management framework (DRMF) model	Baas et al. (2008)	The three stages in this model are risk reduction, response, and recovery.
Wheel-shape disaster management model	Rowshandel Arbatani et al. (2008)	This model, which is based on the disaster cycle, is one of the complete disaster management models. It is made up of a mix of logical and integrated models.
Risk management model	Zimmermann and Stössel (2011); BPDMP (2013)	This model aims to improve community resilience and reduce the risk of the disaster by combining logical and integrated models.
Saldana-Zorrilla model	Saldana-Zorrilla Saldana (2015)	This model proposes a guideline for risk management and enhancing risk reduction strategies and planning.
Institutional model for collaborative disaster risk management	Tau et al. (2016)	This model combines the theoretical, political, and technological aspects of disaster risk reduction.

Source: Alrehaili, N.R., Almutairi, Y.N., Alghamdi, H.M. & Almuthaybiri, M.S. (2022:97- 99). Adapted.

Therefore, Nojavan et al. (2018:8) have identified the Wheel Shape Model (2008) and the Cuny Model (1998) as being particularly pertinent for fostering skills development in disaster management, a key focus of this section. Additional models that emphasise the significance of skills development in disaster preparedness and response include the Capability Approach, Sendai Framework for Disaster Risk Reduction, Traditional Disaster Theory, Expand-Contract Model, and Disaster Pressure and Release Model (Runga, 2023). Amongst the various disaster management models and frameworks outlined above, only two will receive detailed examination in this review: the Capability Approach and the Sendai Framework for Disaster Risk Reduction (SFDRR). The decision to focus on these models is based on two primary reasons. The former, the Capability Approach, is a well-established model with a significant amount of literature and research supporting its principles. The latter, the Sendai Framework for Disaster Risk Reduction, is an internationally recognized initiative endorsed by the United Nations, and it has been widely embraced and implemented across the globe.

- **Capability Approach**

Proposed by the esteemed economist and philosopher Amartya Sen in 1980, the Capability Approach emphasises the importance of individuals' abilities, well-being, and the improvement of their quality of life by addressing the obstacles that restrict their freedom to lead a fulfilling lifestyle. This approach highlights the necessity of enhancing individuals' capabilities to pursue fulfilling lives, rather than merely focusing on quantifying resources or economic growth (Sen, 1993, Runga, 2023). As a general framework, Sen (1993) emphasised that the Capability Approach can be used in various fields. Thus, in the context of disaster management, Dreze and Sen (2002), as cited by Runga (2023), posit that the emphasis of disaster risk reduction efforts should be on enhancing the capabilities of individuals to anticipate and effectively respond to disasters, rather than solely concentrating on minimizing their risk exposure. The Capability Approach therefore underscores the importance of equipping individuals impacted by disasters with the ability to adequately prepare, manage, recover, and bounce back. These abilities encompass capabilities such as education, awareness, and skills development (Runga, 2023).

- **The Sendai Framework for Disaster Risk Reduction (SFDRR)**

Capacity building, a key component of the Sendai Framework, highlights the importance of strengthening the knowledge and capabilities of individuals, communities and institutions in understanding and managing disaster risk. The Sendai Framework for Disaster Risk Reduction (SFDRR) for the period 2015-2030 aims to increase awareness of disaster risks, improve disaster risk management, and enhance readiness for effective recovery and re-building, with objectives focusing on enhancing disaster risk management; effectively managing disaster risk; and investing in disaster risk reduction to address the complexity of contemporary disaster risks (UNISDR, 2015; Faivre et al., 2018). The four priorities of the SFDRR include understanding disaster risk, enhancing disaster risk governance, investing in disaster risk reduction, and improving disaster preparedness (Bennett, 2020; Runga, 2023). Education and international cooperation are essential for achieving these priorities and building the necessary skills and knowledge for disaster risk reduction and response (Surianto et al., 2019). Hence, it is crucial for individuals and communities to be equipped with the necessary information and skills to analyse and address disaster risks effectively.

For example, in order to address the first priority, a comprehensive understanding of disaster risks requires individuals and communities to possess the necessary knowledge, skills and competencies to identify risks and their contributing factors (Mannan et al., 2021). This empowers decision-makers and communities to analyse, evaluate and incorporate disaster risk information into planning processes (Mizutori, 2020). In terms of the second priority, strengthening disaster risk governance at all levels is essential for prevention, mitigation and response efforts (Mizutori, 2020). This necessitates expertise in developing laws, policies and strategies for risk prevention and reduction, as well as raising awareness and providing training initiatives to implement risk prevention policies within communities.

As part of the third priority of the SFDRR, investing in risk reduction underscores the significance of investing in skills development, innovative approaches and technologies to enhance disaster resilience (UNISDR, 2015). Training in the use of these technologies, which facilitate risk assessment, monitoring and early warning, is thus crucial. The fourth priority of the Sendai Framework highlights that education and training play a critical role in enhancing disaster preparedness and facilitating effective response, recovery and reconstruction efforts following a disaster (UNISDR, 2015).

Skills Development and Disaster Management: Empirical Literature Review

Skills development and disaster and risk management are essential within the context of sustainable development and resilience building. As nations encounter growing complexities and vulnerabilities caused by a range of natural and human-induced disasters, it is becoming imperative to foster relevant skills and competencies (Cvetković et al., 2021). This section explores how skills development programs are depicted in the existing empirical literature on Disaster Management in Mozambique and Japan. Additionally, it outlines the strategies and methods employed to empower individuals and communities in disaster prevention and mitigation efforts. The section begins by examining the case of Mozambique, followed by an analysis of Japan, and concludes with a concise summary of the insights offered by the literature pertaining to both nations.

Mozambique and Skills Development in Disaster Management

This brief overview of the empirical literature comprises five categories of studies, beginning with those that focus on the training components in previous Disaster Management initiatives in Mozambique. The following perspective in this literature review relates to technological considerations, as demonstrated by an analysis of how training is depicted in a study that investigates the use of mobile phones in Mozambique's early warning systems for disasters. This is followed by an assessment of how training is analysed, firstly in a scholarly publication on Mozambique's Disaster Management policies, and secondly in a research paper on Disaster Management in rural Mozambique. Finally, the section concludes with an evaluation of extant literature on the training aspects of Disaster Management in Mozambique following Cyclone Idai.

- **Mozambique Disaster Management Projects**

Although developing countries like Mozambique are still in need of substantial enhancements in their capacities pertaining to the prevention, preparedness and mitigation of risks and disasters, various projects, awareness and educational initiatives have been launched to equip the communities and individuals with the requisite knowledge and skills. Some of these projects include the CARE post-Idai shelter, CIDMA project, Safer Schools project and PRO-GRC, as expanded below (Few et al., 2016; Ferreira Nogueira, `2019; Yordanov et al., 2020; Deprez and Labattut, 2020).

Furthermore, a study conducted by Deprez and Labattut (2020) on experts, managers, and local communities aimed to evaluate the emergency shelter work conducted by CARE Mozambique's office and the lessons learnt from their response to Cyclone Idai. The study revealed that the expertise of the Emergency Shelter Team highlighted the significance of prior training in preparedness for efficient and prompt responses. Therefore, recommendations were made for the development of operational guidelines to support shelter recovery.

In addition, Safer Schools and PRO-GRC Disaster Management projects also emphasised the importance of capacity building in Disaster Management efforts. Few et al. (2016) presented these projects as part of thirteen case studies showcasing the use of participatory techniques in engaging community leaders and stakeholders in Disaster Management concepts and practices. These projects aim to strengthen ownership and collaboration amongst all stakeholders, focusing on preparedness, response, recovery, prevention, and long-term mitigation efforts.

The use of technology devices for training in Disaster Management initiatives in Mozambique has also been explored. Ferreira Nogueira (2019) conducted a study on the use of mobile technology in the country's disaster early warning systems, emphasizing the training of volunteers in utilizing coded forms and sending SMS messages in response to disasters. The study also highlights the importance of training volunteers in compiling post-disaster data and transmitting alerts via SMS. Likewise, Yordanov et al. (2020) discussed one of the courses offered through the CIDMA project, which focuses on the use of geospatial information technology in disaster management efforts. The course, delivered in a blended format, aimed to provide practical experience to students on popular GIS platforms such as OpenStreetMap, Mapillary, QGIS, and Copernicus. Exercises based on Mozambique's data were included in the course to empower students to utilize GIS for identifying the causes and impacts of natural disasters and implementing mitigation strategies.

In the aftermath of Cyclone Idai, studies by Baltazar and Rossetto (2020), Kubo et al. (2022) and Budimir et al. (2022) assessed the impact of training on cyclone preparedness. These studies emphasise the importance of training in enabling effective emergency responses, reducing errors, and streamlining information dissemination during disasters.

However, scholars have identified weaknesses in Mozambique's National Disaster Management policy, particularly in terms of training and disaster preparedness. Studies by scholars highlight the need for the continuous training of skilled human resources in Disaster Management and DRR policies, emphasising the uneven distribution of training and knowledge amongst local decision-makers and stakeholders (Koivisto & Nohrstedt, 2017; Lundgren & Strandh, 2022).

Skills Development in Disaster Management in Japan

This sub-section briefly overviews extant literature on the role of skills development or training in Disaster Management in Japan. It is worth noting that all the overviewed papers, except for the one on life-long preparedness and adaptive politics, were written with the purpose of sharing with readers the lessons that can be learned from Japan with regard to their Disaster Management experiences. It is also worth noting that the papers included in this brief overview were published between 2013 and 2021.

- **Japan's Disaster Management life-long preparedness and adaptive politics**

Japan has developed advanced disaster management systems and strategies with a strong focus on skills development. Due to the country's susceptibility to disasters, preparedness education is deeply ingrained in Japanese culture. Terms such as 'kojo', 'jijo', 'gojo' and 'Kyojo' have been developed and are utilized to enhance understanding and promote self-reliance and mutual support within communities (Kitagawa, 2021). This cultural approach extends to schools and can be compared to the concept of 'Ubuntu' in terms of disaster preparedness and assistance. Disaster preparedness and education are integrated into the school curriculum to ensure that students are informed about potential risks and have the necessary knowledge to respond during emergencies (Suindramedhi, 2015; Kitagawa, 2016; Kitagawa, 2021). Furthermore, specialized training programs are offered for professionals in different fields related to disaster management, including firefighters, medical personnel, engineers, and volunteers. In addition, the

Japanese government ensures that regular training and simulations are conducted to familiarize individuals and organizations with their roles and responsibilities in case of emergencies (Suindramedhi, 2016) & Ishiwatari et al., 2021). These training programs often incorporate technology to simulate disaster scenarios and teach responders how to effectively use specialized equipment.

Kitagawa (2016) conducted an analysis of the policy discourse behind Japan's four forms of aid for Disaster Management preparedness. These four forms of aid are public aid (kojo), self-help (jijo), neighborhood mutual aid (gojo), and distant mutual aid (kyojo). Kitagawa (2016) highlights the example of MEXT (Ministry of Education, Culture, Sports, Science and Technology) to illustrate how a public organization, such as a ministerial department, contributes to the public aspect of Disaster Management preparedness in Japan. Kitagawa (2016) describes how MEXT provides curriculum guidelines and advice information to schools to assist them in developing their disaster operations and drills manuals, in addition to enhancing the compulsory 'enrichment of Safety Education' course introduced after the 2011 earthquake and tsunami. The aspect of self-aid in Disaster Management in Japan is presented by Kitagawa (2016) in terms of how citizens and organizations maintain an emergency arsenal with stocked food and water, and how *tendeko* teachings have saved lives. According to Kitagawa (2016), *tendeko* teaches people not to rely on others when facing a tsunami, but instead to independently seek higher ground, as evidenced by the successful evacuation of Kamaishi Elementary School students during the 2011 tsunami. Regarding distant mutual aid (kyojo), Kitagawa (2016) discusses it within the context of community engagement and partnership initiatives aimed at disaster preparedness, including those led by philanthropic organizations such as trade unions, the Red Cross, and religious groups.

- **Training Lessons Learned from Japan's Disaster Management Experiences**

This sub-section provides a brief overview of studies by Arora (2013), Suindramedhi (2016) and Ishiwatari et al. (2021). Each of these studies aimed to analyze the lessons that can be derived from Japan's Disaster Management experiences. Arora (2013) conducted field research in conjunction with surveys of academics, Disaster Management Experts and various participants, including graduate students and employees of non-governmental organizations.

The results of Arora's (2013) study revealed that, except for one training question, a majority of participants responded neutrally, with slightly less than half of the participants providing such responses. This was particularly evident in questions regarding the effectiveness and perceived utility of Disaster Management training.

Interestingly, nearly two-thirds of respondents agreed on the effectiveness of community disaster training and simulated drills in mitigating casualties. On the contrary, as noted by Ishiwatari et al. (2021), certain lessons in Disaster Management derived from the 2011 Great East Japan Earthquake include the global coordination of Disaster Management workshops and virtual dialogues, along with initiatives by the Japan International Cooperation Agency (JICA). The authors also highlight JICA programs as a method to enhance Disaster Management capabilities through the sharing of knowledge, providing examples such as the organization of training courses on Disaster Management internationally.

CONCLUSION

In summary, the literature review on skills development in both aforementioned countries indicates that both countries recognize the importance of skills development and have established mechanisms to prepare for and mitigate disasters. It is worth noting, however, that the training strategies employed by each country in preparing communities and individuals differ somewhat. Disaster awareness and preparedness are ingrained in Japanese culture, even being integrated into the school curriculum. While both countries utilize technology in their disaster mitigation strategies, Japan is notably more advanced in its technology for disaster management. This suggests that developing countries such as Mozambique have yet to make significant strides in attaining the level of technological sophistication seen in more developed countries like Japan.

This study explores the impact of skills development on disaster management in a broader context. With limited existing research on the connection between Human Resource Management and disaster management, future studies could focus on the Human Resource perspective of skill development in disaster management.

REFERENCES

Adams, R. M., Evans, C., Wolkin, A., Thomas, T., & Peek, L. (2022). Social vulnerability and disasters: Development and evaluation of a CONVERGE training module for researchers and practitioners. Disaster Prevention and Management. *International Journal (Toronto, Ont.)*, *31*(6), 13–29.

Alrehaili, N. R., Almutairi, Y. N., Alghamdi, H. M., & Almuthaybiri, M. S. (2022). A Structural review on disaster management models and their contributions. *International Journal of Disaster Management*, *5*(2), 93–108. doi:10.24815/ijdm.v5i2.27087

Arimoto, M. (2023). *Preliminary Cross-Cultural Research on Resilience Based on the Sendai Framework for Disaster Risk Reduction: Toward extending questionnaire survey by combining ethnography after flood\ndamage in a local town of Japan* (Doctoral dissertation, Tohoku University).

Arora, P. (2013). *Leveraging Human Resource for Effective Disaster Management*. Academic Press.

Asghar, S., Alahakoon, D., & Churilov, L. (2006). A comprehensive conceptual model for disaster management. *Journal of Humanitarian Assistance*, *1360*(222), 1–15.

Baltazar, C. S., & Rossetto, E. V. (2020). Mozambique Field Epidemiology and Laboratory Training Program as responders workforce during Idai and Kenneth cyclones: A commentary. *The Pan African Medical Journal*, *36*(1). Advance online publication. doi:10.11604/pamj.2020.36.264.21087 PMID:33088393

Banerji, P., & Singh, N. (2013). Comparative analysis of disaster management between Japan & India. *IOSR Journal of Business and Management*, *13*(6), 62–74. doi:10.9790/487X-1366274

Bennett, D. (2020). Five years later: Assessing the implementation of the four priorities of the Sendai framework for inclusion of people with disabilities. *International Journal of Disaster Risk Science*, *11*(2), 155–166. doi:10.1007/s13753-020-00267-w

Birkmann, J., Cardona, O. D., Carreño, M. L., Barbat, A. H., Pelling, M., Schneiderbauer, S., & Welle, T. (2013). Framing vulnerability, risk and societal responses: The MOVE framework. *Natural Hazards*, *67*(2), 193–211. doi:10.1007/s11069-013-0558-5

Birkmann, J., Sorg, L., & Welle, T. (2017). Disaster vulnerability. The Palgrave handbook of unconventional risk transfer, 329-356.

Busayo, E. T., Kalumba, A. M., Afuye, G. A., Ekundayo, O. Y., & Orimoloye, I. R. (2020). Assessment of the Sendai framework for disaster risk reduction studies since 2015. *International Journal of Disaster Risk Reduction*, *50*, 101906. doi:10.1016/j.ijdrr.2020.101906

Celse, J., & Kensen, M. (2023). Can positional concerns be a threat to disaster management? Assessing the prevalence of positional concerns among socially vulnerable populations in Trinidad & Tobago. *Environmental Hazards*, *22*(2), 152–176. doi:10.1080/17477891.2022.2096556

Cvetković, V. M., Tanasić, J., Ocal, A., Kešetović, Ž., Nikolić, N., & Dragašević, A. (2021). Capacity development of local self-governments for disaster risk management. *International Journal of Environmental Research and Public Health*, *18*(19), 10406. doi:10.3390/ijerph181910406 PMID:34639706

Deprez, S., & Labattut, E. (2020). *Lessons Learnt from CARE's Shelter Responses to Cyclone Idai in Malawi.*

Dreze, J., & Sen, A. (2002). *India: Development and participation.* OUP Oxford.

Druker, M. J. (1974). Development and structure of public attitudes toward the multiple levels of American government. The University of Wisconsin-Madison.

Faivre, N., Sgobbi, A., Happaerts, S., Raynal, J., & Schmidt, L. (2018). Translating the Sendai Framework into action: The EU approach to ecosystem-based disaster risk reduction. *International Journal of Disaster Risk Reduction*, *32*, 4–10. doi:10.1016/j.ijdrr.2017.12.015

Felbermayr, G., & Gröschl, J. (2014). Naturally negative: The growth effects of natural disasters. *Journal of Development Economics*, *111*, 92–106. doi:10.1016/j.jdeveco.2014.07.004

Ferreira Nogueira, D. (2019). *Mobile-Based Early Warning Systems in Mozambique: An exploratory study on the viability to integrate Cell Broadcast into disaster mitigation routines*. Academic Press.

Few, R., Scott, Z., Wooster, K., Avila, M. F., & Tarazona, M. (2016). Strengthening capacities for disaster risk management II: Lessons for effective support. *International Journal of Disaster Risk Reduction*, *20*, 154–162. doi:10.1016/j.ijdrr.2016.02.005

Glago, F. J. (2021). Flood disaster hazards; causes, impacts and management: a state-of-the-art review. *Natural hazards-impacts, adjustments and resilience*, 29-37.

Goto, K., Goff, J., & Paris, R. (2021). Ten years since the 2011 Tohoku-oki tsunami-Progress in paleotsunami research. *Earth-Science Reviews*, *216*, 103598. doi:10.1016/j.earscirev.2021.103598

Ishiwatari, M., Ranghieri, F., Taniguchi, K., & Mimura, S. (2021). Learning from megadisasters in Japan: Sharing lessons with the world. *Journal of Disaster Research, 16*(6), 942–946. doi:10.20965/jdr.2021.p0942

Kim, S. J., & Bostwick, W. (2020). Social vulnerability and racial inequality in COVID-19 deaths in Chicago. *Health Education & Behavior, 47*(4), 509–513. doi:10.1177/1090198120929677 PMID:32436405

Kitagawa, K. (2016). Disaster preparedness, adaptive politics and lifelong learning: A case of Japan. *International Journal of Lifelong Education, 35*(6), 629–647. doi:10.1080/02601370.2016.1231230

Kitagawa, K. (2021). Conceptualising 'disaster education'. *Education Sciences, 11*(5), 233. doi:10.3390/educsci11050233

Koivisto, J. E., & Nohrstedt, D. (2017). A policymaking perspective on disaster risk reduction in Mozambique. *Environmental Hazards, 16*(3), 210–227. doi:10.1080/17477891.2016.1218820

Kong, T. M., de Villiers, A. C., Ntloana, M. B., Pollard, S., & Vogel, C. (2020). Implementing capacity development for disaster risk reduction as a social learning system. *International Journal of Disaster Risk Reduction, 50*, 101740. doi:10.1016/j.ijdrr.2020.101740

Koshimura, S., Hayashi, S., & Gokon, H. (2014). The impact of the 2011 Tohoku earthquake tsunami disaster and implications to the reconstruction. *Soil and Foundation, 54*(4), 560–572. doi:10.1016/j.sandf.2014.06.002

Kubo, T., Chimed-Ochir, O., Cossa, M., Ussene, I., Toyokuni, Y., Yumiya, Y., & Salio, F. (2022). First Activation of the WHO Emergency Medical Team Minimum Data Set in the 2019 Response to Tropical Cyclone Idai in Mozambique. *Prehospital and Disaster Medicine, 37*(6), 727–734. doi:10.1017/S1049023X22001406 PMID:36325992

Lundgren, M., & Strandh, V. (2022). Navigating a double burden–Floods and social vulnerability in local communities in rural Mozambique. *International Journal of Disaster Risk Reduction, 77*, 103023. doi:10.1016/j.ijdrr.2022.103023

Mannan, S., Haque, D. M. E., & Sarker, N. C. D. (2021). A study on national DRR policy in alignment with the SFDRR: Identifying the scopes of improvement for Bangladesh. *Progress in Disaster Science, 12*, 100206. doi:10.1016/j.pdisas.2021.100206 PMID:34632365

Mizutori, M. (2020). Reflections on the Sendai Framework for disaster risk reduction: Five years since its adoption. *International Journal of Disaster Risk Science, 11*(2), 147–151. doi:10.1007/s13753-020-00261-2

Mugabe, V. A., Gudo, E. S., Inlamea, O. F., Kitron, U., & Ribeiro, G. S. (2021). Natural disasters, population displacement and health emergencies: Multiple public health threats in Mozambique. *BMJ Global Health*, *6*(9), e006778. doi:10.1136/bmjgh-2021-006778 PMID:34489329

Mwale, F. D., Adeloye, A. J., & Beevers, L. (2015). Quantifying vulnerability of rural communities to flooding in SSA: A contemporary disaster management perspective applied to the Lower Shire Valley, Malawi. *International Journal of Disaster Risk Reduction*, *12*, 172–187. doi:10.1016/j.ijdrr.2015.01.003

Nojavan, M., Salehi, E., & Omidvar, B. (2018). Conceptual change of disaster management models: A thematic analysis. *Jamba (Potchefstroom, South Africa)*, *10*(1), 1–11. doi:10.4102/jamba.v10i1.451 PMID:29955258

Noy, I., & Yonson, R. (2018). Economic vulnerability and resilience to natural hazards: A survey of concepts and measurements. *Sustainability (Basel)*, *10*(8), 2850. doi:10.3390/su10082850

Perera, I., Meedeniya, D., Banerjee, I., & Choudhury, J. (2013, December). Educating users for disaster management: An exploratory study on using immersive training for disaster management. In *2013 IEEE International Conference in MOOC, Innovation and Technology in Education (MITE)* (pp. 245-250). IEEE. 10.1109/MITE.2013.6756343

Rahmani, M., Muzwagi, A., & Pumariega, A. J. (2022). Cultural factors in disaster response among diverse children and youth around the world. *Current Psychiatry Reports*, *24*(10), 481–491. doi:10.1007/s11920-022-01356-x PMID:35953637

Runga, A. (2023). *An Investigation into the effectiveness of climate-related policies on disaster preparedness and response in Zimbabwe. The Case of Cyclone Idai in Chimanimani District* (Master's thesis, University of Agder).

Salvucci, V., & Santos, R. (2020). Vulnerability to natural shocks: Assessing the short-term impact on consumption and poverty of the 2015 flood in Mozambique. *Ecological Economics*, *176*, 106713. doi:10.1016/j.ecolecon.2020.106713

Scott, Z., Wooster, K., Few, R., Thomson, A., & Tarazona, M. (2016). Monitoring and evaluating disaster risk management capacity. *Disaster Prevention and Management*, *25*(3), 412–422. doi:10.1108/DPM-01-2016-0002

Sen, A. (1980). Equality of What? In S. McMurrin (Ed.), *Tanner Lectures on Human Values* (Vol. I). University of Utah Press.

Sen, A. (1993). Capability and well-being. *The Quality of Life, 30*, 270-293.

Shi, P., Xu, W., Ye, T., He, C., & Li, N. (2011). Developing Disaster Risk Science–Discussion on the Disaster Reduction Implementation Science. *Journal of Natural Disaster Science, 32*(2), 79–88. doi:10.2328/jnds.32.79

Suindramedhi, S. (2015). Disaster Management: Lesson Learned from Japan. *GMSARN International Journal, 9*(3), 83–92.

Surianto, S., Alim, S., Nindrea, R. D., & Trisnantoro, L. (2019). Regional policy for disaster risk management in developing countries within the sendai framework: A systematic review. *Open Access Macedonian Journal of Medical Sciences, 7*(13), 2213–2219. doi:10.3889/oamjms.2019.614 PMID:31456854

Tasri, E. S., Karimi, K., & Muslim, I. (2022). The effect of economic variables on natural disasters and the impact of disasters on economic variables. *Heliyon, 8*(1), e08678. doi:10.1016/j.heliyon.2021.e08678 PMID:35028461

Thompson, D. (2010). *A Framework for Making Multistate Disaster Management Systems Perform: Insights from the Caribbean Emergency and Disaster Response Agency*. Now CDEMA.

Tselios, V., & Tompkins, E. L. (2019). What causes nations to recover from disasters? An inquiry into the role of wealth, income inequality, and social welfare provisioning. *International Journal of Disaster Risk Reduction, 33*, 162–180. doi:10.1016/j.ijdrr.2018.10.003

Tsuji, Y., Satake, K., Ishibe, T., Harada, T., Nishiyama, A., & Kusumoto, S. (2014). Tsunami heights along the pacific coast of northern Honshu recorded from the 2011 Tohoku and previous great earthquakes. *Pure and Applied Geophysics, 171*(12), 3183–3215. doi:10.1007/s00024-014-0779-x

UNISDR (United Nations International Strategy for Disaster Reduction). (2015). *Sendai framework for disaster risk reduction 2015–2030*. UNISDR.

Ward, P. S., & Shively, G. E. (2017). Disaster risk, social vulnerability, and economic development. *Disasters, 41*(2), 324–351. doi:10.1111/disa.12199 PMID:27174613

Yordanov, V., Brovelli, M. A., Carrion, D., Barazzetti, L., Francisco, L. J. A., Comia, H. R., & Caravela, M. I. (2020). Capacity building for disaster management in Mozambique through teaching public participatory GIS and spatial data infrastructure. *The International Archives of the Photogrammetry, Remote Sensing and Spatial Information Sciences*, *44*, 151–158. doi:10.5194/isprs-archives-XLIV-3-W1-2020-151-2020

Zorn, M. (2018). Natural disasters and less developed countries. *Nature, tourism and ethnicity as drivers of (de) marginalization: Insights to marginality from perspective of sustainability and development*, 59-78.

Chapter 2
Socio-Economic Variables and Disaster Incidents in KwaZulu-Natal, South Africa

Maliga (Mal) Reddy
Durban University of Technology, South Africa

Zamokuhle Mbandlwa
Durban University of Technology, South Africa

Strinivasan Soondrasan Pillay
Durban University of Technology, South Africa

Genius Murwirapachena
https://orcid.org/0000-0003-0767-932X
Durban University of Technology, South Africa

Sibongiseni Soondrasan Ngema
Department of Cooperative Governance and Traditional Affairs, South Africa

ABSTRACT

South Africa constantly experiences disaster incidents linked to extreme weather, fires, structural collapse, and drowning. While the prevalence of these incidents varies in each region, KwaZulu-Natal, which is one of South Africa's coastal provinces, usually records several cases. For example, between 2018 and 2022 the province reported 5428 disaster incidents that affected 66026 households, damaged 73433 houses, and caused 805 fatalities. The total estimated costs of damages from these incidents over the period were more than R12.21 billion (about US$666.7 million). This chapter examines the effects of selected socio-economic variables on disaster incidents in the KwaZulu-Natal province of South Africa. Panel data for the province's district and metropolitan municipalities over the period 2018 to 2022 are estimated using fixed and random effects regression models. Results from the chapter shed light on the policy direction necessary to mitigate the prevalence and impact of disaster incidents both in the province and other similar environments.

DOI: 10.4018/979-8-3693-2721-0.ch002

INTRODUCTION

South Africa, a nation characterized by diverse landscapes and climatic conditions, grapples with a persistent challenge – the occurrence of various disaster incidents. These incidents, ranging from extreme weather events like strong winds, lightning, heavy rains, hailstorms, and flooding to fires, structural collapses, and drownings, have far-reaching consequences on the lives and livelihoods of its citizens. Among the provinces, KwaZulu-Natal, situated along the country's eastern coastline, emerges as a region frequently affected by such calamities. The occurrence of overwhelming and devastating incidents that continue to ravage communities is a clear indication of the impact of climate change in the province. The province will continue to suffer and all events shall be devastating and thereby put all communities at risk. This requires more effort to be exerted to create awareness and develop_strategies to mitigate and adapt to its effects. The concern is the state of capacity to adapt, prepare, and respond effectively to prevent and reduce the severity impact of natural hazards through collaborative efforts by various organs of state, non-state actors, scientific bodies, and other stakeholders that will produce credible information to enhance decision making.

The period between 2018 and 2022 witnessed a staggering 5,428 reported disaster incidents in KwaZulu-Natal, impacting 66,026 households, causing damage to 73,433 houses, and tragically resulting in 805 fatalities. The economic toll of these incidents is equally profound, with estimated damages exceeding R12.21 billion (approximately US$666.7 million). These statistics paint a sobering picture of the challenges faced by the province and underscore the urgent need for effective mitigating strategies.

Considering these circumstances, this chapter aims to delve into the intricate relationship between socio-economic variables and the occurrence of disaster incidents in KwaZulu-Natal. By analyzing panel data spanning the years 2018 to 2022, encompassing district and metropolitan municipalities within the province, the study employs fixed and random effects regression models to unravel the underlying patterns and influences. The insights derived from this analysis are crucial not only for understanding the dynamics of disaster incidents in KwaZulu-Natal but also for formulating targeted policy interventions that can mitigate their prevalence and impact.

This chapter contributes to the existing body of knowledge by offering a nuanced exploration of the socio-economic factors that either exacerbate or alleviate the risk and consequences of disaster incidents. By shedding light on the specific variables at play, the findings presented herein aim to guide policymakers, political office bearers, public practitioners, emergency responders, and community stakeholders toward evidence-based strategies that can enhance resilience and reduce vulnerabilities in KwaZulu-Natal and similar environments.

As the global community grapples with the increasing frequency and severity of climate-related events, the outcomes of this study hold broader relevance, providing a template for proactive disaster management strategies that can be adapted and applied in diverse geographical contexts. In doing so, this chapter seeks to contribute meaningfully to the ongoing discourse on disaster risk reduction, sustainable development, and resilience-building, emphasizing the imperative for collaborative efforts in safeguarding communities from the impact of natural and human-induced hazards.

OVERVIEW OF THE STUDY SITE

The geographical location of KwaZulu-Natal (KZN), as enunciated in Figure 1, is a coastal province situated along the east coast of South Africa, with a majestic shoreline on the Indian Ocean. The province has three international borders: Eswatini (formerly Swaziland) and Mozambique to the north and Lesotho to the west. There are also three provincial borders: Mpumalanga to the northwest, Free State to the west, and Eastern Cape to the south. More significantly, KZN is home to the country's two busiest ports: these being Durban situated on the east coast within the eThekwini Metropolitan area and Richard Bay on the Zululand coast in the north of KwaZulu-Natal province.

In essence, the study site within the KZN province constitutes the following District and Metro Municipalities: Amajuba, UThukela, UMgungundlovu, UMzinyathi, UMkhanyakude, ILembe, Zululand, UGu, Harry Gwala, King Cetshwayo and EThekwini Metropolitan.

The climate in KwaZulu-Natal varies from subtropical in the coastal areas experiencing warm and humid weather to more temperate conditions inland, enduring much cooler temperatures. The province is susceptible to various weather-related events, including heavy rains, storms, and heat waves that can lead to flooding, landslides, and other disasters. Historically, the province has been experiencing various disaster incidents, including floods, storms, drought and fires. Vulnerabilities are often exacerbated by socio-economic factors, creating complex challenges for disaster risk reduction and response measures. The frequency and impact of disasters underscore the need for targeted interventions to enhance resilience.

Figure 1. Position of KZN on the South African map and the extent of disaster damages
Source: Authors' own diagram using maps from KZN COGTA (2022)[1]

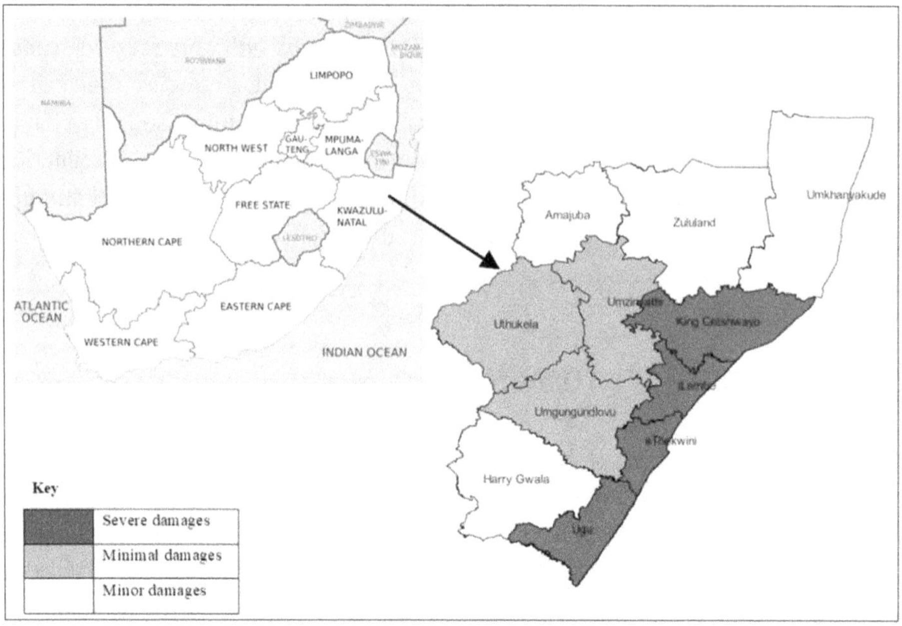

LITERATURE REVIEW ON DISASTER MANAGEMENT THEORIES THAT EXIST IN DISASTER MANAGEMENT

A literature review on disaster management involves exploring various theories and concepts related to the field. The following key theories in disaster management underpin the Disaster and Risk Management:

Disaster Risk Reduction (DRR)

Disaster risk reduction focuses on preventing the emergence of new risks, diminishing existing risks, and effectively managing residual risk. These efforts contribute significantly to enhancing resilience and, consequently, advancing sustainable development (Twigg, 2015).

The policy objective of disaster risk management is encapsulated in disaster risk reduction strategies and plans. These strategies delineate specific goals and objectives over various timeframes, accompanied by tangible targets, indicators, and time frames. Aligned with the Sendai Framework for Disaster Risk Reduction 2015-2030, these goals are directed towards averting the creation of disaster risk, mitigating existing risk, and fortifying resilience in economic, social, health, and environmental domains (Izumi, Shaw, Djalante, Ishiwatari and Komino, 2019).

A universally endorsed disaster risk reduction policy is outlined in the United Nations-backed Sendai Framework for Disaster Risk Reduction 2015-2030, ratified in March 2015. The envisioned outcome over the 15-year period is a substantial reduction in disaster risk and associated losses in terms of lives, livelihoods, health, and the assets of individuals, businesses, communities, and nations, across economic, physical, social, cultural, and environmental dimensions. DRR focuses on minimizing vulnerabilities and increasing resilience to disasters. Seddiky, Giggins and Gajendran (2020) argued that DRR involves strategies to reduce the impact of hazards and prevent disasters from occurring. Vulnerability, resilience, mitigation, preparedness, and adaptation are central to DRR. This is what could have been initiated by the South Africa government and this theory could have been implemented to secure a disaster-prone area.

Social Vulnerability Theory

This theory emphasizes the social factors that contribute to vulnerability in the face of disasters. It explores how socio-economic, cultural, and political factors influence a community's ability to cope with and recover from disasters.

Social vulnerability arises when the extent of risk to an individual's life and well-being from a specific and identifiable event in health, nature, or society is influenced by the hardships imposed by unfavorable social conditions. One prevalent approach to assess social vulnerability involves consolidating various social factors into an index (Zakour, Gillespie, Zakour and Gillespie, 2013). This scoping review had a broad objective of surveying the literature on social vulnerability indices. Our primary goals included delineating the characteristics of social vulnerability indices, comprehending their composition, and elucidating how these indices are employed in scholarly works. Social capital, community cohesion, inequality, and social networks are crucial in understanding vulnerability (Spielman, Tuccillo, Folch, Schweikert, Davies, Wood and Tate, 2020).

Disaster Management Cycle

This theory divides the disaster management process into different phases, including mitigation, preparedness, response, and recovery. It provides a structured approach to managing disasters. Disasters not only bring emotional devastation but also impose significant financial burdens on both individuals and organizations. For instance, the most expensive hurricane in U.S. history, Katrina in 2005, resulted in approximately $190 billion in damages. Recent data from the National Centers for Environmental Information reveals that the three costliest hurricanes occurred within the last decade (Sawalha, 2020).

The U.S. Geological Survey predicts an increase in the frequency and severity of natural hazards in the future due to climate change. This underscores the growing importance of trained disaster management professionals as communities nationwide brace themselves for and respond to unprecedentedly intense natural disasters (Malilay, Heumann, Perrotta, Wolkin, Schnall, Podgornik, Cruz, Horney, Zane and Roisman, 2014). Although preventing natural disasters is impossible, professionals can adopt measures to minimize their impact. One crucial measure involves using the disaster management cycle as a framework to prevent or alleviate the consequences of disasters. In collaboration with communities, local and national governments, and organizations, knowledgeable professionals employ this cycle to devise optimal strategies for reducing the damage and costs inflicted by disasters.

Community-Based Disaster Risk Management (CBDRM)

CBDRM involves the active participation of local communities in identifying and managing their own risks. It recognizes the importance of local knowledge and community engagement in disaster resilience. Disaster management typically relies on external experts and costly or hard-to-reach resources. Andrew (2018) view Community-Based Disaster Risk Management (CBDRM) as a model that represents an approach to risk analysis and disaster management which was initiated and organized by local communities. The success of CBDRM depends on sustaining community-level interest and motivation in disaster risk management (DRM) and integrating CBDRM strategies into national policy. In the realm of CBDRM, Hazard, Vulnerability, and Capacity Assessments (HVCAs) play a crucial role. HVCAs enable the identification of vulnerable groups, understanding the factors contributing to their vulnerability and impact, assessing their needs and capacities, and ensuring that projects, programs, and policies effectively address these needs (Van Niekerk, Nemakonde, Kruger and Forbes-Genade, 2018).

Children hold a unique position characterized by both vulnerability and capability in the context of disasters and disaster risk. Despite often being viewed as passive victims, various international treaties recognize that children possess distinct power, abilities, and rights. To enhance the effectiveness of CBDRM and HVCAs, it is essential to explore ways of (i) actively involving children and youth and (ii) adapting tools and frameworks to specific contexts (Sarabia, Kägi, Davison, Banwell, Montes, Aebischer and Hostettler, 2020).

Social-ecological Systems Theory

This theory explores the dynamic interactions between social and ecological components in a system. It emphasizes the interconnectedness of human and natural systems in the context of disasters. A socio-ecological system comprises a bio-geo-physical unit interconnected with its associated social actors and institutions. These systems, marked by complexity and adaptability, are defined by spatial or functional boundaries surrounding specific ecosystems and contextual issues. The term socio-ecological system (SES) encompasses several key features: A cohesive system involving both biophysical and social factors interacting consistently and resiliently. Definition across various spatial, temporal, and organizational scales, often hierarchically linked. Critical resources (natural, socio-economic, and cultural) regulated by a combination of ecological and social systems. A perpetually dynamic, complex system with continuous adaptation (Cumming and Allen, 2017).

Scholars utilize the concept of socio-ecological systems to underscore the integration of humans into nature, challenging the artificial separation between social and ecological systems. Despite differences in the meaning of resilience in social and ecological contexts, the SES approach emphasizes the interconnectedness of these systems through feedback mechanisms, acknowledging their shared characteristics of resilience and complexity. The theoretical foundations of socio-ecological systems rest on the idea that humans are an integral part of nature, rejecting the notion of a clear boundary between social and natural systems. This perspective, initially proposed by Berkes and Folke, has been further developed, highlighting social-ecological keystones and the significance of biocultural diversity for system resilience (Partelow, 2018).

In the past, the intersection between social and natural sciences in addressing socio-ecological systems was limited. However, this changed in the 1970s and 1980s with the emergence of subfields such as environmental ethics, political ecology, environmental history, ecological economics, common property, and traditional ecological knowledge. Each of these subfields serves as a bridge connecting natural science and social science thinking. Elinor Ostrom and her collaborators developed the Social-Ecological Systems (SES) framework, incorporating common-pool resource theory and collective self-governance. This framework draws on systems ecology and complexity theory, addressing societal concerns like equity and human well-being. SES theory encompasses concepts from resilience, robustness, sustainability, and vulnerability, aiming to provide a comprehensive understanding of SES dynamics and attributes (Van Assche, Verschraegen, Valentinov and Gruezmacher, 2019).

Approaching SESs from a complex system perspective seeks to integrate various disciplines to address environmental challenges effectively. The management of complex systems involves adaptive and flexible processes capable of handling uncertainty and change. SESs, characterized by complexity and adaptability, require continuous learning, testing, and knowledge development to navigate uncertainty and change, differing from simple systems in their nonlinearity, uncertainty, emergence, scale, and self-organization (Preiser, Biggs, De Vos and Folke, 2018).

Institutional Analysis and Development (IAD) Framework

The IAD framework focuses on the role of institutions in shaping how communities and societies manage disasters. It examines the rules, norms, and organizations that influence decision-making. Elinor Ostrom, the first woman to be awarded the Nobel Memorial Prize in Economic Sciences in 2009, developed the Institutional Analysis and Development framework (IAD). This framework consists of a set of concepts aimed at analyzing commons, such as fisheries and woodlands. Ostrom's focus was on identifying institutional structures that support sustainable management of resource stocks, striking a balance between individual usage and the broader public interest. In contrast to rational choice models, the IAD was created to formally explore and document governance structures, actor positions, and both formal and informal rules governing resource extraction from common resources (Ostrom, 2011).

Described as a "multi-level conceptual map" by Ostrom, the IAD allows for zooming in and out of hierarchical aspects within social system governance structures. The framework aids in understanding complex collective action problems by breaking them down into 'action arenas,' which are more manageable components. The analyst assumes a fixed structure in the short term, requiring "actors in positions" for an action situation to exist, with actors making choices within the existing rule structure. The study of outcomes in collective choice situations considers the influence of institutional arrangements, socio-economic conditions, and the physical environment on actors. The IAD categorizes institutional arrangements into seven rule types. Overall, the IAD serves as a systematic method for documenting policy analysis functions, akin to techniques commonly used in physical and social sciences, shedding light on how institutions operate and evolve over time (Whaley, 2018).

These theories provide a foundation for understanding the complexities of disaster management and can guide research, policy development, and practical interventions in the field. It's essential to note that these theories are often interconnected, and their application may vary based on the specific context and nature of the disaster. Even though these theories are fundamentally important for Disaster and Risk Management but this chapter specifically derive from two theories which are Hazard, Vulnerability, and Capacity Model and Social-ecological Systems Theory because the main focus of these theories is on the Resilience, adaptive capacity, and the feedback loops between human and ecological systems are critical elements (Cole, 2017).

Resilience Theory

Resilience theory, in the context of socio-economic variables and disaster incidents, provides a framework for understanding how communities and individuals can effectively respond to and recover from various shocks and stressors, including natural disasters. When applied to the specific case of KwaZulu-Natal, South Africa, resilience theory becomes particularly relevant due to the region's vulnerability to a range of environmental hazards, including floods, droughts, and storms. Resilience is commonly defined as the capacity of a system to absorb disturbances, reorganize, and continue functioning while undergoing change. In the context of socio-economic and disaster resilience, the theory emphasizes the interplay between social, economic, and environmental factors. Resilient communities are those that can anticipate, adapt to, and recover from disruptive events (Moosa, 2021).

In the case of KwaZulu-Natal, several socio-economic variables play a crucial role in determining the resilience of the population to disasters. High levels of poverty and economic inequality exacerbate the impact of disasters. Vulnerable populations lack the resources and infrastructure to cope with and recover from disasters. According to Chabana (2020), education levels influence the understanding of disaster risks and the ability to implement effective response strategies. Awareness campaigns and education programs can enhance community resilience. The quality of infrastructure and overall development levels significantly affect how well a community can withstand and rebound from disasters. Access to proper housing, transportation, and healthcare facilities is crucial.

Strong social networks and community cohesion contribute to resilience. Communities with strong social ties often collaborate more effectively during crises, providing support and resources to each other. KwaZulu-Natal, located in a geographically diverse region of South Africa, experiences a range of natural disasters, including floods, droughts, and storms. Climate change further intensifies these risks. Recent incidents, such as floods or drought-induced water scarcity, have highlighted the need for a comprehensive approach to disaster management.

Linking Resilience Theory With Socio-Economic Variables and Disasters

Resilience theory suggests that communities with higher adaptive capacity can adjust and respond effectively to changing circumstances. Socio-economic variables like education, infrastructure, and social cohesion directly influence adaptive capacity in KwaZulu-Natal. Addressing socio-economic vulnerabilities, such as poverty and inadequate infrastructure, can reduce the overall vulnerability of communities to disasters. Resilience-building initiatives should target these variables to enhance the ability of communities to withstand shocks. Resilience is not solely a top-down process but requires active community participation. Engaging with local communities, considering their unique socio-economic context, and incorporating their knowledge into disaster preparedness plans can strengthen overall resilience (Madubula, 2021).

Policymakers should design and implement interventions that address socio-economic disparities and enhance the resilience of vulnerable populations. This includes investments in education, healthcare, and infrastructure development. Applying resilience theory to the socio-economic variables and disaster incidents in KwaZulu-Natal involves understanding the dynamic interactions between these factors. Strategic planning, community engagement, and targeted interventions can contribute to building a more resilient society capable of effectively responding to and recovering from various disasters (Thulare and Moyo, 2021).

Disaster Risk and Vulnerability

The conceptualization of disasters as a convergence of hazards and vulnerabilities has been extensively explored in the works of Blaikie et al. (2003, 2014) in "The Challenge of Disasters and our Approach" and "At Risk: Natural Hazards, People's Vulnerability, and Disasters." This foundational text posits that disasters are not merely a consequence of natural events but are shaped by social, economic, and environmental dynamics. Cutter et al. (2009, 2010, 2012) contribute to this discourse with their exploration of social vulnerability, emphasizing the interplay between socio-economic factors and the impact of environmental and climate related hazards. Another intriguing approach postulated by Emrich et al. (2011) focuses on assessments of hazards vulnerability by describing and integrating hazard zone information on climate-sensitive hazards with socioeconomic and demographic data to create an index showing both the areal extent of hazard exposure and social vulnerability. Equally important, is the reflection on the development and practice of disaster risk management and climate change policy integration (Del Lama Soares, 2023) in promoting adaptable, relevant, and targeted strategic disaster risk interventions.

Climate Change and Extreme Weather

The Intergovernmental Panel on Climate Change (IPCC) has been instrumental in revealing the interconnectedness between climate change, extreme weather events, and disaster risk. The IPCC's reports, particularly "Climate Change 2014: Impacts, Adaptation, and Vulnerability," (Tong, et al., 2014), highlights a global perspective on the implications of climate change on disaster occurrences. Field et al. (2012) further delves into the nuanced relationship between climate change and disaster adaptation, offering insights into the challenges and opportunities presented by a changing climate.

Southern Africa is now considered to be a climate change 'hot spot', with rising temperatures, droughts, flooding, and unpredictable weather systems set to become the new normal. Over the last six decades, significant changes in climate have been observed in South Africa. Consequently, climate zones across the region are shifting, ecosystems and landscapes are degrading due to fires, droughts and heat waves. This has, in some parts, translated into natural disasters, vector-borne diseases, as well as food and water insecurity, which consequently threaten livelihoods (World Bank, 2022). Recent research on environmental migration in the region suggests that environmental factors such as the severe drought of 2015–2016 or the loss of ecosystem services contribute to migration across the region, often from other countries into South Africa (Bega, 2022). In 2017, it was estimated that 15 000 South Africans were internally displaced, that is, forced to leave their homes, because of natural or human-inflicted hazards, staying within the country's borders. In 2018 and 2019, more than 2.6 million and 3.4 million weather-related displacements occurred in sub-Saharan Africa (IPCC, n.d.). Experts at the South African Weather Service (SAWS) say severe and extreme weather events are becoming more frequent and intense as a result of the changing climate (Siva, et al., 2021).

The 28th Conference of the Parties (COP 28) held in Dubai during December 2023, consisted of the 28th meeting of the Conference of the Parties (COP) to the UN Framework Convention on Climate Change (UNFCCC), the 18th meeting of the Conference of the Parties serving as the Meeting of the Parties to the Kyoto Protocol (CMP 18), the 5th session of the Conference of the Parties serving as the Meeting of the Parties to the Paris Agreement (CMA 5), and the 59th sessions of the Subsidiary Body for Scientific and Technological Advice (SBSTA 59) and the Subsidiary Body for Implementation (SBI 59).

At the COP 28, various declarations aimed at building Africa's climate resilience were highlighted. In particular, the establishment of green banks, greenhouse gas trading systems, public-private partnerships, transformative financing measures such as green bonds and sustainability-linked instruments, together with a host of new funding pledges. Some of the crucial announcements impacting Africa (Houston, 2024, Maslin, et al., 2023, and Obergassel, et al., 2022), are as follows:

- the African Group of Negotiators emphasised their continued commitment to energy transition, climate finance and the urgent need for funding for climate change adaptation. The Group noted that the continent needs between USD 200 and 400 billion a year by 2030 for loss and damage and USD 400 billion a year for adapting to climate change.
- some countries pledged contributions to the Loss and Damage Fund's start-up phase. Countries pledging contributions include the European Union member states (USD 245 million), the United Kingdom (USD 50.5 million), the United States (USD 17.5 million) and Japan (USD 10 million).
- South African President Cyril Ramaphosa welcomed the setting up of the loss and damage fund but said that global banking reform and debt relief to enable funding for developing countries needed to be on a far larger scale.
- Multilateral development banks (MDBs) and development finance institutions (DFIs) are increasingly anchoring climate funding in Africa because they can shoulder political risk, access government protections, enter markets that others cannot and are uniquely capable of facilitating innovative financing structures and long-term lending. However, the amount of capital needed to build climate resilience is significant and private equity, debt finance and specialist funds are entering the market. Multi-finance and blended solutions are growing in popularity to de-risk deals and support a broader ecosystem of lenders and investors. To mobilise private sector finance, investor-friendly policies and incentives are needed to reduce barriers such as uncertain licensing processes, a lack of market liquidity and bankability.
- Africa and Middle Eastern leaders announced the launch of the USD 10 billion Global Green Growth Institute's (GGGI) SAFE initiative at COP 28, a public-private partnership intended to transform food systems for millions facing food insecurity due to climate change.
- MDBs, including the African Development Bank (AfDB), Islamic Development Bank and the International Fund for Agricultural Development, announced a joint investment of USD 1.6 billion to assist in the development of Special Agro-Industrial Processing Zones (SAPZs) with a focus on climate-resilient food systems, as part of the SAFE initiative.
- South Africa presented its finalised Just Energy Transition Investment Plan (JET-IP) at COP 28. The JET IP is aligned with the Cabinet-approved National Just Transition Framework and outlines the investments required to achieve the country's decarbonisation commitments while promoting sustainable development and ensuring a just transition for affected workers

and communities. The JET IP covers electricity, new energy vehicles and green hydrogen and identifies USD 98 billion in financial requirements over the next five years, to come from both the public and private sectors.

Stemming from the above, it is evident that in the pursuit of fortifying African countries against the impact of climate change, significant strides have been made through collaborative efforts and financial commitments from global stakeholders. With substantial financing pledged by major players worldwide, coupled with the announcement of numerous climate adaptation initiatives, the participation of Multilateral Development Banks (MDBs) and Development Finance Institutions (DFIs) in facilitating and innovating the financing process has been instrumental. Furthermore, the increasing mobilization of the private sector in climate resilience projects demonstrates a promising trend towards bolstering Africa's resilience to climate-related challenges.

However, despite these positive developments, there remains a relentless need for further action and investment to adequately address the multifaceted challenges posed by climate change in Africa. While access to funding is improving, more comprehensive strategies and sustained efforts are required to ensure that vulnerable communities are adequately protected and that adaptation measures are effectively implemented. Moreover, the equitable distribution and utilization of financial resources, as well as the alignment of initiatives with local needs and priorities, are critical considerations for maximizing the impact of climate resilience efforts.

In addition, some of the other outcomes of the conference related to disaster risk management includes (Stocktake, 2023 and IISD, 2023):

- the adoption of the framework for the Global Goal on Adaptation (GGA), established in the Paris Agreement, which aims to guide the implementation of the goal and, among other things, establishes impact, vulnerability, and risk assessment (by 2030), multi-hazard early warning systems (by 2027), climate information services for risk reduction and systematic observation (by 2027), and country-driven, gender-responsive, participatory, and transparent national adaptation plans (by 2030);
- the designation of the consortium of the UN Office for Disaster Risk Reduction and the UN Office for Project Services as the host of the Santiago Network on loss and damage;
- the launch of the implementation of the work programme on just transition pathways, with at least two hybrid dialogues to held prior to the two annual sessions of the Subsidiary Bodies;
- the decision to continue and strengthen the dialogue to exchange views on and enhance understanding of the scope of Article 2.1(c) of the Paris

Agreement (on aligning finance flows with low-GHG climate resilient development) and its complementarity with Article 9 of the Paris Agreement (on climate finance).

Overall, these recent developments at COP 28 offer valuable opportunities to enhance climate adaptation and resilience-building efforts across the globe, in Africa and South Africa extending into the province of KwaZulu-Natal. By integrating the principles and objectives outlined in these initiatives, stakeholders can work towards a more sustainable and climate-resilient future for the continent and country.

Community Resilience and Adaptive Capacity

The profound interrogation of vulnerability and resilience articulated by Manyena (2006) on the resilience theory is pivotal for understanding the dynamics of community resilience. The discussion on resilience revisited serves as a conceptual framework for analyzing how socio-economic factors influence adaptive capacity in KwaZulu-Natal. Cutter, et al. (2014), in the study *"The Geographies of Community Disaster Resilience"*, extends these arguments further within the local and spatial context. The research explores the topographies of community disaster resilience, offering insights into the spatial dimensions and local factors influencing resilience. Other scholars (Norris, et al., 2008) project community resilience as a metaphor and theory, offering insights into the various capacities that contribute to readiness in the face of disasters. Such deliberations are crucial and informs the examination of adaptive capacity in KwaZulu-Natal, aligning with the chapter's focus on understanding how socio-economic variables interact with geographical aspects in KwaZulu-Natal. The multidimensional insights derived from these sources contribute to a nuanced investigation of the socio-economic factors that either exacerbate or alleviate the risk and consequences of disaster incidents, guiding evidence-based strategies for resilience-building in KZN.

Methodology and Empirical Model Specification

Panel regression models are adopted to examine the effects of selected socio-economic variables on disaster incidents in the KwaZulu-Natal province of South Africa. The use of panel data allows for the analysis of several municipalities over a longer period. This generally increases sample size and provides insights into questions that can neither be addressed by cross-sectional nor time-series data alone (Reed and Ye, 2011). Several advantage of using panel data are documented in the econometrics literature (Anton and Nucu, 2020; Croissant and Millo, 2019; Hill et al., 2020; Hsiao, 2022). Amongst these advantages is the ability of panel regression models to allow for variables that vary both over time and between municipalities as well as those that vary in predictable ways. This makes panel regression models more dynamic and preferable in examining the link between disaster incidents and socio-economic variables in the context of a developing economy. The fixed effects (FE) and random effects (RE) are the commonly used panel regression estimators in the econometric literature (Bell, Fairbrother and Jones 2019; Hill et al., 2020). Although these two estimators have different assumptions, the basic mathematical formulation for panel regression modelling is specified as:

$$y_{it} = \alpha + x'_{it}\beta + v_i + \epsilon_{it} \tag{1}$$

where y_{it} is the dependent variable for municipality i in time period t (i.e., year t in the context of this study); α is the constant/intercept; x'_{it} are the independent variables; β is the coefficient of each independent variable; v_i is the municipal specific error term differing between municipalities but having a constant value for any particular municipality; and ϵ_{it} is the common error term (this has the usual properties of mean 0, uncorrelated with itself, uncorrelated with x, uncorrelated with v, and homoscedastic). The common error term (ϵ_{it}) is then decomposed to $\epsilon_{it} = v_t + \omega_{it}$, where ω_{it} is a conventional error term that better describes v_t. Irrespective of the properties of v_t and ϵ_{it}, if equation 1 is true, then it is also true that:

$$\bar{y}_i = \alpha + \bar{x}_i\beta + v_i + \bar{\epsilon_i} \tag{2}$$

where $\bar{y}_i = \sum_t y_{it} / T_i$ and $\bar{x}_i = \sum_t x_{it} / T_i$; while $\bar{\epsilon_i} = \sum_t \epsilon_{it} / T_i$. Econometric literature further suggests that when equation 2 is subtracted from equation 1, then it must also be true that:

$$\left(y_{it} - \bar{y}_i\right) = \left(x_{it} - \bar{x}_i\right)\beta + \left(\epsilon_{it} - \bar{\epsilon_i}\right) \tag{3}$$

Thus, equations 1, 2 and 3 generally provide the theoretical basis for estimating β in panel regression modelling (Reed and Ye, 2011; Wooldridge, 2016). The FE estimator (i.e., the within estimator) would then imply estimation using the OLS to estimate equation 3, while the RE estimator is a weighted average of estimates produced by the between and within estimators, and is equivalent to:

$$\left(y_{it} - \theta \bar{y}_i\right) = \left(1 - \theta\right)\alpha + \left(x_{it} - \theta \bar{x}_i\right)\beta + \left\{\left(1 - \theta\right)v_i + \left(\epsilon_{it} - \theta \bar{\epsilon}_i\right)\right\} \quad (4)$$

where θ is a function of σ_v^2 and σ_ϵ^2. Therefore, if $\sigma_v^2 = 0$ it means v_i is 0, then $\theta = 0$ implying that equation 2 can be directly estimated using the OLS. Equally, if $\sigma_\epsilon^2 = 0$, it means ϵ_{it} is 0, then $\theta = 1$, suggesting that the within estimator returns all the information available, thus the regression will have an R^2 of 1. Given the above, the empirical model estimated in this study is specified as follows:

$$Incidents_{it} = \alpha_{it} + \beta_1 Density_{it} + \beta_2 Indigent_{it} + \beta_3 Equitable_{it} + \beta_4 Councillors_{it} + \beta_5 Coastal_{it} + v_i + \epsilon_{it} \quad (5)$$

where $Incidents_{it}$ is the number of recorded disaster incidents in municipality i and year t, $Density_{it}$ is the population density (i.e., number of people per square kilometre), $Indigent_{it}$ refers to the reported number of indigent households receiving free basic services, $Equitable_{it}$ is the total equitable share grant received from the national government, $Councillors_{it}$ refers the total number of councillors, $Coastal_{it}$ is a dummy where 1 indicates a coastal municipality and 0 otherwise (coastal municipalities are more susceptible to severe disaster and risk incidents)[2].

Density, Indigent, and Equitable are the key independent variables in the model estimated in this study. However, correlation tests revealed that both Density and Indigent were highly correlated with Equitable. This was expected given that the national government also considers each municipality's population and the number of poor households when determining its equitable share of the national revenue. Addressing multicollinearity usually includes eliminating one of the correlated variables (Kalnins, 2018; Lindner, Puck and Verbeke, 2020). While this approach is commonly used in the literature, it is not the only solution to the problem of multicollinearity. In cases where multicollinearity exists amongst key variables, the correlated variables may be combined into a single variable, dimensionality reduction may be adopted, or variables may be standardise/ centred (Ismaeel, Midi and Sani, 2021; Kalnins, 2018). Alternatively, analysts may convert one or more of the correlated variables into natural logarithms and the coefficients of the converted variables will be interpreted as elasticities (Allam et al., 2020; Khalid, Usman, and Mehdi, 2021; Xie, Wang, and Cong, 2020). This approach was adopted in this study and the problem of multicollinearity significantly reduced after Density and Indigent were converted into natural logarithms.[3] The rest of the variables specified in equation 5 maintained their original absolute values and the empirical model then assumed the following semi-translog function:

$$Incidents_{it} = \alpha_{it} + \beta_1 lnDensity_{it} + \beta_2 lnIndigent_{it} + \beta_3 Equitable_{it} + \beta_4 Councillors_{it} + \beta_5 Coastal_{it} + v_i + \epsilon_{it} \quad (6)$$

where $lnDensity_{it}$ is the natural logarithm of population density, while $lnIndigent_{it}$ is the natural logarithm of the number of indigent households in each municipality. Further, the data was tested for heteroscedasticity using the Breusch-Pagan/Cook-Weisberg. Test results produced a Chi^2 whose p-value was statistically significant indicating the presence of heteroscedasticity, implying that the RE model should have robust standard errors. Subsequently, both the FE and RE estimators were used estimate the model specified in equation 6. These estimators have different assumptions and the most efficient estimator given the dataset should always be adopted (Bell, Fairbrother and Jones 2019; Hill et al., 2020; Hsiao, 2022). As such, the Hausman test which is commonly used to compare between FE and RE in the econometric literature was used to determine the most efficient estimator in the context of this study. The Hausman test produced a Chi^2 whose p-value was statistically significant indicating that FE as the most efficient technique. Therefore, the results estimated using a FE panel data regression model were adopted and interpreted in this study.[4]

Data and Descriptive Statistics

The study uses annual panel data for the 11 large municipalities in the KwaZulu-Natal province of South Africa for the period 2018 to 2022. Thus, the panel data used consists of cross-sections (N = 11 district/metropolitan municipalities) and time series (T = 5 years). The KwaZulu-Natal Provincial Department of Cooperative Governance and Traditional Affairs (CoGTA – KZN) has a Provincial Disaster Management Centre that monitors and reports on disaster and risk incidents in the province. Incidents in each local municipality are consolidated into a single summarised report for each district municipality. Thus, CoGTA – KZN provides consolidated data for each district municipality on the number of incidents recorded, households affected, houses destroyed, number of people affected, and fatalities recorded. The department further categorises incidents according to type per each municipality, where common incident types are strong winds, lightning, fires, heavy rains, structural collapse, hailstorm, flooding, and drowning (CoGTA – KZN, 2022). Additionally, data for independent variables were obtained from the Non-Financial Census of Municipalities (P9115) published by Statistics South Africa, and the Local Government Equitable Share output published by the South African National Treasury.[5] Descriptive statistics for the key variables used in this study are given in Table 1.

Table 1 show an average of 99 disaster incidents with higher standard deviations (overall and between) that indicate there are larger variations from the sample mean over the investigated period. This variation is further confirmed by the minimum and

Table 1. Summary statistics of key variables (N = 55; n = 11; T = 5)

		Mean	SD	Minimum	Maximum
Incidents (Number)	Overall	99	126	3	590
	Between		63	36	270
	Within		111	-165	419
Population (Number)	Overall	1 082 557	949 172	531 191	4 152 175
	Between		985 598	552 537	3 997 885
	Within		38 481	931 188	1 236 847
Area (km^2)	Overall	8 578	3 831	2 556	14 800
	Between		3 981	2 556	14 800
	Within		0	8 578	8 578
Density (Population per km^2)	Overall	235	427	51	1 624
	Between		444	53	1 564
	Within		13	176	295
Households (Number)	Overall	274 865	296 698	117 256	1 247 391
	Between		307 980	121 169	1 188 251
	Within		14 299	212 381	334 005
Indigent (Number)	Overall	66 137	180 213	397	768 258
	Between		185 212	1 431	622 403
	Within		26 722	9 035	211 992
Equitable (Rands)	Overall	1 200 000 000	749 000 000	515 000 000	4 060 000 000
	Between		760 000 000	602 000 000	3 430 000 000
	Within		161 000 000	668 000 000	1 830 000 000
Councillors (Number)	Overall	156	36	103	222
	Between		38	103	220
	Within		3	153	169

Notes: SD = Standard Deviation; N = Observations; n = Municipalities; T = Timeframe in years
Source: Authors' own table, 2024

maximum statistics which show that disaster incidents varied from as little as 3 to as high as 590 during the period. The average population for the sampled municipalities was 1 082 557 people. However, population statistics for the sample varied from 531 191 to 4 152 175 people. On the other hand, the average population density for the sample was 235 people per square kilometre (km^2), with density varying from 51 to 1 624 people per km^2. The average population density is consistent with what is reported in the literature, making KwaZulu-Natal the second densely populated province in South Africa, after Gauteng which in 2022 had a population density of

609 people per km^2. The Northern Cape province had the least population density with an average of 3 people per km^2, while the rest of the provinces had less than 68 people per km^2. South Africa's average population density in 2022 was 50 people per km^2, which was relatively lower compared to other emerging economies like Morocco (86), Kenya (89), Ghana (134), Nigeria (238), India (433), China (150), and Turkey (108).[6] In this study, we hypothesise a positive relationship between population density and disaster incidents.

South African municipalities are expected by law to identify indigent households and provide them with basic public services. Therefore, the number of indigent households is used as a proxy for poverty, where a positive relationship is hypothesised between poverty and disaster incidents. Table 1 shows an average of 66 137 indigent households for the sample, with higher standard deviations indicating larger variations from the sample mean, and one municipality had as high as 768 258 indigent households. We further examine the impact of the equitable share grant on disaster incidents. The equitable share is an unconditional grant allocated to municipalities according to Section 214(1) of the South African Constitution. This grant enables municipalities to deliver free basic services and subsidises the cost of core services for municipalities with the least potential to cover costs from their own revenues (CoGTA, 2023). In this study, the equitable share grant refers to grants received by the district and all local municipalities within its jurisdiction. The average grant for the sample was R1 200 000 000 (about, US$65 580 149) and the highest grant received was R4 060 000 000 (about, US$221 879 505).[7] The study also controls for the number of councillors in each municipality, where we expect more councillors to provide swift responses that aid to mitigate disaster incidents. Population per councillor or households per councillor variables would have been ideal but were dropped from analysis due to high correlation. Correlation was tested using the variance inflation factor (VIF) and the correlation values are presented in Table 2.

Table 2. Correlation values

	VIF	1/VIF
Equitable	7.33	0.136
lnDensity	5.43	0.184
Councillors	2.38	0.421
lnIndigent	2.05	0.489
Coastal	1.74	0.576
Mean VIF	3.78	

Source: Authors' own table, 2024

VIF measures how much the variance of the estimated regression coefficient is inflated if the independent variables are correlated. Generally, a VIF value of 1 indicates no correlation between a given explanatory variables, while a value between 1 and 5 indicates moderate correlation (Lavery et al., 2019; Salmerón Gómez et al., 2016; Shrestha, 2020). While VIF values that are generally less than 5 are commonly recommended in the literature, several studies including Lavery et al. (2019) and Shrestha (2020) suggest that VIF values that are less than 10 can be acceptable. In the context of the variables used in this study, the overall VIF value was 3.78, although equitable share had a VIF value of 7.33. The overall VIF of 3.78 suggests that the variance of the regression coefficients would be 3.78 times greater than it would have been if the predictor variables had been linearly independent with each other. This is generally acceptable since the mean VIF value is moderate (Lavery et al., 2019; Salmerón Gómez et al., 2016). Thus, the VIF values for the variables are low enough to warrant the reliability of the regression models estimated.

Results

As explained earlier in the methodology, various pre-estimation tests were performed before adopting the most efficient model between the FE and RE.[8] Heteroscedasticity was tested using the Breusch-Pagan/Cook-Weisberg test which produced a Chi^2 of 25.01 whose p-value of 0.000 was statistically significant at 1% suggesting the presence of heteroscedasticity. Therefore, the FE and RE models were expressed with robust standard errors to address the confirmed heteroscedasticity problem. Subsequently, the Hausman test was used to determine the most efficient estimate between FE and RE. This test measures whether FE or RE is consistent or biased using the null hypothesis that RE is appropriate, against the alternative hypothesis that FE is appropriate (Henningsen and Henningsen, 2019; Hsiao, 2022; Papke and Wooldridge, 2023). The decision rule of this test is that the alternative hypothesis that FE is appropriate is accepted if the Chi^2 produces a statistically significant p-value. In this study, the Hausman test produced a Chi^2 of 9.30 whose p-value of 0.026 was statistically significant at 5%, suggesting FE as the most efficient estimator.

When FE is adopted as the most efficient estimator, it is important to test whether time-fixed effects are needed when estimating the FE model (Murwirapachena and Kabange, 2023). Testing for time-fixed effects is very important as it allows the model to account for individual effects unique to each municipality (Callaway and Karami, 2023; Imai and Kim, 2021). Therefore, a joint test that all coefficients on the indicators for time (years in the context of this study) were 0 in all equations was performed. Thus, the study assessed whether the dummies were equal to zero for all years. The decision rule is that if the dummies were equal to 0, then no time-fixed effect is needed, otherwise time-fixed effects are required. The test produced and *F* score of 11.77 with a statistically significant p-value of 0.001, confirming the need for time-fixed effects in the FE model. Therefore, an FE model with robust standard errors and time-fixed effects was estimated in this study. Results from estimation are presented in Table 3.

Estimates presented in Table 3 are interpreted in terms of the statistical significance, sign, and magnitude of each variable's coefficient. In the context of this study, a positive and statistically significant coefficient suggests the variable has a positive effect on the prevalence of disaster incidents (i.e., the variable increases the chance of disaster incidents). On the other hand, a statistically significant negative coefficient suggests the variable has an inverse effect on disaster incidents (i.e., lowers the chance of disaster incidents). From the selected explanatory variables expressed in our FE model, only Equitable and Councillors were statistically significant at 10%

Table 3. Estimation results

	Coefficient	**Robust SE**	**p-Value**
lnDensity	427.554	904.962	0.647
lnIndigent	-3.198	7.693	0.686
Equitable	9.21e-08*	4.37e-08	0.061
Councillors	-6.705*	3.311	0.070
Coastal	0	omitted	
Year			
2019	-193.994***	42.404	0.001
2020	-271.278***	62.146	0.001
2021	-286.481***	80.976	0.005
2022	-235.993**	91.412	0.027
_cons	-778.239	4410.533	0.863
N	55		
Prob>F	0.000		
R^2			
Within	0.659		
Between	0.428		
Overall	0.232		
Rho	0.943		
Sigma_u	322.277		
Sigma_e	79.423		

Note: ***, ** and * = significance at 1%, 5%, 10% level, respectively. SE = Standard error.
Source: Authors' own table, 2024

significance level. Density and Indigent were statistically insignificant, suggesting that they were not important determinants of disaster incidents in the context of the FE model estimated in this study. On the other hand, Coastal which was a dummy for whether the municipality is located on the coast or not was omitted due to multicollinearity. This is usually the case with FE which commonly omits one variable, especially dummy variables when some multicollinearity exists (Breuer and deHaan, 2023; Kropko and Kubinec, 2020).

The positive and statistically significant coefficient of equitable share suggests that municipalities receiving more equitable share grants from the national government also have higher disaster incidents. While this is not consistent with our *a priori* expectations, this could be because the equitable share grant is given to municipalities so they can be able to provide basic services. As such, municipalities with limited own-revenue generation opportunities as well as larger numbers of poorer households receive higher equitable share grants (CoGTA, 2023). Therefore, the positive coefficient of the equitable share variable can be interpreted to mean that poorer and overly populated municipalities are more vulnerable to the prevalence of disaster and risk incidents.

Differently, the negative coefficient for councillors suggests that having more councillors reduces the prevalence of disaster incidents. Thus, the higher the number of councillors, the lower would be the prevalence of disaster incidents, *ceteris paribus*. This revelation is consistent with our *a priori* expectations where an inverse relationship between disaster incidents and the number of councillors was expected. The reasoning behind our expectations were that having more councillors in a municipality allows for swift responses to actions that mitigate disaster incidents. The average number of councillors for the sample was 156, where the municipality with the least number of councillors in the sample had 103 councillors, while the one with the highest had 222 councillors. The average population per councillor in the sample was 6 451 people, while the average households per councillor is 1 597. Thus, each councillor in the sample served 6 451 people or 1 597 households. Therefore, the negative coefficient for councillors reported in the estimated model suggests that reducing the population/households per councillor by increasing the number of councillors reduces the prevalence of disaster incidents, *ceteris paribus*.

CONCLUSION

The findings of this study underscore the significant impact of socio-economic variables on the occurrence of disaster incidents in KwaZulu-Natal, South Africa. Through the analysis of panel data spanning from 2018 to 2022, the research has shed light on crucial insights that can inform policy interventions aimed at mitigating the prevalence and severity of such incidents.

Firstly, the study reveals a strong correlation between certain socio-economic factors and the frequency of disaster incidents. Variables such as income level, housing quality, and access to infrastructure emerge as significant predictors of vulnerability to disasters. These findings emphasize the importance of addressing socio-economic disparities in disaster preparedness and response strategies.

Secondly, the research highlights the role of governance and urban planning in mitigating disaster risks. Factors such as land-use planning, building regulations, and emergency management infrastructure significantly influence the resilience of communities to disaster events. Effective governance structures and proactive urban planning can play a pivotal role in reducing the vulnerability of populations to disasters.

Lastly, the study underscores the need for multi-sectoral collaboration and community engagement in disaster risk reduction efforts. By involving local communities, civil society organizations, and government agencies in decision-making processes, it is possible to develop context-specific and culturally appropriate strategies for disaster resilience. Engaging stakeholders from diverse sectors fosters a holistic approach to disaster risk management, ensuring that interventions are inclusive and sustainable.

Building upon the insights gleaned from this study, several recommendations can be proposed to enhance disaster risk reduction efforts in KwaZulu-Natal and similar regions. The following salient recommendations encompass addressing socio-economic disparities, strengthening governance structures, and fostering community resilience.

Invest in Early Warning Systems and Community Education

- ❖ Invest in early warning systems and climate information services to enhance climate change response and adaptation measures
- ❖ Implement and enhance early warning systems to provide timely alerts about potential disasters. Early warning systems must actively involve communities at risk to ensure a constant state of preparedness and enable early action.
- ❖ Conduct regular community education programs to ensure that residents understand the risks and know how to respond appropriately.

Infrastructure Development

- ❖ Invest in resilient infrastructure to withstand natural disasters, such as flood-resistant buildings, reinforced bridges, and stormwater drainage systems.
- ❖ Upgrade and maintain critical infrastructure, including roads, water supply systems, and energy grids, to minimize damage during disasters.

Community Engagement and Empowerment

- ❖ Foster community engagement and empower local communities to actively participate in disaster preparedness and response efforts.
- ❖ Establish community-based organizations focused on disaster risk reduction and response, ensuring representation from diverse socio-economic groups.

Social Safety Nets

- ❖ Develop and strengthen social safety nets to provide financial assistance and support to vulnerable populations affected by disasters.
- ❖ Ensure that existing social programs are adaptable to emergency situations and can quickly reach those in need.

Insurance and Risk Financing

- ❖ Encourage the use of insurance and risk financing mechanisms to help individuals, businesses, and local governments recover more quickly after a disaster.
- ❖ Promote community-based insurance schemes that consider the specific socio-economic context of KwaZulu-Natal.

Climate-Smart Agriculture

- ❖ Promote climate-smart agricultural practices that enhance resilience to climate-related disasters, such as drought-resistant crops and sustainable water management techniques.
- ❖ Provide training and support for farmers to adopt resilient farming practices.

Education and Skills Development

- ❖ Invest in education and skills development programs to improve the socio-economic status of the population, making them more resilient to the impacts of disasters.
- ❖ Focus on education about sustainable practices and environmental conservation.

Foster Collaborations With Scientific Bodies and Institutions

- ❖ The need to foster collaborations with scientific bodies and institutions that are involved in knowledge generation and management is imperative to monitor the development of hazards and provide long-lasting solutions that are informed by scientific evidence and trends.
- ❖ It involves a knowledge cycle of access to, interpretation, communication and use of relevant, accurate and reliable disaster risk information by targeted user communities and their feedback on how the information is used.

Government Policy and Planning

- ❖ Integrate disaster risk reduction and climate change adaptation into regional development plans and policies.
- ❖ Ensure that land-use planning takes into account the vulnerability of different areas to disasters and guides responsible development.

International Collaboration

- ❖ Collaborate with international organizations and neighbouring regions to share knowledge, resources, and best practices in disaster risk reduction and response.
- ❖ Engage in joint research and development initiatives to address common challenges.

Monitoring and Evaluation

- ❖ Establish a robust monitoring and evaluation system to assess the effectiveness of disaster risk reduction strategies and socio-economic development initiatives.
- ❖ Use feedback from communities to continuously improve and adapt interventions.

By implementing these recommendations, KwaZulu-Natal can enhance its resilience to disasters and promote sustainable development, taking into account the unique socio-economic context of the region.

In summary, the findings of this study have significant implications for policy and practice in disaster risk reduction. By addressing underlying socio-economic vulnerabilities, strengthening governance structures, and fostering community resilience, it is possible to enhance the capacity of populations to withstand and recover from disaster events. Moving forward, concerted efforts must be made to translate these research findings into actionable policies and interventions, ultimately building more resilient and sustainable communities in KwaZulu-Natal and beyond.

REFERENCES

Allam, A. S., Bassioni, H. A., Kamel, W., & Ayoub, M. (2020). Estimating the standardized regression coefficients of design variables in daylighting and energy performance of buildings in the face of multicollinearity. *Solar Energy*, *211*, 1184–1193. doi:10.1016/j.solener.2020.10.043

Andrew, M. (2018). Revisiting community-based disaster risk management. In *Reducing Disaster Risks* (pp. 42–52). Routledge.

Anton, S. G., & Nucu, A. E. A. (2020). The effect of financial development on renewable energy consumption. A panel data approach. *Renewable Energy*, *147*, 330–338. doi:10.1016/j.renene.2019.09.005

Bega, S. (2022, March). Southern Africa's environment migrants are the forgotten refugees. Mail and Guardian. *Environment*, *19*. https://mg.co.za/environment/2022-03-19-southernafricas-environment-migrants-are-the-forgotten-refugees/

Bell, A., Fairbrother, M., & Jones, K. (2019). Fixed and random effects models: Making an informed choice. *Quality & Quantity*, *53*(2), 1051–1074. doi:10.1007/s11135-018-0802-x

Blaikie, P., Cannon, T., Davis, I., & Wisner, B. (2003). The challenge of disasters and our approach. *Environmentalism: Critical Concepts*, *4*, 233.

Blaikie, P., Cannon, T., Davis, I., & Wisner, B. (2014). *At risk: natural hazards, people's vulnerability and disasters*. Routledge. doi:10.4324/9780203714775

BreuerM.deHaanE. (2023). Using and interpreting fixed effects models. *Available at* SSRN. doi:10.2139/ssrn.4539828

Callaway, B., & Karami, S. (2023). Treatment effects in interactive fixed effects models with a small number of time periods. *Journal of Econometrics*, *233*(1), 184–208. doi:10.1016/j.jeconom.2022.02.001

Chabana, D. (2020). *Family influence on youth disaster risk and vulnerability perceptions in a South African context* (Doctoral dissertation, North-West University (South Africa)).

CoGTA. (2023). *Municipalities received their first tranche of the Local Government Equitable Share.* Available at: https://www.cogta.gov.za/index.php/2023/07/13/municipalities-received-their-first-tranche-of-the-local-government-equitable-share/

CoGTA – KZN. (2022). *April 2022 inclement weather conditions report. Report No: 1.22 of May 2022*. Provincial Disaster Management Centre.

Cole, D. H. (2017). Laws, norms, and the Institutional Analysis and Development framework. *Journal of Institutional Economics*, *13*(4), 829–847. doi:10.1017/S1744137417000030

Croissant, Y., & Millo, G. (2019). *Panel data econometrics with R*. Wiley.

Cumming, G. S., & Allen, C. R. (2017). Protected areas as social-ecological systems: Perspectives from resilience and social-ecological systems theory. *Ecological Applications*, *27*(6), 1709–1717. doi:10.1002/eap.1584 PMID:28618079

Cutter, S. L. (2010). *Social science perspectives on hazards and vulnerability science*. Springer Netherlands.

Cutter, S. L., Ash, K. D., & Emrich, C. T. (2014). The geographies of community disaster resilience. *Global Environmental Change*, *29*, 65–77. doi:10.1016/j.gloenvcha.2014.08.005

Cutter, S. L., Boruff, B. J., & Shirley, W. L. (2012). Social vulnerability to environmental hazards. In *Hazards vulnerability and environmental justice* (pp. 143–160). Routledge. doi:10.4324/9781849771542

Cutter, S. L., Emrich, C. T., Webb, J. J., & Morath, D. (2009). Social vulnerability to climate variability hazards: A review of the literature. *Final Report to Oxfam America, 5*, 1-44.

Del Lama Soares, F. (2023). Climate Change and Disaster Risk Management Policy Integration in Pacific Island Countries: Trajectories and Trends. In *Climate Change Strategies: Handling the Challenges of Adapting to a Changing Climate* (pp. 389–410). Springer Nature Switzerland. doi:10.1007/978-3-031-28728-2_18

Emrich, C. T., & Cutter, S. L. (2011). Social vulnerability to climate-sensitive hazards in the southern United States. *Weather, Climate, and Society*, *3*(3), 193–208. doi:10.1175/2011WCAS1092.1

Field, C. B. (Ed.). (2012). *Managing the risks of extreme events and disasters to advance climate change adaptation: special report of the intergovernmental panel on climate change*. Cambridge University Press. doi:10.1017/CBO9781139177245

Henningsen, A., & Henningsen, G. (2019). Analysis of panel data using R. In *Panel data econometrics* (pp. 345–396). Academic Press. doi:10.1016/B978-0-12-814367-4.00012-5

Hill, T. D., Davis, A. P., Roos, J. M., & French, M. T. (2020). Limitations of fixed-effects models for panel data. *Sociological Perspectives*, *63*(3), 357–369. doi:10.1177/0731121419863785

Houston, L. J. H. (2024). Climate action and the pursuit of environmental sustainability in Africa: An analysis of Africa's involvement in COP26 discussions. In *Human Rights and the Environment in Africa* (pp. 375–395). Routledge.

Hsiao, C. (2022). Analysis of panel data (No. 64). Cambridge University Press.

IISD (International Institute for Sustainable Development). (2023). The Reporting Service for Environment and Development Negotiations. *Earth Negotiations Bulletin*, *12*(842).

Imai, K., & Kim, I. S. (2021). On the use of two-way fixed effects regression models for causal inference with panel data. *Political Analysis*, *29*(3), 405–415. doi:10.1017/pan.2020.33

Interparliamentary Panel on Climate Change (IPCC). (n.d.). *The IPCC's Sixth Assessment Report Impacts, adaptation options and investment areas for a climate-resilient southern Africa*. https://cdkn.org/sites/default/files/2022-%2003/IPCC%20 Regional%20Factsheet_Southern%20Africa_Web.pdf

Ismaeel, S. S., Midi, H., & Sani, M. (2021). Robust multicollinearity diagnostic measure for fixed effect panel data model. *Malaysian Journal of Fundamental and Applied Sciences*, *17*(5), 636–646. doi:10.11113/mjfas.v17n5.2391

Izumi, T., Shaw, R., Djalante, R., Ishiwatari, M., & Komino, T. (2019). Disaster risk reduction and innovations. *Progress in Disaster Science*, *2*, 100033. doi:10.1016/j.pdisas.2019.100033

Kalnins, A. (2018). Multicollinearity: How common factors cause Type 1 errors in multivariate regression. *Strategic Management Journal, 39*(8), 2362–2385. doi:10.1002/smj.2783

Khalid, K., Usman, M., & Mehdi, M. A. (2021). The determinants of environmental quality in the SAARC region: A spatial heterogeneous panel data approach. *Environmental Science and Pollution Research International, 28*(6), 6422–6436. doi:10.1007/s11356-020-10896-9 PMID:32997249

Kropko, J., & Kubinec, R. (2020). Interpretation and identification of within-unit and cross-sectional variation in panel data models. *PLoS One, 15*(4), e0231349. doi:10.1371/journal.pone.0231349 PMID:32315338

Lavery, M. R., Acharya, P., Sivo, S. A., & Xu, L. (2019). Number of predictors and multicollinearity: What are their effects on error and bias in regression? *Communications in Statistics. Simulation and Computation, 48*(1), 27–38. doi:10.1080/03610918.2017.1371750

Lindner, T., Puck, J., & Verbeke, A. (2020). Misconceptions about multicollinearity in international business research: Identification, consequences, and remedies. *Journal of International Business Studies, 51*(3), 283–298. doi:10.1057/s41267-019-00257-1

Madubula, N. P. (2021). *A comparative analysis of the vulnerability of informal and formal households to disaster risks: The case of Bekkersdal* (Doctoral dissertation).

Malilay, J., Heumann, M., Perrotta, D., Wolkin, A. F., Schnall, A. H., Podgornik, M. N., Cruz, M. A., Horney, J. A., Zane, D., Roisman, R., Greenspan, J. R., Thoroughman, D., Anderson, H. A., Wells, E. V., & Simms, E. F. (2014). The role of applied epidemiology methods in the disaster management cycle. *American Journal of Public Health, 104*(11), 2092–2102. doi:10.2105/AJPH.2014.302010 PMID:25211748

Manyena, S. B. (2006). The concept of resilience revisited. *Disasters, 30*(4), 434–450. doi:10.1111/j.0361-3666.2006.00331.x PMID:17100752

Maslin, M. A., Lang, J., & Harvey, F. (2023). A short history of the successes and failures of the international climate change negotiations. *UCL Open Environment, 5*, 5. doi:10.14324/111.444/ucloe.000059 PMID:37475957

Moosa, B. (2021). *A policy framework towards fostering urban resilience: the case of KwaZulu-Natal, South Africa* (Doctoral dissertation, University of Johannesburg).

Murwirapachena, G., & Kabange, M. M. (2023). Can commercialisation address consumer debt in local government? A case study of South African metropolitan municipalities. *Canadian Journal of Development Studies. Canadian Journal of Development Studies*, 1–25.

Norris, F. H., Stevens, S. P., Pfefferbaum, B., Wyche, K. F., & Pfefferbaum, R. L. (2008). Community resilience as a metaphor, theory, set of capacities, and strategy for disaster readiness. *American Journal of Community Psychology*, *41*(1-2), 127–150. doi:10.1007/s10464-007-9156-6 PMID:18157631

Obergassel, W., Arens, C., Beuermann, C., Elsner, C., Hermwille, L., Kreibich, N., ... Spitzner, M. (2022). Last Call for 1.5 Degrees. *Carbon & Climate Law Review*, *16*(4), 225–242. doi:10.21552/cclr/2022/4/4

Ostrom, E. (2011). Background in institutional analysis and development framework. *Policy Studies Journal: the Journal of the Policy Studies Organization*, *39*(1), 7–27. doi:10.1111/j.1541-0072.2010.00394.x

Papke, L. E., & Wooldridge, J. M. (2023). A simple, robust test for choosing the level of fixed effects in linear panel data models. *Empirical Economics*, *64*(6), 2683–2701. doi:10.1007/s00181-022-02337-y

Partelow, S. (2018). A review of the social-ecological systems framework. *Ecology and Society*, *23*(4), art36. doi:10.5751/ES-10594-230436

Preiser, R., Biggs, R., De Vos, A., & Folke, C. (2018). Social-ecological systems as complex adaptive systems. *Ecology and Society*, *23*(4), art46. doi:10.5751/ES-10558-230446

Salmerón Gómez, R., García Pérez, J., López Martín, M. D. M., & García, C. G. (2016). Collinearity diagnostic applied in ridge estimation through the variance inflation factor. *Journal of Applied Statistics*, *43*(10), 1831–1849. doi:10.1080/02664763.2015.1120712

Sarabia, M. M., Kägi, A., Davison, A. C., Banwell, N., Montes, C., Aebischer, C., & Hostettler, S. (2020). The challenges of impact evaluation: Attempting to measure the effectiveness of community-based disaster risk management. *International Journal of Disaster Risk Reduction*, *49*, 101732. doi:10.1016/j.ijdrr.2020.101732

Sawalha, I. H. (2020). A contemporary perspective on the disaster management cycle. *Foresight*, *22*(4), 469-482.

Seddiky, M. A., Giggins, H., & Gajendran, T. (2020). International principles of disaster risk reduction informing NGOs strategies for community based DRR mainstreaming: The Bangladesh context. *International Journal of Disaster Risk Reduction*, *48*, 101580. doi:10.1016/j.ijdrr.2020.101580

Shrestha, N. (2020). Detecting multicollinearity in regression analysis. *American Journal of Applied Mathematics and Statistics*, *8*(2), 39–42. doi:10.12691/ajams-8-2-1

Sivakumar, V., & Fazel-Rastgar, F. (2021, February). Heavy rainfall resulting from extreme weather disturbances in eastern coastal parts of South Africa: 11 April 2022. In *Proceedings of the Earth and Environmental Sciences International Webinar Conference* (pp. 161-186). Cham: Springer International Publishing.

Spielman, S. E., Tuccillo, J., Folch, D. C., Schweikert, A., Davies, R., Wood, N., & Tate, E. (2020). Evaluating social vulnerability indicators: Criteria and their application to the Social Vulnerability Index. *Natural Hazards*, *100*(1), 417–436. doi:10.1007/s11069-019-03820-z

Stocktake, P. (2023). The Reporting Service for Environment and Development Negotiations. *Earth Negotiations Bulletin*, *12*(837), 2.

Thulare, M. H., & Moyo, I. (2021). COVID-19 and street traders in the City of uMhlathuze, KwaZulu-Natal, South Africa: On responses and adaptation mechanisms. *Cogent Social Sciences*, *7*(1), 2006392. doi:10.1080/23311886.2021.2006392

Tong, J., Xiucang, L., Qingchen, C., Jiashuang, Y., & Erda, L. (2014). Highlights and understanding of climate change 2014: Impacts, adaptation, and vulnerability. *Advances in Climate Change Research*, *10*(3), 157.

Twigg, J. (2015). *Disaster risk reduction*. Academic Press.

Van Assche, K., Verschraegen, G., Valentinov, V., & Gruezmacher, M. (2019). The social, the ecological, and the adaptive. Von Bertalanffy's general systems theory and the adaptive governance of social-ecological systems. *Systems Research and Behavioral Science*, *36*(3), 308–321. doi:10.1002/sres.2587

Van Niekerk, D., Nemakonde, L. D., Kruger, L., & Forbes-Genade, K. (2018). Community-based disaster risk management. Handbook of disaster research, 411-429.

Whaley, L. (2018). The critical institutional analysis and development (CIAD) framework. *The International Journal of the Commons*, *12*(2). doi:10.18352/ijc.848

World Bank. (2022). *South Africa Overview*. https://www.worldbank.org/en/country/southafrica/overview

Xie, Q., Wang, X., & Cong, X. (2020). How does foreign direct investment affect CO2 emissions in emerging countries? New findings from a nonlinear panel analysis. *Journal of Cleaner Production*, *249*, 119422. doi:10.1016/j.jclepro.2019.119422

Zakour, M. J., Gillespie, D. F., Zakour, M. J. & Gillespie, D. F. (2013). Vulnerability theory. *Community Disaster Vulnerability: Theory, Research, and Practice*, 17-35.

ENDNOTES

1 The South African map was extracted from https://en.m.wikipedia.org/wiki/File:Map_of_South_Africa_with_English_labels.svg

2 Detailed discussions of these variables are given in the next section.

3 Initially, only Equitable was converted to natural logarithms but correlation remained high.

4 Results for all pre-estimation tests performed in this study are presented as Appendix 1.

5 The datasets are publicly accessible online at: https://www.statssa.gov.za/?page_id=1859 and http://mfma.treasury.gov.za/Media_Releases/LGESDiscussions/Pages/default.aspx

6 The 2022 population densities for South African provinces are published on http://www.geo-ref.net/en/zaf.htm

7 The Rand is South Africa' local currency and as of 2 January 2024 the exchange rate was US$1 = R18.30.

8 Results for all pre-estimation tests performed in this study are presented in Appendix 1.

APPENDIX

Table 4. Other pre-estimation tests

	Test	Chi^2	F	p-Value
Heteroscedasticity	Breusch-Pagan/Cook-Weisberg	25.01	-	0.000
FE versus RE	Hausman	9.30	-	0.026
Time-fixed effects in FE	Testparm i	-	11.77	0.001

Chapter 3
Building Resilient Communities Through Social Capital in Cities:
Reflections From City of Harare, Zimbabwe

Elias Mabaso
Independent Researcher, USA

ABSTRACT

This chapter explores community resilience in urban environments focusing on communities in the city of Harare in Zimbabwe. Despite its increased use in research and practical application for disaster risk management, resilience is a complex and contested issue. In this work, resilience is examined using a concurrent nested mixed method approach for data collection and analysis through the use of random and purposive sampling and questionnaire surveys. Some of the key findings include the diverse understanding of resilience by communities and urban local authorities leading to interventions mostly by urban planners that do not holistically promote community resilience. Urban informality is not always a hazard, but can promote community resilience if well managed, and social capital is one of the most important pillars for supporting community resilience building. The chapter concludes by emphasizing the importance of risk localization and a holistic understanding of resilience at community levels if city authorities are to reduce disasters and strengthen community resilience.

DOI: 10.4018/979-8-3693-2721-0.ch003

INTRODUCTION

With the increasing magnitude and impacts of disasters the world over, governments and humanitarian organisations have implemented various strategies to mitigate impacts and reduce loss of lives. Urban communities have not been spared either from the impacts of disasters. By 2050, about 68% of the global population is projected to live in urban areas with the fastest growth likely to be in urban centres of developing countries (UN DESA, 2018). The disproportionate population living in cities is likely to lead to increased hazards and disasters. With such a high likelihood of increased urban risks, strategies that empower affected urban communities to effectively cope, respond, recover, and move forward in a sustainable way need to be implemented.

Despite contestations on the meaning of resilience, it is considered one of the strategies that can minimise the negative impacts of disasters on communities living in urban areas (Banica, 2020). A resilient community moves forward as they respond, adapt, recover, and transform livelihoods to sustainable levels (Kamara, et al., 2018). There is a need for a shift in disaster risk management strategies in urban areas if the unprecedented impacts of hazards are to be effectively addressed. It is imperative to implement strategies that promote adaptation, absorption, and transformation in the context of possible increase in disaster impacts. The strategies should not only focus on common resilience measures that include absorptive, responsive, and transformative capacities (Kamara, et al 2018), but also non-traditional measures and traits such as social capital and its role in promoting community resilience. Social capital as a resilience enabler is characterised by important resilience process measures that include networking, collaboration, learning, diversity of strategies and innovation (Rotarangi and Stephenson, 2015). The processes and or resilience pathways tend to be overlooked in urban resilience research and yet should be integral for a holistic understanding of resilience.

The chapter presents different resilience-enabling options for urban communities particularly in the face of climate change. Climate change is likely to cause a major shift in the livelihood patterns of urban communities as they grapple with increasing hazards and disaster impacts. In Zimbabwe, despite being regarded as one of the urban hazards, by urban local authorities (Kamete, 2004), informal urban agriculture has become the 'new normal' and a major urban land use depicting pronounced utilisation of social capital which promotes community resilience. The question to be addressed is whether social capital as reflected by communities who practise informal urban agriculture in Harare city of Zimbabwe plays a significant role in promoting disaster risk reduction and subsequently leading to community resilience.

Harare, Zimbabwe's largest city has one of the most widespread urban informality practices in southern Africa, particularly informal urban agriculture. Understanding the local drivers of risk and community resilience is pertinent if urban disaster risk reduction and planning interventions are to promote community resilience. There has been an overemphasis and reliance on international city resilience measures and approaches that have mostly been developed and applied in Western world cities. Some of the internationally developed tools and approaches for assessing resilience such as the United Nations Disaster Risk Reduction Office's Disaster Resilience Scorecard for cities, the Rockefeller Foundation City Resilience Index, and the World Bank's City Strength Diagnostic have limitations when applied at the community level. The limitations partly emanate from limited consideration of local risk drivers. The socio-economic, physical, and cultural local risk drivers can be resilience inhibiting or promoting factors hence the need for a good understanding of their significance at community level. A wholesale application of international measures without consideration of risk localisation, may lead to negative consequences and worsening of risks (Barnes, 2020).

Research on resilience tends to focus on the macro city level and less on various communities and individuals that constitute the city's diverse socio-economic and demographic profiles. Conclusions drawn from aggregated findings that are based on wider city studies on resilience may hide differences that characterise individuals and communities. Community resilience focused studies consider that people's adaptation, coping, transformative capacities, and processes are not the same, but determined by multiple drivers that can be social, physical, and political in nature (Barnes et al., 2020).

Important urban community resilience characteristics that include social capital have not received much research attention and analysis and yet they play an important role in the determination of overall resilience and community well-being. Most research on urban resilience does not systematically link urban resilience to urban risks and planning beyond natural hazard-induced disasters and engineering designs. City resilience studies also tend to be strongly biased towards engineering and ecological resilience with little attention to social and political resilience traits (Meerow et al, 2016),. The chapter's objectives include identifying hazards that normally lead to disasters and negatively affect communities in the city of Harare. This is considered in the context of the types and processes of social capital and its role in urban community resilience strengthening. The chapter advances possible disaster risk reduction and urban planning options that integrate social capital as an important planning ingredient in promoting community resilience.

The chapter is divided into 6 sections. The first section outlines the research context and objectives. The second section presents the research methodology highlighting the processes and tools used for data collection and analysis. The third section is a literature review on community resilience and social capital. The section explores the theoretical framework and conceptual analysis of resilience and social capital from global and local perspectives. Section four brings together the results of the analysis of data to establish trends and possible themes. Section five is an interpretation of the findings based on research objectives while the conclusion provides a synthesis of the key findings.

METHODOLOGY.

To enable a comprehensive coverage of the complex and diverse issues which constitute disaster risks, social capital, community resilience, a concurrent nested mixed methods approach was adopted throughout the processes from the data collection stage up to analysis. Concurrent nested mixed methods focus on collecting, analysing, and mixing quantitative and qualitative data in a single study with the data being collected and analysed simultaneously (Findlay and Li, 1999). This provided a better understanding and insight into the research issues under investigation as well as strengthening data reliability and validity.

A case study of Cold Comfort in Harare, a residential area that practices extensive and diverse informal urban agriculture was adopted. This enabled in-depth analysis and possible generalisations to the wider context of urban informality, social capital and planning as well as community resilience in Harare and Zimbabwe. Secondary data sources on disaster risk management, resilience, and social capital were reviewed and analysed. Primary data were collected through household survey and interviews with key informants, and opinion leaders following a random sampling of participants from the selected residential area. Structured and semi-structured questionnaires were used for the diverse targeted groups. Semi structured questions were used to collect data from professionals who included urban planners, university lecturers and NGO officials. Structured questionnaires were used to collect data from communities mostly involved in informal urban agriculture activities through focused group discussions and individual household interviews. Data analysis was integrated with data collection throughout the process as different themes emerged during the analysis.

The Importance of Community Resilience in Cities

The 21st century has experienced an escalation of urban challenges which in some cases have accentuated urban risks. Some of the urban risks include high incidences of diseases mostly linked to poor sanitation and contaminated drinking water and food insecurity particularly challenges of food access and utilisation leading to malnutrition and in some cases death (Lehmann, 2028; Elmqvist et al., 2021). Urban areas also experience increased flooding which in some cities has been worsened by increased rainfall associated with climate change (Tabari,2020; Shamsuddin, 2020).

As traditional disaster management has been mostly dominated by response, resilience thinking and processes continue to be strongly advocated for in urban planning to enable disaster affected people to prepare for, respond, absorb and adapt to disasters impacts. Meerow et al., (2016) observed that urban resilience despite its conceptual fuzziness has become increasingly a favoured concept as cities continue to grow and grapple with uncertainties and challenges.

Manyena (2014) highlighted the complexity surrounding resilience when he refers to resilience as ubiquitous with multiple faces and multiple spaces.

Despite the seemingly increased focus on promoting community resilience as a solution to addressing some of the urban disaster impacts, the diverse conceptualisation, and interpretations of resilience by scholars and practitioners reveal the loaded nature of resilience. This conceptual haziness makes application and measurement of community resilience in urban settings problematic (Meerow et al., (2016). Resilience will remain a critical consideration in urban disaster risk reduction and urban planning particularly if urban authorities are to effectively address some of the myriad of urban risks likely to be worsened by climate change. It is therefore pertinent to explore and understand the current and possible future etymological and practical application of resilience in urban environments and how it is promoted or hindered by social capital.

Resilience and Disaster Risk Reduction

Despite the seemingly clear link between resilience and disasters, unpacking the relation between Disaster Risk Reduction (DRR) and resilience has generated a lot of debate among disaster scholars. According to UNISDR (2015), DRR is one of the pre-requisites for promoting resilience. However, resilience is more than DRR as it focuses on what enables individuals and households to function in adverse or disruptive conditions now and in the future (UNDRR, 2016). DRR on the other hand tends to focus on minimising impact of future disasters and not much on how communities will continue to function thereafter (UNISDR, 2015). The limitation of UNISDR 's approach to DRR and resilience seems to be its increased focus on resilience to large scale natural hazard induced disasters with little or no consideration of resilience issues in conflict related disasters. What seems to be common among disaster scholars and practitioners is that resilience is normally conceptualised within the framework of DRR, with resilience being an outcome of effective DRR policies and practices.

An understanding of disaster causes, and philosophical foundations is pertinent for better conceptualisation of the link between resilience, disasters, and disaster risks. In the early history of mankind, disasters were predominantly understood as Acts of God. The theistic view perceived disaster causes from a supernatural perspective with human beings having no control on what was considered divine retribution on humans for their evil deeds (Wisner, et al., 2012). The theistic view has been criticised for making man subservient to some force outside his control hence taking no objective measures to counter what was perceived as supernatural forces beyond human control. This approach to addressing disaster impacts had a bearing not only on preparedness and response mechanisms, but also resilience measures which were centred on pleasing the deity so that he would not punish his subjects by disasters.

A counter perspective to the theistic view of disaster causes was the objectivist view which was propounded in the 1930s up to 1960s (Chipangura et al., 2017). This was influenced by the deterministic philosophical perspective which stipulated that human actions and outcomes are determined by their environment. Disasters were viewed as acts of nature or mainly caused by natural hazards which could be quantified and addressed through scientific approaches. Response mechanisms tended to focus on technical solutions such as construction of flood defence systems with the belief that man could easily control disaster impacts by controlling the hazard (Chipangura, et al., 2017). The hazard or objectivist view has been challenged for emphasizing technical solutions without considering socio-political issues that may exacerbate disaster causes and effects (Chipangura et al., 2017).

The possibilist school of thought challenged the deterministic or hazard focused construct by viewing disasters from a social perspective. From a possibilist or constructivist perspective, disasters are a result of the intersection of hazards and vulnerability with social as well as economic issues accentuating vulnerability instead of being determined by natural hazards (Blaikie, et al., 2004). The Pressure and Release model (PAR) proposed by Blaikie et al (2004) illustrates disaster cause and effects from a constructivist perspective. The model explains occurrence of disasters as an interaction between natural hazards and vulnerable people (Blaikie, et al., 2004). There is need to understand the connections between the impact of hazards on people and social factors as well as processes that generate vulnerability to understand issues of disaster occurrence and possible trajectories.

UNISDR (2015) defines DRR as interventions that are aimed at prevention of new and reducing existing disaster risk and managing residual risk, all of which contribute to strengthening resilience and therefore to the achievement of sustainable development. The conclusion to be drawn from the definition is that DRR is not an end but should lead to resilience and finally contribute to sustainable development.

Resilience Typologies and Contestations

The difficulty in identifying a common understanding of resilience seems to stem from the original use and application of the term. Scholarly and practitioners' definitions of resilience tend to reflect disciplinary focus and bias. Some of the definitions of resilience are based on conceptualisation of resilience from a natural sciences perspective. Some of the natural sciences disciplines that have influenced definition of resilience include ecology, engineering, and physics. Other resilience definitions stem from social sciences that include sociology, psychology, and political sciences.

A seemingly common understanding among scholars is that the term resilience originated from Latin '*resilire*' which means to 'leap back' or bouncing back to a steady state (Merilainen, 2019).

Some selected definitions of community resilience are provided below.

While there are several resilience typologies, the above definitions show four broad resilience groups which are engineering, ecological, socio-ecological and evolutionary resilience.

Engineering view of resilience emphasises stability of the system as well as single equilibrium after a disturbance, that is how fast the disturbed system returns to its original state (Folke,2006). The disturbance could be a disaster such as an earthquake or social related challenges such as financial crisis or social upheavals. In addition to the emphasis on equilibrium or stable state after disturbance, other attributes of the engineering resilience construct include persistence, resistance, and predictability (Davidson et al. 2016).

Table 1. Resilience definitions

	Resilience Definition	Author	Disciplinary Focus
1	Ability of systems to resist against external influences and return to a state of equilibrium or stable state.	Folke, (2006)	Engineering
2	Ability of communities to withstand external shocks to their social infrastructure.	Adger (2000)	Sociological
3	The ability of the community to bounce back, respond to, recover from, and absorb the impacts and cope with adverse circumstances whether economic, environmental, or social.	Ainuddin and Routray (2012)	Multi-disciplinary (integrated)
4	The ability of an urban system and all its constituent socio-ecological; and socio-technical networks across spatial scales-to maintain or rapidly return to desired functions in the face of a disturbance, to adapt to change and to quickly transform systems that limit current or future adaptive capacity.	Merrow et al., (2016)	Multi-disciplinary
5	The ability of at-risk individuals, households, communities and systems to anticipate, cushion, adapt, bounce back better and move on from the effects of shocks and hazards in a manner that protects livelihoods and recovery gains, and supports sustainable transformation.	UNDP, 2015	Socio-ecological
6	The ability of a system, community, or society exposed to hazards to resist, absorb, accommodate to, and recover from the effects of a hazard promptly and efficiently by preserving and restoring essential basic structures.	UNISDR, 2016)	Integrated (multi-disciplinary)

Source: Classifications developed by author based on textual analysis (2020)

Unlike engineering resilience which emphasises stable equilibrium, ecological resilience approach acknowledges the possibility of systems shifting to a new equilibrium state if it is to retain its pre-disaster functionality (Sharif and Yamagata, 2016). The ecological view of resilience attempts to address the weaknesses inherent in the single stable state and equilibrium by including multiple stable states which may not lead to equilibrium (Davidson, et al., 2016).

Due to the emphasis on stability, the engineering and ecological resilience approaches have been criticised for being too conservative with a tendency of focusing much on endogenous (internal) stressors and little attention on external stressors (Davoudi, 2013). In addition, Davoudi (2013) points out that the approach fails to account for political and power relations and social dynamics within complex systems that tend to characterise urban systems. Brown (2014) notes that the ecological approach to resilience tend to promote a technical approach-imposed rationality that is alien to the practice of ordinary people. The technical approach seems to be influenced by the hazard conceptualisation of disasters which focuses more on scientific solutions to disaster impacts without considering the diversity of causes and trajectories. Some of the causes could be political or socio-cultural and may not be easily amenable to technical solutions. The limitations of the engineering and ecological approach to the understanding of resilience, particularly failure to consider issues of political ecology, which emphasizes role played by the human elements that include politics and culture in determining change of any system led to the socio-ecological approach as well as the evolutionary and inter-disciplinary approaches. These approaches tend to depict several common traits with a wide range of common indicators that can be used to assess community resilience. The socio-ecological typology analyses issues of ecology and infrastructure concerns in resilience and how they relate to psychological, sociological, and political factors (UNDP, 2015).

Socio-ecological conceptualisation of resilience emphasizes adaptive and dynamic nature of urban resilient systems which constantly undergo transformations (Coaffee and Lee, 2016). The system may not return to an equilibrium or stable state after the disturbance but is characterised by self-organising capacity and learning which enables the resilient system to not only bounce back, but also bounce forward (Manyena et al., 2009; Sharifi and Yamagata, 2016; Coaffee and Lee, 2016). Miller et al., (2010) contend that the challenge with a socio-ecological approach to understanding systems is the usual disconnect between social issues and ecological variables. Their view is that there can be tension between social resilience and ecological resilience since social resilience may be realised at different scale, such as at community level while ecological resilience through extended feedbacks may be operational at a larger or extended scale beyond the community level and may even transcend national borders.

The evolutionary approach to resilience considers the nature of the constantly changing non-equilibrium systems to embrace changes and adaptations that ensure continuous system functionality (Coaffee and Lee, 2016). The approach attempts to address limitations of the engineering and ecological resilience approaches which are deeply rooted in the 'bouncing back' conceptualisation of resilience. The bouncing back understanding of resilience has been criticised as being too narrow and seemingly advocating for return of affected communities to the same state that led them to being negatively affected by a disaster. Shaw (2012) advocates for a more proactive conception of resilience as leaping forward, which seemingly embraces the evolutionary resilience approach. Evolutionary resilience emphasises institutional and governance issues particularly how organisational flexibility and learning can be promoted together with public awareness (Coaffee and Clarke, 2015).

While there is no single agreed approach to conceptualisation and measurement of community resilience, the common aspect in most of the definitions of community resilience is a focus on capacity or ability to continue functioning after some disturbance that may cause change or a shift from the original state of the individual or community circumstances.

Resilience and Social Capital

Social capital which has also been narrowly considered as social cohesion is one of the core resilience traits that needs great attention if resilience of urban communities is to be comprehensively understood. From a resilience perspective, social capital looks at existence of unifying forces or linkages that can promote resilience (Wilson et al.,2020). In addition, social capital considers how communities collaborate and apply learning, social networks, collective efforts when dealing with hazards and disasters to help them absorb, adapt, transform, and realise sustainable wellbeing (Tammar, Shongi and Rahaman, 2020). Social capital has also been defined as the value embedded in relationships which include social networks, cohesion or connectedness, norms, trust and reciprocity (Aldrich and Meyer, 2015). Table 1.2 provides more definitions of social capital.

Social capital can be bonding, which refers to personal relations that are based on a sense of collective identity such as family or close friendship (Mpanje et al., 2018). Examples of bonding capital include assistance or help that households or individual can receive from people within their communities. In dealing with hazards and disasters, bonding capital is useful for helping to absorb the impacts of hazards and or disasters as affected people receive support from or provide support to affected neighbours. Bridging social capital on the other hand is when individuals or communities connect with others who though different may have similar political views, religion or from the same race (Kyne and Aldrich, 2019). Bridging capital

Table 2. Social capital definitions

	Definition	Author
1	"The sum of the resources, actual or virtual that accrues to an individual or a group by virtue of possessing a durable network of more or less institutionalised relationships of mutual acquaintance and recognition."	Bourdieu and Wacquant, (1992 p.119)
2	"The web of social relationships that influence individual behaviour and thereby affect economic growth."	Pennar (1997p.154)
3	"Social capital refers to social values (norms) networks and trust that facilitate a group's purposeful action."	Falk (2000 p.1)
4	Networks of relationships among people who live and work in a particular society, enabling that society to function effectively.	Hellerstein and Neumark, (2020)
5	Social capital refers to networks with bonding, bridging and linking capacity to facilitate cooperative, collective, inclusive action and reduce opportunistic behaviour. It represents a range of key resources that exist in social relationships, networks, links connections, associations, customs, and norms.	Neal (2019)

Source: Compiled by author based on textual analysis of various secondary data sources (2020)

can promote diversity and innovation which are important for resilience building and or strengthening.

What can clearly be deduced from the various definitions is the focus on the strengthening of relationships and collaboration of the community which is important for resilience building. Social capital can enhance networks, and standards (Tammar, Abosuliman and Rahaman, 2020) which can lead to the strengthening of collaboration, partnerships and information sharing, which constitute some of the important social capital resilience measures. Tammar, Abosuliman and Rahaman (2020) further identify creativity as one of the main elements of social capital which is also important for resilience strengthening. They define creativity as the ability of adapting to new circumstances and learning from the disaster experience for communities to improve their situation during the recovery stages. However, this definition is limited since creativity as a resilience element is not only limited to recovery but, can also be important in preparedness and disaster response. Increasing a community's social capital both bonding and bridging can help to strengthen and enhancing community well-being. Communities lacking social cohesion tend to reflect higher rates of social disorder, anxiety, and depression which is likely to have a negative impact on the community's resilience (Tammar, Abosuliman and Rahaman, 2020).

Enhancing community resilience requires approaches well beyond just technical or infrastructural interventions which tends to be the focus of ecological and engineering resilience but should also include consideration of diverse social and psychological factors. This is important considering that most disasters are a social construct being primarily human caused and central to some of the causes are socio-political issues

(O'Keefe et al., 1976). Social capital is closely linked to other resilience approaches and measures. Wilson, et al (2020) advanced a similar notion by retorting that adaptation, can be moderated by social structures and dynamics that may include, shared social norms and strong social networks which are elements of bridging social capital. These can help in information sharing and other reciprocal relationships with important implications for individuals and community resilience. Strong social ties can help to ameliorate negative health outcomes from natural hazard events, thus enhancing community resilience. This shows the significance of understanding the various components of social capital and how it can impact different resilience processes and outcomes. As indicated by Wilson et al. (2020) community cohesion and connectedness which are some of the major elements of social capital can build a sense of collective memory that reduces the uncertainties inherent in decisions pertaining to infrequent hazards. They also conclude that community cohesion can heighten the sense that preparation for a hazard is important, improve the process of identifying relevant resources or information, and foster more rapid knowledge transfer and risk-appropriate social re-organization. This contributes to strengthening of resilience elements for cohesive members of community.

Social capital also plays an important role in risk reduction which can contribute to resilience through knowledge sharing, clear communication, social learning and people–place connections (Wilson et al., 2020). However, some researchers have indicated the negative effects of social capital for promoting adaptation to risks of climate change. For example, high bonding social capital strengthened the misperception among elderly residents in the United Kingdom that they were not at risk from heat waves, further increasing their vulnerability (Wolf et al, 2010).

Whilst research on resilience has tended to focus on ecological resilience, the issue of soft resilience traits that include social capital cannot be ignored and play a significant role in holistically understanding community resilience (Panday et al.,2021).

COMMON HAZARDS AND DISASTERS IN COLD COMFORT, HARARE

Understanding what community perceives or considers as the main hazards and disasters provides a good basis for determining resilience measures of coping, adaptation, and transformative strategies the communities have been employing over the years. It also helps to understand the types and traits of social capital that promote resilience. Without an understanding of the common hazards, the suggested resilience measures may not relate to the lived realities of the communities. Figure

1.1 shows the main hazards and disasters reported by communities in Cold Comfort residential area of Harare City, Zimbabwe. The findings corroborate with the Zimbabwe Vulnerability Assessment Committee (2020) (ZIMVAC) assessment of hazards in other parts of Harare city. Drought, veld fires, theft and financial crisis were reported as hazards of greater concern.

Most communities identified financial crises as the highest-ranking hazard affecting their agriculture activities and general livelihoods. Drought and dry spells were ranked second with 92% of the sampled communities identifying it as a major hazard that affect their agriculture and other livelihoods. There were notable differences on prioritisation of hazard types which seemed to be determined by location of the community in the residential area. Most community members practising informal urban agriculture and with pieces of land for cultivation located on wetland areas did not prioritise drought but floods as the main hazard affecting them. Communities on the hill slopes ranked drought as one of the most important hazards to be addressed. A very small number of community members, 5% considered soil erosion as a hazard of concern.

Community members practising informal urban agriculture along the stream highlighted floods as a major hazard. The wetland area is subject to seasonal flooding especially during years with normal to above normal rainfall. As expected, community members utilising hill slopes for crop cultivation did not consider flooding as a hazard but blamed those who practised farming in wetlands and along the stream of having settled in harm's way.

Asked whether it was not a problem to be farming on hill slopes and under electrical lines, the community members indicated that this did not create any environmental challenges since the land was not being used for any other purpose. They justified by indicating that if they did not utilise the open land for crop cultivation, tall grasses would grow posing risks of fires and become habitat for poisonous snakes that will be a risk to the family since the local authority or city council did not regularly cut grass.

One of the challenges which some community members considered a disaster was theft of farm produce particularly green maize mealies and vegetables grown outside the residential premises. This corroborates with findings by ZIMVAC (2020) assessment that identified theft of agriculture produce as a major problem for most urban communities living in the high-density residential areas of Harare city.

Communities' Perspective of Resilience and Disaster

When asked what they understood by resilience, respondents provided a wide range of definitions some shown on Table 1.3. The definitions indicate how perceptions on disasters and resilience from a community and individual perspective differ from

Table 3. Community definitions of resilience and disaster

Resilience	Disaster
Doing something that helps you.	An unexpected happening that leads to suffering including loss of life and requires government or NGOs to help.
Ability to respond to disasters without help from NGOs and government.	Something that causes death or suffering of the family or community.
Protecting oneself from hunger.	Something that comes from God as punishment for sins.
Having enough food for the family.	What causes death or suffering to a community or family.
Access to resources to meet family needs during times of disasters.	National problem that affects everybody and no one seems to be able to deal with it.
Having a good life meeting all your family needs (well-being).	When there is shortage of food or water.
Something done to reduce climate change.	An accident or danger that one encounters when not prepared.
Living within your means.	This means floods in other countries, but we have no disaster in Zimbabwe.
Staying for a long time without being affected by disasters.	Some events or process that cause massive destruction and suffering.
Helping each other during times of challenges.	

Source: Author from focus group discussions (2019)

conventional views normally held by the academic community and organisations that work with disaster affected people. This can have major implications on implementation of interventions to promote community resilience.

Most of the definitions consider disasters from a natural hazard perspective while a few consider the human agent. Of interest are the broad perspectives on resilience which in the widely spoken local Shona language is not easy to translate into a single word unlike disaster which communities understood as *"njodzi"* which is closer in meaning to "danger" or "hazard". There is no distinction between hazard, risk and disaster in Shona language and all are referred to as *njodzi*. Community perspectives of resilience highlight both the narrow and broad scope of issues related to one's immediate experiences and responses. Some of the community definitions of disasters and resilience consider the broader issues of resilience and aspects of social capital, for example, helping each other during times of challenges and working with relatives and friends to withstand effects of disasters.

Community Adaptation and Response to Common Disasters

This section highlights the diversity of mechanisms and strategies employed by communities to adapt and respond to common disasters which in some cases translate into resilience outcomes. Some of the strategies reflect resilience traits of adaptative, absorptive, and transformative processes. While residents of Cold Comfort who practice informal urban agriculture prioritised several hazards that may lead to disasters (Figure 1.1), this section focuses on two hazards that most communities indicated as being of high frequency and normally cause disasters in Cold Comfort residential area. The two hazards discussed are food insecurity and financial crises. These hazards are however not mutually exclusive and impacts of one may lead to worsening of the other hazard and or lead to a disaster of a different type. For example, financial crises can lead or worsen food insecurity as communities fail to get adequate financial resources to buy agricultural inputs and or supplement their food requirements. This therefore means that some of the adaptation and disaster response mechanisms that communities implement are common for the two hazards.

Adapting and Responding to the Impacts Food Insecurity

Food security refers to a situation when all people at all times access sufficient and nutritious food to maintain a healthy and active life (International Food Research Institute, 2002). Communities in Cold Comfort residential area of Harare responded to the impact of food insecurity through informal trading (81%), borrowing (24%) reducing meals (37%), and skipping meals (36%). Figure 1.2 shows some of the community adaptation and response strategies. This also confirms findings by ZIMVAC (2020) assessments in surrounding residential areas of Harare city where major mechanisms of reducing impacts and coping with food insecurity included skipping meals, reducing quantity of meals eaten and relying on less expensive food stuffs.

While food availability is not a major challenge since food is available on the market including the informal market, the major challenge is the limited access due to the high food prizes that are beyond the reach of some community members. Communities indicated that food was easily available particularly in the informal markets even during drought, but normally beyond the reach of many due to the high costs. Unlike high-density residential areas, the residents of Cold Comfort do not receive government or Non-Governmental Organisations assistance since the assumption is that the residents have the means to meet their food requirements as can be deduced from a comment by one of the community members.

Assistance is given to those who are very poor in areas such as Mbare and Mufakose residential areas. People who live in Cold Comfort are not poor and most can afford to meet their family requirements without assistance.

However, some community members indicated that because of financial difficulties and many unemployed community members, food needs and vulnerability assessments must cover all residential areas and people in need should receive the same assistance provided to people in the high-density residential areas. They argued that where one stays should not be used as a criterion for determining vulnerability and needs.

Responding to Financial Crises

Community members in Cold Comfort indicated financial crisis as the most common hazards that affected them. Community members have devised several strategies to reduce impact and respond to financial crisis as highlighted in Figure 1.3. Some of the common financial challenges that have led to a financial crisis include shortage of cash both local currency and other acceptable currencies since Zimbabwe uses a multi-currency system, declining value of the local currency when compared to the major currencies such as the US dollar and declining formal employment opportunities which contributed to loss of earnings. Most informal traders do not use electronic money transactions such as debt or credit cards but prefer cash transactions. Vegetable and tomato venders also prefer buyers to use cash rather than plastic money or debit cards transactions as well as cash transfers using mobile phones.

Some of the strategies communities employ to reduce risks and adapt as well as respond to financial crises as shown in Figure 1.3 include borrowing (41%), support from relatives and friends (30%) and informal trade (31%). There are no NGOs offering support in Cold Comfort unlike neighbouring areas as highlighted in the ZIMVAC 2020 report. However, like the ZIMVAC (2020) findings in other surrounding areas of Harare showed that remittances are also an important source of support in Cold Comfort with 9% of interviewed community members indicating that remittances have helped them to mitigate impacts of financial crises. Sending money from outside the country helps community members to get disposable cash for daily needs when compared to accessing money through the local banking system which always experiences cash shortages.

Community members do not only borrow money to address challenges of financial crises, but some have borrowed inputs such as maize seeds, fertilisers, and food with an anticipation of returning when conditions improved especially when they are involved in informal trade to augment their earnings. A response by one of the community members summed up the significant role informal trade plays to alleviate financial challenges.

Everyone has turned into a businessperson now and everything can be sold to get money. There is no family that is not involved in selling something and the whole Cowie Street in Cold Comfort has turned into a business centre where one can buy almost everything from basic food items to even household electrical goods and furniture.

This was also observed by this researcher during transact walks with almost every road intersection having a market stall selling an assortment of commodities that included tomatoes, vegetables, maize grain, fruits, clothing, plates, and mobile airtime, Figure 1.4. A visit to Harare city centre also showed that informal trading was widespread during evening when council law enforcement officials have gone to their houses. Some of the streets in the city centre were turned into market centres with some totally blocked from traffic as informal traders from different residential areas converge to sell products taking advantage of the high population concentration in the city centre.

Social Capital in Cold Comfort

The common social networks or binding social capital identified by communities include money lending clubs, church groups, informal networks for information sharing (WhatsApp group). Communities indicated that they actively participated in informal networks that support agriculture activities particularly social media chat groups for information sharing. There were no community members who practised informal urban agriculture who indicated that they belonged to organised formal or official government established farming groups. Communities indicated that some social groups particularly church affiliated groups were important for information sharing and providing support during harvesting of crops and field preparation. Collaboration through social networks was highlighted as important for crop and animal management information sharing. Community members used social network both within and outside their residential area to exchange information on disease control, crop varieties and weather conditions and diseminating warning alerts to other community members when council officials were spotted in some areas raiding farm produce being sold in undesignated areas. Some also used the social networks for marketing produce such as chicken and mushrooms. Figure 1.6 shows common community social capital processes for communities that practice informal urban agriculture in Cold Comfort.

Figure 1.5 shows some of the major lessons that communities who practise informal urban agriculture highlighted as important for promoting resilience in the context of recurrent hazards and disasters. Importance of support networks which are a form of binding social capital was highlighted as an important lesson by 9% of the interviewed community members. The various support network included religious associations, money lending clubs and farming groups. Some of the support networks are not formal, for example, community members indicated that they belonged to a WhatsApp social media group for sharing agriculture information. The support networks were considered the most important source of agriculture information including warning group members of pending destruction of crops by city council officials. Second to social networks was the importance of diversification with 97% of the sampled communities indicating that experiences in conducting urban agriculture had taught them the importance of diversifying crop production and other income generating activities beside agriculture.

Importance of saving was indicated by 72% of the communities practicing urban agriculture as necessary to strengthen financial status and responding to other future disasters. This does not only refer to saving money but farming produce for use during the following agriculture season. Despite the seemingly significant understanding of the laws that govern land use in urban areas including urban farming, learning urban by-laws was the least important for Cold Comfort farming communities. Only 4% indicated that they had learned about requirements for animal production such as minimum chickens to be kept without a need for a permit from the urban local authority. Some of the communities who indicated that they had learned about urban by-laws considered some of the urban laws as restrictive and not good for informal urban agriculture, for example, restrictions on the number of chickens to be kept even if one has adequate space. Other lessons learned by 44% of the interviewed community members included the importance of collaborating with other neighbours for protection of crops from thieves. Community members emphasised the importance of networking with other farmers to learn how to address challenges.

Urban Resilience Not a One Size Fits All

Community resilience is not a permanent trait that is acquired once and remains static, but changes due to changing socio-economic, environmental, and physical factors. Some disasters are not recurrent, for example, floods in some of the residential areas in the City of Harare, Zimbabwe. After periods of continuous unreliable rainfall interspersed with drought for almost five years, some communities in Harare showed limited understanding of flood risk reduction and response mechanisms. This is likely to have a bearing on resilience trait to be developed or strengthened over time. Similarly, resilience traits depicted for one type of disaster are not easily transferable or applied to another type of disaster. While some social capital components may promote resilience processes or pathways for responding to financial crises and possibly drought elements, for example, networking processes, collaboration and innovation, some would not be suitable for applying to flood response. In this case, communities may still need to employ diversity of strategies for flood response or limited collaboration which may not be the same collaborative process applied when responding to food insecurity or financial crises.

While informal urban agriculture may reduce disaster risk by enhancing food security for some urban communities during a particular season (Warren et al., 2015), that should not be translated as an indication of well-developed resilience for every community member. Resilience should go beyond episodic or temporary coping with a hazard and should show long term ability to absorb, adapt and cope leading to sustainable well-being (Manyena, 2006). Resilience is also context specific and place specific, as determined by the predominant hazards (Manyena, 2014). In Cold Comfort residential area of Harare, communities who settled along the river valley and in wetlands had different set of skills on diversity and flexibility compared to communities on dryland areas who relied on seasonal rainfall. The communities who practised informal urban agriculture activities along the river valley prioritised flood disaster risk reduction and response strategies while communities living on hill slopes did not consider flood as a hazard hence did not prioritise flood risk reduction strategies. With time this is likely to affect development of flood resilient traits for communities living on hill slopes if they were to be exposed to floods. Communities or individuals may show certain resilience traits such as adaptation or coping and yet may not show transformative resilience traits.

Urban communities who practice informal urban agriculture in Cold Comfort residential area of Harare city, utilise bonding social capital mostly collaboration to fight hazards of wildfires. This also applies when there are diseases and pests outbreak affecting their maize plants. Communities show increased social networking by use of social media to share information on pest control. Disasters of low magnitude with less impact such as periodic dry spells lead to less application of diverse response strategies and less utilisation of binding social capital. In such cases, urban communities can easily absorb and adapt to the impacts without radically transforming their daily activities such as by reducing the number of meals consumed per day.

Resilience characteristics are embedded in other local processes and strategies such as culture, knowledge, and experiences which are common traits of social capital as shown on Figure 1.6. These characteristics make the building blocks for disaster resilience enabling communities to bounce forward after a disaster.

Social Capital as a Resilience Enabler

Social capital can be a resilience node (outcome) and not solely a pathway/enabler or (process). This means that urban communities who show strong social capital traits are likely to be resilient to a wide range of disasters. The indicators of social capital include social networks, collaboration, response support systems and shared values (Kyne and Aldrich, 2019). Figure 1.6 shows common indicators of social capital depicted by residents of Cold Comfort in Harare. While these are not a complete list of all possible measures of social capital for assessing resilience, they however provide a comprehensive coverage to make valid conclusions on social capital and resilience.

Communities in Cold Comfort residential area of Harare employed bonding social capital traits to respond to drought and financial crisis. The bonding social capital practices included money lending clubs and agriculture input support groups. For example, sharing traditional/indigenous seed varieties not easily obtainable from the formal markets and collaboration for fighting fires. Communities in Cold Comfort residential area have several social networks both formal and informal which they have activated over the years during and after disasters. The informal networks that include WhatsApp agriculture information sharing groups are used to share information on pests and diseases as well as available markets for agriculture products. The informal networks have helped farmers sustain operations during disasters and addressing impacts of hazards that include drought, financial crisis as well as pests and diseases.

Without proper understanding of resilience enablers such as bonding social capital, interventions to strengthen resilience outcomes such as adaptive and absorptive capacity may fall short and or undermine strengthening of resilience. Bridging capital refers to whether households can get support from people living outside the community (Kyne and Aldrich, 2019). These could be relatives, government service providers such as Agricultural Extension workers, city planners and council officials who provide information on urban land use and financial institutions that provide loans. For communities in Cold Comfort that practise informal urban agriculture, there are no well-developed systems and structures that reflect operations of bridging capital. Communities who practise informal urban agriculture do not have access to agriculture extension services neither do they get assistance from government and NGO's during times of crisis such as food insecurity. Communities in some high-density residential areas of Kuwadzana, Dzivarasekwa and Warren Park received assistance to address food insecurity during the lean season. Some of the major social capital components that are significant pathways to facilitate resilience of communities in Cold Comfort as highlighted in Figure 1.6 include social networks in the form of bonding and bridging capital, participation in agriculture related social activities, respecting values and norms that hold the people of Cold Comfort together such as not stealing from neighbours and helping neighbours when they face challenges.

Findings from communities in Cold Comfort are in line with Twigg's conclusions that a focus on resilience should put more emphasis on what communities can do for themselves and how to strengthen their capacities rather than concentrating on their vulnerabilities to disasters or their needs in emergencies Twigg (2007).

The capacities in the case of Cold Comfort are depicted in the various types of social capital that farmers employ such as bonding social capital for information sharing to address challenges paused by crop pests and diseases, and to inform other community members on market trends. Community members also show capacities to respond to fire hazards and have developed resilience traits to deal with recurrent fire hazards through strengthening collaboration to put out fires. Urban communities also employ social capital particularly collaboration to protect their produce from thieves who were reported to be a menace in most areas of the city especially during harvest times.

Risk Informed Urban Planning for Community Resilience

Research on urban agriculture tend to limit analysis to role of agriculture in promoting food security as well as economic and environmental impact particular for the urban poor. This limited analysis which has also tended to inform urban planning misses a critical component of the role of informal urban agriculture in building resilience particularly in the context of increasing urban risks due to climate change induced hazards and vulnerabilities. Resilience for urban communities practising informal agriculture goes beyond meeting food security needs which can be temporarily achieved using outside intervention such as humanitarian aid by NGOs and government.

Urban planning has tended to focus on physical components and technical solutions through land use planning and infrastructure development (Davoudi, et al., 2012). Ironically, agriculture land use has not been seriously considered as an important land use for urban planning in Zimbabwe despite it occupying a large portion of urban land use, in Harare city especially during the rainy season. Urban planners need to balance issues of physical planning with social vulnerability analysis considering that disasters are a social construct (O'Keefe, et al. 1976). Lack of consideration of the social vulnerabilities in planning compromises people's ability to cope with and respond to hazards which may negatively affect resilience building. The challenge for urban planners has been considering informal urban agriculture as a hazard and not compatible with city aesthetic value (Kamete, 2004). There has not been much effort in investing and taking consideration of the possible positive role of informal urban agriculture in promoting resilience from a social and even environmental perspective. For example, urban farming could be used to reduce runoff in some bare surfaces which may lead to increased soil erosion in some of the non-vegetated areas and hill slopes common in most residential areas of Harare city. With increased build up areas and tarmac surfaces, issue of urban heat islands become a hazard and increases risks which may have negative health effects on the city residents (Memon, et al., 2008). Promoting urban agriculture can therefore be one of the ways of greening the city and contribute to reducing temperatures of exposed areas of the city (Humaida, et al., 2023).

Urban planners should move away from the 'ritualistic' planning practices strongly driven by obsession with beautifying the city and by so doing undermining issues of social capital which greatly contributes to people's resilience. Resilience and risk informed urban planning approach should consider issues of community participation including cultural values as integral in urban planning which can contribute to community resilience building. City resilience should balance issues of 'soft' resilience measures with 'hard' resilience which tends to be strongly inclined towards infrastructure and not the social measures such as networking skills, social cohesion, flexibility and efficiency. This integrated approach to resilience should be informed by consideration of diversity of socio-cultural and economic variables that are at play at local level. The starting point should be an understanding of local drivers of risks particularly local hazards, Figure 1.1, vulnerabilities, and capacities. This risk localisation helps to inform community and individual's disaster response ability as well as ability to bounce forward in the context of recurrent disasters.

Applying international urban planning approaches and tools for addressing challenges experienced by urban communities who practice informal agriculture and other urban informality for livelihood sustenance has several limitations and challenges. Some of the international tools and approaches fail to contextualise the various local drivers of change that determine risk levels and resilience building. Hazards can be context specific, and this tends to also shape the response mechanisms and how these can strengthen resilience. For Cold Comfort residential area in Harare, hazards such as veld fires and theft of agriculture produces that have become a recurrent challenge have led farmers to device response mechanisms for several years and some of the mechanisms indicate resilience traits, for example, social networks and collaborative mechanisms, Figure,1.2 and Figure 1.6. Most of the international measures that include UNDDR's disaster resilience score card, tend to be applicable at macro city level and pose challenges when applied at community or micro levels. There can also be local hazards, see Figure 1.1, and drivers of risks that may not easily fit in the adopted international framework due to their complexity, for example, local political drivers. While planers may want to use rationality and follow the book in line with international norms and standards, the issue of 'political correctness' may override these considerations. In this case, the politicians who may be interested in short term gains and attracting votes tend to encourage people to settle on wetlands which are then subsequently utilised for agriculture. Communities who settle in such zones will then adjust some of their adaptive and or absorptive capacity to reflect local realities which may not be easy to capture if one applies international tools to assess risks and resilience without consideration of local dynamics or risk localisation. Therefore, trying to understand resilience from a limited technical view can create challenges.

Risk informed urban planning should adopt an integrated risk analysis which considers not only structural or physical risk drivers but also non -structural interventions that include early warning and promoting social capital for urban risk management. Understanding of hazards and vulnerability analysis for planning intervention must go beyond natural hazards and include political and social drivers of change that may have impact on people's resilience. Some of the drivers of change can be important pathways to resilience nodes, for example, social networks common in Cold Comfort as a form of bonding capital can be an important pathway to coping and risk absorption over time leading to resilience.

City residents may show high levels of resilience to one disaster and yet very low to another, Figure 1.2 and Figure 1.3, hence the importance of considering the multi-dimensional and multi-sector nature of resilience not only as 'bouncing back' but 'leaping forward' in case of any disaster that may affect the urban community (Manyena, et al., 2009). A holistic understanding of risks should not only be informed by focusing on vulnerability of infrastructure to disasters, but how these interact with the human elements particularly the socio-political drivers of change. An approach that factors in a broad understanding of these factors can strengthen integrated urban planning for resilience. In some cases, socio-cultural and political factors may be more determinant factors of city changes including city residents' response to various disasters which in the long run may have a bearing on their resilience.

With climate change likely to worsen hazards and disasters that include floods and food insecurity in urban areas (Tabari, 2020; Shamsuddin, 2020) planners need to become more innovative and change their rigid and at times too ritualistic planning approaches that may contribute to urban disaster risk creation and compromise community resilience.

CONCLUSION

Understanding urban community resilience requires a holistic approach to the assessment of a diverse range of hazards and local urban risk drivers. Imported notions of resilience which in most cases tend to be Eurocentric and imposed on communities by humanitarian organisations may not always be useful to promote resilience since they tend to be divorced from local cultural values and norms. Understanding some of the socio-cultural norms and values can be significant in promoting resilience while disregarding the cultural norms and values in trying to promote resilience may lead to resistance and possibly accentuate disaster risk creation.

Urban land use planning has tended to focus on physical components and technical solutions to urban challenges. Urban planners need to balance issues of physical planning with social vulnerability analysis considering that disasters are a social construct. Lack of consideration of the social vulnerability drivers in urban planning tends to compromise people's ability to cope with and respond to hazards which negatively affect resilience building.

Urban planners should move away from the 'ritualistic' planning heavily driven by obsession with beautifying the city and by so doing undermining issues of social capital which is an important enabler of community resilience.

Applying international measures of resilience for understanding resilience of communities who practice informal urban agriculture tend to have limitations since some fail to contextualise the various parameters that determine resilience. Hazards can be context specific which tend to also shape the response mechanisms and how these can strengthen resilience. Therefore, trying to understand resilience from a limited technical view can create or even worsen hazards leading to disaster risk creation rather than promoting community resilience.

ACKNOWLEDGMENT

Some of the findings of this research are based on data collected for my unpublished Doctor of Philosophy (PhD) thesis. I acknowledge the support and guidance provided by my two supervisors, Professor Desmond Manatsa of Bindura University of Science Education and Professor Andrew Collins of Northumbria University in the United Kingdom

REFERENCES

Adger, W. N. (2000). Social and ecological resilience: Are they related? *Progress in Human Geography*, *24*(3), 347–364. doi:10.1191/030913200701540465

Ainuddin, S., & Routray, J. K. (2012). Earthquake hazards and community resilience in Baluchistan. *Natural Hazards*, *63*(2), 909–937. doi:10.1007/s11069-012-0201-x

Aldrich, D. P., & Meyer, M. A. (2015). Social capital and community resilience. *The American Behavioral Scientist*, *59*(2), 254–269. doi:10.1177/0002764214550299

Banica, A, Kourtit, K. & Nijkamp, P. (2020) Natural disaster as a development opportunity: a spatial economic resilience interpretation. *Review of Regional Research, 40*(3).

Barnes, M. L., Wang, P., Cinner, J. E., Graham, N. A., Guerrero, A. M., Jasny, L., Lau, J., Stutcliffe, S. R., & Zamborain-Mason, J. (2020). *An Invitation to Reflexive sociology*. University of Chicago Press.

Brown, K. (2014). Global environmental change I: A social turn for resilience? *Progress in Human Geography, 38*(1), 107–117. doi:10.1177/0309132513498837

Chipangura, P., van Niekerk, D., & Van der Waldt, G. (2017). Disaster Risk Policy Problem Framing: Insights from Societal Perceptions in Zimbabwe. *International Journal of Disaster Risk Reduction*, *22*, 317–324. doi:10.1016/j.ijdrr.2017.02.012

Coaffee, J., & Clarke, J. (2015). On securing the generational challenge of urban resilience. *The Town Planning Review, 86*(3), 249–255. doi:10.3828/tpr.2015.16

Coaffee, J., & Lee, P. (2016). *Urban resilience: planning for risk, crisis and uncertainty*. Palgrave. doi:10.1007/978-1-137-28884-4

Davidson, J., Jacobson, C., Lyth, A., Dedekorkut-Howes, A., Baldwin, C., Ellison, J., Holbrook, N. J., Howes, M. J., Serrao-Neumann, S., Singh-Peterson, L., & Smith, T. F. (2016). Interrogating resilience: Toward a typology to improve its operationalization. *Ecology and Society*, *21*(2), 27. doi:10.5751/ES-08450-210227

Davoudi, S., Brooks, E., & Mehmood, A. (2013). Evolutionary resilience and strategies for climate adaptation. *Planning Practice and Research, 28*(3), 307–322. doi:10.1080/02697459.2013.787695

Davoudi, S., & Porter, L. (2012). Applying the resilience perspective to planning: Critical thoughts from theory and practice. *Planning Theory & Practice, 13*(2), 299. doi:10.1080/14649357.2012.677124

Elmqvist, T., Andersson, E., McPhearson, T., Bai, X., Bettencourt, E., Brondizo, J., Colding, G., Daily, G., Folke, C., Grimm, N., Haase, D., Ospina, D., Parnell, S., Polasky, S., Seto K.C. and Van Der Leeuw, S. (2021) Urbanization in and for the Anthropocene. *Nature Partner Journals/Urban Sustainability, 1*(6), 1-6. . doi:10.1038/s42949-021-00018-w

Falk, I., & Kilpatrick, S. (2000). What Is Social Capital? A Study of Rural Communities. *Sociologia Ruralis*, *40*(1), 87–110. doi:10.1111/1467-9523.00133

Findlay, A. M., & Li, F. L. N. (2004). Methodological issues in researching migration. *The Professional Geographer*, *51*(1), 50–59. doi:10.1111/0033-0124.00144

Folke, C. (2006). Resilience, The emergence of a perspective for social-ecological systems analyses. *Global Environmental Change*, *16*(3), 253–267. doi:10.1016/j.gloenvcha.2006.04.002

Hellerstein, J. K., & Neumark, D. (2020). Social Capital, Networks, and Economic Wellbeing. IZA DP No. 13413, Bonn, Germany.

Humaida, N., Saputra, M. H., & Hadiyan, Y. (2023). Urban gardening for mitigating heat island effect. *IOP Conference Series. Earth and Environmental Science*, *1133*(1), 012048. doi:10.1088/1755-1315/1133/1/012048

Kamara, J. K., Wali, N., Agho, K., & Renzaho, A. M. N. (2018). Resilience to climate induced disasters and its overall impact on well-being in Southern Africa: A mixed methods systematic review. *International Journal of Environmental Research and Public Health*, *15*(11), 2375. doi:10.3390/ijerph15112375 PMID:30131077

Kamete, A. (2004). Home industries and the formal city in Harare, Zimbabwe, Reconsidering informality: *Perspectives From Urban Africa*, 120–138.

Kyne, D., & Aldrich, D.P. (2019). Capturing bonding, bridging and linking social capital through publicly available data. *Risks, Hazards and Crisis in Public Policy*, 1-26.

Lehmann, S. (2018). Implementing the urban nexus approach for improved resources-efficiency of developing cities in Southeast-Asia, City. *Cultura e Scuola*, *13*, 46–56. doi:10.1016/j.ccs.2017.10.003

Manyena, B. (2009). *Disaster resilience in development and humanitarian interventions* [PhD Thesis]. Northumbria.

Manyena, B. (2014). Disaster resilience: A question of 'multiple faces' and 'multiple spaces? *International Journal of Disaster Risk Reduction*, *8*, 1–9. doi:10.1016/j.ijdrr.2013.12.010

Manyena, B. (2016). After Sendai: Is Africa Bouncing Back or Bouncing Forward from Disasters? *International Journal of Disaster Risk Science*, *7*(1), 41–53. doi:10.1007/s13753-016-0084-7

Manyena, B., O'Brien, G., O'keefe, P., & Rose, J. (2011). Disaster resilience: A bounce back or bounce forward ability? *Local Environment*, *16*(5), 417–424. doi: 10.1080/13549839.2011.583049

Manyena, S. B. (2006). The concept of resilience revisited. *Disasters, 30*(4), 433–450. doi:10.1111/j.0361-3666.2006.00331.x PMID:17100752

Meerow, S., Newell, J. P., & Stults, M. (2016). Defining urban resilience: A review. *Landscape and Urban Planning, 147*, 38–49. doi:10.1016/j.landurbplan.2015.11.011

Meerow, S. J. P., & Newell, J. P. (2016). Urban resilience for whom, what, when, where, and why. *Urban Geography, 40*(3), 309–329. doi:10.1080/02723638.2016.1206395

Memon, R. A., Leung, D. Y. C., & Chunho, L. (2008). A review on the generation, determination and mitigation of urban heat island. *Journal of Environmental Sciences (China), 20*(1), 120–128. doi:10.1016/S1001-0742(08)60019-4 PMID:18572534

Merilainen, E. (2019). The dual discourse of urban resilience: Robust city and self-organised neighbourhoods. *Disasters, 44*(1), 125–151. doi:10.1111/disa.12367 PMID:31231816

Miller, F., Osbahr, H., Boyd, E., Thomalla, F., Bharwani, S., Ziervogel, G., & Nelson, D. (2010). Resilience and vulnerability: Complementary or conflicting concepts? *Ecology and Society, 15*(3), 1–11. doi:10.5751/ES-03378-150311

Mpanje, D., Gibbons, P., & McDermott, R. (2018). Social capital in vulnerable urban settings: An analytical framework. *Journal of International Humanitarian Action, 3*(4), 4. Advance online publication. doi:10.1186/s41018-018-0032-9

Neal, J., & Neal, Z. (2019). Implementation capital: Merging frameworks of implementation outcomes and social capital to support the use of evidence-based practices. *Implementation Science : IS, 14*(16), 16. Advance online publication. doi:10.1186/s13012-019-0860-z PMID:30764850

O' Keefe, P., Westgate, K., & Wisner, B. (1976). Taking the naturalness out of natural disasters. *Nature, 260*(5552), 566–567. doi:10.1038/260566a0

Panday, S., Rushton, S., Karki, J., Balen, J., & Barnes, A. (2021). The role of social capital in disaster resilience in remote communities after the 2015 Nepal earthquake, International. *Journal of Disaster Reduction, 55*, 102112.

Pennar, K. (1997). *The tie that leads to prosperity: The economic value of social bonds is only beginning to be measured.* Business Weekly.

Rotarangi, S. J., & Stephenson, J. (2014). Resilience pivots: Stability and identity in a social-ecological-cultural system. *Ecology and Society*, *19*(1), 28. doi:10.5751/ES-06262-190128

Shamsuddin, S. (2020) Resilience resistance: The challenges and implications of urban resilience implementation. *Journal of Cities*. . doi:10.1016/j.cities.2020.102763

Sharifi, A., & Yamagata, Y. (2016). Urban Resilience Assessment: Multi Dimensions Criteria and Indicators. In Y. Yamagata & H. Maruyama (Eds.), *Urban Resilience* (pp. 259–276). Springer. doi:10.1007/978-3-319-39812-9_13

Shaw, K., & Maythorne, L. (2012). Managing for local resilience: Towards a strategic approach. *Public Policy and Administration*, *28*(1), 43–65. doi:10.1177/0952076711432578

Tabari, H. (2020). Climate change impact on flood and extreme precipitation increase with water availability. *Scientific Reports*, *10*(1), 13768. doi:10.1038/s41598-020-70816-2 PMID:32792563

Tammar, A., Abosuliman, S. S., & Rahaman, K. R. (2000). Social Capital and Disaster resilience nexus: A study of flash floods recovery in Jeddah City. *Sustainability (New Rochelle, N.Y.)*, *12*, 1–16.

Twigg, J. (2007). *Characteristics of a disaster resilient community*. http://www.proventionconsortium.org/themes/default/pdfs/characteristics/commu nity_characteristics_en_lowres.pdf

UN DESA. (2018). *Revision of World Urbanization Prospects*. United Nations.

UNDP. (2015). *Building Resilience in Zimbabwe: Towards a Resilience Strategic Framework. Zimbabwe Resilience Building Fund (ZRBF)*. UNDP Zimbabwe.

UNISDR. (2011). *Hyogo Framework for Action 2005-2015, Building the Resilience of Nations and Communities to Disaster, Mid-term review. 2010-2011*. UNISDR.

UNISDR. (2015). *Sendai framework on disaster risk reduction 2015–2030*. UN Office for Disaster Risk Reduction.

Warren, E., Sophie Hawkesworth, S., & Knai, C. (2015). Investigating the association between urban agriculture and food security, dietary diversity, and nutritional status: A systematic literature review. *Food Policy*, *53*, 54–66. doi:10.1016/j.foodpol.2015.03.004

Wilson, R. S., Herziger, A., Hamilton, M., & Brooks, J. S. (2020). From incremental to transformative adaptation in individual responses to climate-exacerbated hazards. *Nature Climate Change*, *10*(3), 200–208. doi:10.1038/s41558-020-0691-6

Wisner, B., Blaikie, P., & Cannon, T. (2004). *At risk: Natural Hazards, People's vulnerability and disasters*. Routledge, Elsevier.

Wisner, B., Gaillard, J. C., & Kelman, I. (2012). *Handbook of Hazards and Disaster Risk Reduction and Management*. Routledge. doi:10.4324/9780203844236

Zimbabwe Vulnerability Assessment Committee. (2020). *Zimbabwe Urban Vulnerability*. ZIMVAC.

Chapter 4
Re-Imagining Community-Based Disaster Risk Reduction Towards Building Resilient Rural Communities

Francis Ziwanai
Durban University of Technology, South Africa

ABSTRACT

Zimbabwe is prone to a variety of natural and climate-related hazards, which disproportionately affect communities. While community-based disaster risk reduction (CBDRR) initiatives exist in the country, their effectiveness and sustainability in building resilient communities remain limited. There is a need to re-imagine CBDRR approaches in Zimbabwe to address the existing gaps and challenges and enhance community resilience. The aim of the research was to investigate innovative approaches and strategies that can enhance the effectiveness of CBDRR initiatives, particularly in rural areas that are prone to various natural hazards. The research employed a qualitative approach where primary and secondary data sources were used to gather comprehensive insights on the topic. Research findings identified gaps and challenges existing in CBDRR practices which include ineffective community-based early warning systems, centralization of disaster management, inadequate trainings, lack of preparedness, inadequate resources, reactive intervention, and policy gaps. Proposed frameworks and interventions to address these issues were establishment of trained local disaster risk management committee, education and training, preparedness, capacitation of supporting institution, funding, and proactive disaster management intervention in CBDRR efforts. Overall, this research contributes to the ongoing discourse on disaster risk reduction and resilience-building, offering valuable insights and practical recommendations for improving the outcomes of

DOI: 10.4018/979-8-3693-2721-0.ch004

community-based initiatives in rural contexts. The research provides recommendations and guidelines for policymakers, practitioners, and community leaders to implement these innovative approaches and build resilient rural communities in the face of increasing disaster risks.

INTRODUCTION

Disasters pose significant challenges to communities worldwide, and Zimbabwe is no exception. According to Gogo (2020), the country faces a range of natural and climate-related hazards, including droughts, floods, cyclones, and landslides, which have severe impacts on the lives, livelihoods, and infrastructure of its population. In response to these challenges, Community-Based Disaster Risk Reduction (CBDRR) has emerged as a critical approach to building resilience at the local level (Oxfam 2019). CBDRR recognizes the importance of community participation, local knowledge, and context-specific interventions in reducing vulnerabilities and enhancing disaster preparedness (Gogo, 2020).

Chatiza (2020) provided that, while CBDRR initiatives exist in Zimbabwe, there is a need to re-imagine and strengthen these approaches to effectively address the complex and evolving risks faced by communities. The re-imagining of CBDRR involves envisioning innovative strategies, policies, and practices that prioritize community empowerment, foster collaboration among stakeholders, and leverage emerging technologies and best practices. By doing this, it seeks to ensure sustainable development and protect the welfare of Zimbabwean citizens by constructing resilient communities that can endure and recover from disasters (Gogo, 2020).

The chapter explored and proposed a re-imagined CBDRR framework in Zimbabwe that is tailored to the local context and aligns with international best practices. By addressing the existing gaps and challenges in CBDRR implementation, the analysis undertaken seeks to contribute to the development of effective strategies for building resilient communities in Zimbabwe. Through a comprehensive analysis of existing CBDRR practices, policies, and programs in Zimbabwe, this chapter identified areas for improvement and propose evidence-based recommendations for enhancing community resilience.

Ultimately, this study aims to add to the conversation around CBDRR in Zimbabwe by offering useful perspectives and doable suggestions for practitioners, politicians, and community leaders. Zimbabwe can make great progress toward creating resilient communities that are more equipped, more adaptable, and able to lessen the effects of disasters on their lives, livelihoods, and infrastructure by rethinking CBDRR and bolstering its implementation.

RESEARCH METHODOLOGY

For this qualitative study, primary and secondary sources of data were used. Primary data denotes first time collected data whilst secondary data is the already collected and produced data by other researchers (Ajayi, 2023). Primary data is recent, reliable and objective. For primary data, interviews were used for data collection. Secondary data is published data with a degree of validity and reliability (Martins et al., 2018). It is a profound and informed data by expertise and professional researchers making it reliable and valid data. Employing both primary and secondary data provided an increased validity and reliability of collected data (Ajayi, 2023).

For primary data collection, interviews were employed with rural community members in Chirumhanzu, researchers and academics in the field of disaster management and government officials dealing with disaster management. This approach helped in capturing a wide range of perspectives and experiences related to disaster risk reduction in rural areas. Interviews were important in augmenting secondary data as well as juxtaposing people's experiences and perspectives with already published data on disaster risk reduction and building of resilient communities. For secondary sources, published articles, journals, books, records and databases were used.

Research Context

Zimbabwe, located in Southern Africa, is a country that faces numerous natural and climate-related hazards, posing significant risks to its communities. The nation is vulnerable to frequent cyclones, floods, droughts, and other extreme weather disasters that have a catastrophic effect on the infrastructure, economy, and people (Oxfam, 2019). These risks worsen pre-existing vulnerabilities, such as food shortages, poverty, and restricted access to essential services, so undermining community resilience (Gogo, 2020).

In response to these challenges, Community-Based Disaster Risk Reduction (CBDRR) has gained recognition as a crucial approach to building resilience at the local level (Ebba, 2022). CBDRR emphasizes the active involvement of communities in identifying risks, developing mitigation strategies, and implementing disaster preparedness measures. Communities are empowered to take charge of their disaster risk reduction activities by CBDRR, which makes them more resilient and better prepared to respond to and recover from catastrophes (Gogo, 2020). This is achieved by harnessing local knowledge, capacities, and resources.

However, despite the existence of CBDRR initiatives in Zimbabwe, there are several contextual factors that need to be considered when re-imagining and strengthening these approaches (Ebba 2022). The success and sustainability of

CBDRR initiatives are influenced by the socioeconomic and political environment of the nation, which is marked by extreme poverty, inequality, and difficulties with governance. Implementing CBDRR successfully is hampered by a lack of funding, inadequacies in local institutions' capacity, and disjointed coordination between stakeholders (Gogo, 2020).

Additionally, Zimbabwe's diverse geographical and climatic conditions require tailored approaches that account for regional variations in hazards, vulnerabilities, and community dynamics. Due to the urban-rural divide and the large proportion of the people living in rural areas, rural communities confront particular obstacles, such as limited access to infrastructure, services, and information (Nyahunda et al., 2022). As a result, specific solutions are needed to address these issues.

Furthermore, international frameworks and agreements, such as the Sendai Framework for Disaster Risk Reduction and the Sustainable Development Goals (SDGs), provide guidance and targets for disaster risk reduction and sustainable development. Both the Sendai Framework and the SDGs emphasize the importance of community engagement, participatory approaches, and local knowledge in reducing disaster risks and building resilient communities. They call for collaborative efforts among different stakeholders, including communities themselves, to achieve sustainable and resilient development. Aligning CBDRR in Zimbabwe with these global frameworks can ensure coherence, effectiveness, and access to international support.

In this regard re-imagining of CBDRR in Zimbabwe towards building resilient communities is critical calling for the investigation of the distinct requirements, viewpoints, and abilities of nearby communities; appraise current CBDRR procedures and guidelines; investigate novel techniques and technologies; review organizational structures; and promote favorable policy modifications (Nyahunda et al., 2022). The chapter contributes to the creation of context-specific strategies and suggestions that can improve community resilience, lessen vulnerabilities, and advance sustainable development in the face of current and foreseeable dangers by addressing the particular circumstances of Zimbabwe.

Research Problem

Zimbabwe is prone to a variety of natural and climate-related hazards, which disproportionately affect communities. While Community-Based Disaster Risk Reduction (CBDRR) initiatives exist in the country, their effectiveness and sustainability in building resilient communities remain limited. There is a need to re-imagine CBDRR approaches in Zimbabwe to address the existing gaps and challenges and enhance community resilience. Key issues include the inadequate integration of community perspectives, the underutilization of innovative approaches and technologies, weak institutional capacity, limited policy support, and a lack of coordination and collaboration among stakeholders. Thus, there is a pressing need for research to identify and propose context-specific strategies that prioritize community participation, enhance institutional capacity, advocate for policy changes, and foster innovation to ensure the effective implementation of CBDRR initiatives and the development of resilient communities in Zimbabwe.

LITERATURE REVIEW

Resilience Theory

In the context of disaster risk reduction (DRR), the resilient theory focuses on the capacity of communities and systems to anticipate, absorb, adapt to, and recover from shocks and stresses caused by disasters (Gülhan 2021). It acknowledges that catastrophes are unavoidable and that, in the face of persistent hazards, fostering resilience is crucial to reducing their effects and advancing sustainable development.

Resilience theory acknowledges that communities and systems need to be flexible and adaptable to cope with changing conditions and uncertainties caused by disasters (Alexander, 2013). The term "adaptive capacity" describes the capacity to change course and deal with new situations skillfully. Disasters provide distinct difficulties of varying degrees of intensity, necessitating the ability to adapt. In order to enhance readiness and response tactics, it entails growing knowledge and abilities, incorporating new information, and learning from prior experiences (Wenger, 2017).

The theory recognizes that there are multiple pathways to achieve resilience and that diverse strategies and approaches can be effective in reducing risks. Various strategies and approaches can be employed to enhance resilience in the face of disasters (Wenger, 2017). The idea of many routes acknowledges that there isn't a disaster risk reduction strategy that works for everyone. It highlights the value of redundancy, which is the availability of multiple resources, systems, or operations to increase resilience. For example, when it comes to building disaster resilience, there are a number of different approaches that can be taken, such as investing in early warning systems, enhancing social networks and community preparedness, strengthening governance and policy frameworks, and putting money into education and capacity building (Gülhan, 2021). This acts as a safety net and makes it possible for communities to continue carrying out necessary tasks even in the face of disruptions.

Resilience theory acknowledges that communities and systems are interconnected and interdependent. It emphasizes the need to understand the relationships and dependencies between different elements within a system, such as social networks, infrastructure, and ecosystems (Alexander, 2013). All of these need to be utilized in concert with one another to reduce the risk of disasters and create resilient communities. Understanding these relationships makes it easier to spot any cascade effects and create plans that take the resilience of the entire system into account.

As noted by Alexander (2013), resilience theory recognizes that resilience-building efforts should go beyond simply bouncing back to pre-disaster conditions. It highlights how catastrophes have the capacity to impact lives by presenting chances to address underlying weaknesses and encourage constructive transformation. This means that resilience builds the support networks and barriers required for long-term disaster risk reduction, rather than limiting itself to a reactive approach. Stated differently, the goal of resilience-building is to establish circumstances that will promote long-term sustainability, equity, and enhanced well-being, thereby cultivating a community that is more robust to potential threats.

In the context of disaster risk reduction, resilience theory informs approaches that integrate disaster preparedness, response, recovery, and long-term development. It highlights how crucial it is to make investments in policies that improve a community's resilience to shocks and strains (Gülhan, 2021). In addition, it emphasizes the value of community involvement, local expertise, and empowerment in fostering resilience, acknowledging that communities are frequently best suited to recognize their own threats, needs, and solutions.

Establishment of Trained Local Disaster Risk Management Committee

The research pointed that, there must be establishment of trained local disaster risk management committee. At local level there must be trained rural disaster risk management committee or rural meteorological representatives, who could be local volunteers or community members coordinating with the national meteorological representatives pertaining forecasting, early warning systems, evacuation, training and education. In order to have complete and trustworthy data to work with, they are essential in gathering on-the-ground weather observations in their local regions and corresponding with the national representative. A swift intervention strategy can be started by providing precise and fast weather information, allowing rural communities to take the necessary precautions against weather-related hazards.

The representatives for rural meteorology are acquainted with the topography, social dynamics, and potential weak points of the area. Their knowledge can improve early warning systems' precision and efficacy, allowing authorities to promptly notify the public through alerts and advisories (Arifin et al., 2021). In order to make sure that communities are sufficiently ready for approaching dangers, they can assist in assessing the unique risks that rural areas confront and offer tailored forecasts. Additionally, they can identify locations and areas that need to be evacuated (The World Bank, 2019). Locals have a better chance of creating tailored solutions that are pertinent, culturally sensitive, and inclusive of the unique difficulties that various communities experience when they are involved in the decision-making process (Arifin et al., 2021).

The appointment of rural meteorological representatives enables them to actively interact with the community and spread knowledge about the dangers associated with weather and the value of being prepared. To teach locals how to understand weather forecasts, spot warning signals, and take the necessary precautions, they can hold training sessions, workshops, and outreach programs (Manyena, 2006). Through cultivating a preparation culture, they enable rural communities to react to calamities with efficiency.

Education and Training

The study alluded that there must be investment into education and training on disaster management and risk reduction. Trainings must be on-going, so that the locals become abreast with the dynamics in the disaster management and risk reduction. The responsible authorities must select and train the local disaster risk management committee so that they continue facilitating and conducting outreach programs, workshops, and training sessions to educate residents on how to interpret weather forecasts, recognize warning signs, and take appropriate protective measures. It has been established that, education and training programs provide rural communities with essential knowledge about various hazards, potential risks, and strategies for disaster preparedness and response.

Raising awareness makes people more capable of appreciating the significance of risk mitigation strategies and acting appropriately to safeguard both their communities and themselves. Education will improve preparation by enabling rural inhabitants to acquire the knowledge and skills needed to respond to emergencies in a way that is efficient (Manyena, 2006). They study first aid strategies, search and rescue tactics, evacuation protocols, and other crucial response guidelines. This information enables people to respond to catastrophes in a timely and suitable manner, potentially limiting property damage and loss of life (Rangarirai et al., 2023).

Education and training leads to advocacy and policy influence. Education and training initiatives create a platform for rural communities to advocate for their needs and influence disaster management policies at various levels. Well-informed individuals can actively engage with local authorities, government agencies, and NGOs to ensure that disaster risk reduction and resilience building are prioritized in planning, decision-making, and resource allocation processes (Manyena, 2006).

Preparedness

Participants pointed that, the government invest in disaster preparedness among rural dwellers. Disaster preparedness or readiness must be linked to capability of an individual or community to apply disaster knowledge and developed capacities so as to effectively respond and reduce impacts of disasters. This means that rural communities need to be equipped with early warning systems, a disaster emergency plan, knowledge of potential disasters, maps and routes for evacuation, assembly places, high-level areas, and the capacity to respond to emergencies. Effective education and training, the creation of knowledgeable rural disaster risk management committees, cooperation between national and rural disaster risk management committees, the availability of resources, and the capacity of local institutions are the only conditions that may make this possible.

According to Kulig (2000), disaster risk reduction ought to concentrate on creating resilient communities as opposed to just dealing with natural disasters. He contends that in order to create resilient communities that will be able to withstand future disasters, it is imperative that the sources of vulnerability be addressed. According to Trosper (2002), communities need to be able to self-organize prior to, during, and after disturbances, absorb new information, and adjust to them in order to be considered resilient. Strategies for resilience should emphasize that communities should be able to adapt to new priorities that arise from the disaster in addition to being able to cope and recover. This can be in the form of building hazard-resistant structures, adaptive social behaviours, and dissemination of early warnings, among other strategies (Pelling, 2003).

Capacitation of Supporting Institution

There must be capacitation of disaster management supporting institution for example local clinic, hospitals, health centers, police and army and local radio stations. These play a critical role in providing early warming messages, emergence medical response, evacuations and a whole lot in disaster management and risk reduction.

Local radio stations are essential platforms for informing rural populations of impending danger when they are fully operational. When a natural disaster is about to occur, they broadcast notifications urging locals to take immediate precautions to protect their property and lives. Local radio stations provide vital information to the impacted areas during emergencies, including updates on shelter locations, evacuation routes, emergency contact information, and relief services that are available. They can publicize requests for gifts, volunteer labor, and particular humanitarian supplies. Radio stations lessen the effects of disasters and assist save lives by disseminating timely and correct information (Chatiza, 2020). According to Arifin et al., (2021), the information enables people and communities to make well-informed decisions and take the necessary activities to ensure their safety and well-being.

Local health clinics, hospitals, and health centers must be capacitated with human and financial resources. They must have qualified medical personnel on hand who can offer emergency medical care, preserving lives and lessening the severity of injuries. In addition, they can function as shelters or evacuation centers, giving evacuated people and their families a safe haven and vital medical services including first aid, basic healthcare, and prescription drugs (Waugh and Tierney, 2007). In order to improve the general health and resilience of rural populations, they also conduct disease monitoring and control, which involves tracking the spread of diseases both during and after catastrophes (Djalali et al., 2016). Reducing community susceptibility to disasters is largely dependent on local institutions' ability to carry out their duties (Mustai and Billiat, 2015).

Funding

A critical solution for disaster risk reduction lies in the allocation of adequate funding towards building resilient communities. By investing in disaster risk reduction initiatives, such as early warning systems, infrastructure improvements, community education, and capacity building, we can enhance preparedness and response capabilities (Munsaka et al., 2021). Adequate funding enables the implementation of comprehensive strategies that address vulnerabilities and promote resilience at various levels. It supports the development of robust disaster risk reduction programs, facilitates the adoption of innovative technologies, and ensures the sustainability of long-term initiatives. Ultimately, prioritizing funding for disaster risk reduction is crucial for building resilient communities that can effectively mitigate, adapt to, and recover from the impacts of disasters (Rangarirai et al., 2023).

Proactive Disaster Management Intervention

The research alluded that, proactive disaster management and risk reduction are crucial for minimizing the impact of disasters and saving lives and resources. The responsible authorities must invest in the development and implementation of robust early warning systems that can detect and alert communities about impending disasters. These systems can deliver information in a timely manner, enabling individuals to take precautions and, in the case of an emergency, evacuate (Gray et al., 2021).

The government needs to implement training and education initiatives that encourage community involvement and awareness. Local communities need to be empowered with information regarding potential risks, evacuation protocols, and preventative measures (Mavhura and Collins, 2017). Promote the establishment of neighborhood-based organizations capable of actively participating in disaster relief efforts. To pool resources, experience, and technology for proactive disaster management, partnerships between NGOs, government agencies, and businesses in the private sector must be encouraged. More specifically, the private sector needs to be involved in creating cutting-edge technology and solutions that help improve emergency response systems, risk assessment skills, and early warning systems (Mutsau and Billiat, 2015).

By adopting a proactive approach to disaster management and risk reduction, communities can better prepare for potential disasters, mitigate risks, and enhance overall resilience (Rangarirai et al., 2023). This shift from reactive to proactive strategies can save lives, protect livelihoods, and reduce the long-term impact of disasters on rural areas.

CONCLUSION

In conclusion, the research recognizes that, rural communities often encounter unique risks and vulnerabilities due to various factors. Thus, it is crucial to develop tailored risk reduction approaches that empower and engage rural communities while leveraging their inherent strengths. By recognizing the unique challenges faced by rural areas and invest in them, we can create more effective and sustainable strategies for disaster risk reduction. Ultimately, re-imagining community-based disaster risk reduction in rural areas requires a shift from a reactive to a proactive mindset. By investing in education, training, and awareness-raising initiatives, communities can develop a culture of preparedness and risk reduction, ensuring that they are equipped to face future challenges. Sustainable development practices and the integration of disaster risk reduction into broader development plans are also crucial for long-term resilience.

REFERENCES

Ajayi, V. O. (2023). A Review on Primary Sources of Data and Secondary Sources of Data. *European Journal of Education and Pedagogy*, *2*(3), 1–7.

Alexander, D. E. (2013). Resilience and disaster risk reduction: An etymological journey. *Natural Hazards and Earth System Sciences*, *1*, 1257–1284.

Arifin, S., Wicaksono, S. S., Sumarto, S., Martitah, M., & Sulistianingsih, D. (2021). Disaster resilient village-based approach to disaster risk reduction policy in Indonesia: A regulatory analysis. *Jamba (Potchefstroom, South Africa)*, *13*(1), a1021. doi:10.4102/jamba.v13i1.1021 PMID:34230845

Chatiza, K. (2020). *Cyclone Idai in Zimbabwe: An analysis of policy implications for post-disaster institutional development to strengthen disaster risk management.* https://doi.org// doi:10.21201

Djalali, A., Ingrassia, P. L., & Ragazzoni, L. (2016). *Role of Hospitals in a Disaster.* ELSEVIER. doi:10.1016/B978-0-323-28665-7.00006-6

Ebba, E (2022). *Managing disasters in Zimbabwe. A qualitative analysis on Disaster Risk Reduction in Zimbabwe.* Uppsala University.

Gogo, J. (2020). Formulate Disaster Management Policy. *The Herald.* Available online at: https://www.herald.co.zw/formulate-disaster-management-policy/

Gray, B., Eaton, J., Christy, J., Duncan, J., Hanna, F., & Kasi, S. (2021). A proactive approach: Examples for integrating disaster risk reduction and mental health and psychosocial support programming. *International Journal of Disaster Risk Reduction*, *54*, 1–9. doi:10.1016/j.ijdrr.2021.102051 PMID:33732610

Gülhan, S. (2021). An overview of disaster resilience. *Turkish Journal of Health Science and Life*, *4*(3), 106–115.

Kullig, J. (2000). Community resiliency: The potential for community health nursing theory development. *Public Health Nursing (Boston, Mass.)*, *17*(5), 374–385. doi:10.1046/j.1525-1446.2000.00374.x PMID:11013000

Manyena, S. B. (2006). Rural local authorities and disaster resilience in Zimbabwe. *Disaster Prevention and Management*, *15*(5), 810–820. doi:10.1108/09653560610712757

Martins, S.F, Cunha, C.J., & Fernando, S. (2018). Secondary Data in Research – Uses and Opportunities. *Iberoamerican Journal of Strategic Management, 17*(4), 1-4.

Mavhura, E. (2016). Disaster legislation: A critical review of the Civil Protection Act of Zimbabwe. *Natural Hazards*, *80*(1), 605–621. doi:10.1007/s11069-015-1986-1

Mavhura, E., Collins, A., & Bongo, P. P. (2017). Flood vulnerability and relocation readiness in Zimbabwe. *Disaster Prevention and Management*, *26*(1), 41–54. doi:10.1108/DPM-05-2016-0101

Munsaka, E., Mudavanhu, C., Sakala, L., Manjeru, P., & Matsvange, D. (2021). When Disaster Risk Management Systems Fail: The Case of Cyclone Idai in Chimanimani District, Zimbabwe. *International Journal of Disaster Risk Science*, *5*(7), 1–9. doi:10.1007/s13753-021-00370-6

Mutsau, S., & Billiat, E. (2015). Leveraging Schools Systems As A Locus For Disaster Risk Reduction In Zimbabwe. *Journal of Education and Practice*, *6*(29), 163–169.

Nyahunda, L., Tirivangasi, H. M., & Mabila, T. E. (2022). Challenges faced by humanitarian organisations in rendering services in the aftermath of Cyclone Idai in Chimanimani, Zimbabwe. *Cogent Social Sciences*, *8*(1), 1–11. doi:10.1080/23311886.2022.2030451

OECD. (2015). *Disaster Risk Financing: A global survey of practices and challenges*. OECD Publishing.

Oxfam. (2019). *Cyclone Idai in Zimbabwe: An analysis of policy implications, for post-disaster institutional development to strengthen disaster risk management*. Available at: https://oxfamilibrary.openrepository.com/bitstream/handle/10546/620892/bp-impact-response-cyclone-idai-zimbabwe-071119-en.pdf;jsessionid=2110164995CF25C5A08EC8A5E792F09B?sequence=1

Rangarirai, W., Marnani, C., & Rahmat, A. (2023). Analyzing the impact of community-based approaches on disaster preparedness to the risk of tropical cyclone induced flooding in Chimanimani District, Zimbabwe. *3rd International Conference on Disaster Management. Earth Environ. Sci., 1173.*

Trosper, R. L. (2002). Northwest coast indigenous institutions that supported resilience and sustainability. *Ecological Economics*, *41*(2), 329–344. doi:10.1016/S0921-8009(02)00041-1

Waugh & Tierney. (2007). Emergency Management: Principles and Practice for Local Government (2nd ed.). ICMA Press.

Wenger, C. (2017). The oak or the reed: How resilience theories are translated into disaster management policies. *Ecology and Society*, *22*(3), 18. doi:10.5751/ES-09491-220318

World Bank. (2019). *Project Information Document (PID).* Report No. PIDA27189. Washington, DC: World Bank.

Chapter 5
People, Culture, Knowledge, and Skill:
A Philosophical Approach to Disaster and Environment (With Special Reference to the Works of Kenji Miyazawa)

Anjali Sharma
https://orcid.org/0000-0001-6297-3113
Central University of Rajasthan, India

Merry Ray
Central University of Rajasthan, India

ABSTRACT

Disasters and the environment are effects of cultural practices; people's lifestyles and understanding of their relationship with the environment require a deep understanding of ongoing phenomena and developing practical skills to manage them. Miyazawa was an educator, geologist, soil scientist, poet, and storyteller whose prolific writings reflected his life and the situation in Japan during the disaster, helping to understand communities and their people. The vital role of culture, knowledge, and community is to develop the skills to adapt and live in harmony with the environment. Therefore, in this chapter, as the meaning of disaster is explored, communities are framed and discussed in light of existing literature on the influential role that people, culture, knowledge, and skills play in disaster management. In particular, philosophical approaches to disasters will be discussed, and the contribution of Miyazawa's work on resilience, community, and disaster management will be analyzed. The chapter adopts a qualitative approach, including review analysis with secondary data sources.

DOI: 10.4018/979-8-3693-2721-0.ch005

INTRODUCTION

The disasters are scary and dangerous. Disasters are beyond physical or materialistic harm; they affect the environment, mental health, style of living, and the memory of pain and loss of dear ones; they are harsh, ruthless, and fierce. Therefore, it is essential to work on these, being prepared to face the challenges and develop resilience to cope with the long-term effects disaster brings. In challenging times, disaster prevention strategies and resilience development help us protect ourselves from disasters. Awareness of these strategies makes one aware of the self-protection remedies to save; alertness about the danger means understanding the environment's signals, practicing prevention activities, and preparing yourself with the help of the community, i.e., culture, knowledge, and skills to face disaster. Therefore, the conversation focuses on determining people's roles, considering culture, knowledge, and skill in preventing and managing the disaster, and finding out the strategies and challenges of a community for managing disaster.

Philosophers have explored the disaster phenomenon in different personas and concepts of uncertainty, where people use their indigenous knowledge and resources, survival, and various livelihood adaptations for disaster management. (Mishra et al., n.d, pp. 10–13).

Miyazawa Kenji was born in Japan, where natural disasters harshly affected people and their livelihoods. Miyazawa was born on 27th August 1896; two months before his birth, a fiery disaster occurred named the Sanriku earthquake, and a Tsunami also happened; after his birth, there was the Rikku earthquake, which occurred on August 31, 1896. He was concerned about people's suffering, agriculture failure, and natural damage. His poem "Strong in the Rain" is a famous poem that instills resilience, motivation, and understanding of the science of life in challenging times like disaster. The chapter provides insight into the community, stakeholders, and organizations to find different techniques and innovations to prevent disaster (小池直子, 2012, pp. 4–8). In the latter phase, the conversation would include a philosophical discussion on disasters and the role of resilience in managing disasters, including Miyazawa Kenji's work on disaster management.

Overview of the Chapter

The chapter is written in qualitative conversation leading to two phases; firstly, it will discuss the meaning of disaster in general and community constructs and its dimensions: people, culture, knowledge, and skills, including its relationship with the environment as disaster affects both people and environment. The second phase focussed on philosophical explorations, including Miyazawa Kenji's philosophy on disaster management. The chapter was developed by review analysis using secondary data sources.

The discussion begins with the universal perspective of disaster and constructs of community and its dimensions, further explaining the relationship between people, environment, and sustainability. The dialogues lead to the interplay between the role of culture, knowledge, and skills with people and the environment to address the challenges and strategies the community adopts for managing the disaster. The latter phase of the chapter includes a philosophical perspective on disaster and the role of resilience in disasters. The chapter hints at the ideas of the philosopher Miyazawa Kenji that have become the impetus for countries to develop disaster response capabilities.

METHODOLOGY

The chapter adopts a qualitative approach, including review analysis with secondary data sources (i.e., research papers, newspaper articles, and government data). The chapter provides epistemological knowledge by exploring people's roles in preventing and managing disasters and finding out strategies and challenges of community in disaster management. The philosophical explorations are also deduced by interpreting the analysis done.

DATA COLLECTION AND ANALYSIS

Data Collection

Review studies included in the chapter were identified using the Google Scholar database, and specific newspaper articles and government website data were also studied.

All databases were searched using topics and Sub-topics. A list of topics and subtopics are listed in Table 1.

The system used for analysis is Zotero, where, in total, the number of articles reviewed was 90 from 1973 onwards, and the reviewed articles were divided based on the topics.

The final selected review papers were based on the focus on the theme of the title People, Culture, Knowledge, Skill, Community and the Philosopher Miyazawa Kenji and their role in disaster management are listed in Table 2.

Table 1. List of topics and subtopics in the literature search

Topics	Sub-Topics
Disaster	Disaster management, Disaster risk reduction management, emergency care, post-disaster, approaches to disaster, prevention of disaster in Japan, Reconstruction
Community	Community, Conservation, Environment, Resources, Knowledge
People	People, Disaster Management
Sustainable development	Sustainability, climate change, environment
Skill	Disaster Management, Skill
Culture	Culture, Disaster Management
Resilience	Disaster Risk Reduction, Resilience, Disaster Management
Miyazawa Kenji	Miyazawa Kenji, Resilience, Disaster, poem, story

(Google Scholar Search Database and Zotero reference management software are used for analysis from 1973 to 2022)

Table 2. List of number of final selected papers based on topics, year of study

Topics	Year of Study	Number of Results
Disaster	**2002-2021**	**16**
Community	**1973- 2020**	**8**
Sustainable Development	**2004-2021**	**7**
Skill	**2004,2006**	**2**
Culture	**2000,2022**	**2**
Resilience	**2010-2022**	**3**
Miyazawa Kenji	**2011-2021**	**6**
	Total	**44**

(Google Scholar Search Database and Zotero reference management software are used for analysis from 1973 to 2022)

In the final stage, 44 review papers were selected from 1973-2022 for the chapter, and the analysis below is done. Based on the selected review papers, the first phase of the chapter is analyzed, starting with the meaning of disaster. The list of selected data published in Journals, newspapers, Government websites and databases, conferences, Books, Note Series, blogs, and Thesis are listed in Table 3.

DATA ANALYSIS

Universal Perspective on Disaster

The definition of disaster has evolved through warfare, society, and technology. It can also be identified by people in a specific geography and their coping and subsequent survival strategies. There are two approaches to defining disaster (Perry, 2018, pp. 1–41). The classical approach talks about disaster as an action through many dimensions that distort social life. Due to the turbulence in social life, society develops new standard ways with unique personal behavioral attributes and finds new ways of survival in different types of natural disasters and their relationship with ecology. It also talks about how risks come as a natural environmental phenomenon. Moreover, if that hazard affects people in society, it would be called a disaster. The social approach to disaster concerns disasters, which are considered social due to many dimensions. It brings about environmental changes due to human activities and societal ecological approaches. It focuses on society's activities that are not appropriate, the policies, and the official actions that bring negative impact through a disaster. (Perry, 2018, pp. 1–41). According to the emergency events database in the Centre for Research on the Epidemiology of Disasters (CRED), its disaster report published in 2022 data shows that 387 natural hazards and disasters taken place worldwide, which is more than the annual average of 370 disasters occurred last 20 year (2002- 21). The details of natural hazards and disasters and their frequency of occurrence are shown below in Table 4.

Community Constructs and its Dimensions: People, Culture, Knowledge, and Skills

Disasters affect the community and environment and bring drastic changes in people's lives and cultures; both struggle with challenges and develop resilience and survival strategies through knowledge and skills and manage disaster. Therefore, the conversation should first move towards understanding the community constructs and the relationship between People and the Environment.

Community Constructs

There are diverse viewpoints about the community; let us go through a few to develop some understanding. The concept of community involves demographics, ethnicity, profession, orderings, commonness, real world, interaction, society, cultural and space -place, people organization, legal matters, level of goodness, spiritual, economic, gender, and relationships perspective(Walsh & High, 1999, pp. 255–268). Locality,

Table 3. Number of articles included in the review per journal

Publication Name	Number of Articles
GOI-UNDP, Disaster Risk Management Programme (2002)-2009, Evaluation and Review of Lessons Learned	1
GOI, Ministry of Home Affairs, NDMD.	1
International Journal of Strategic Property Management	2
Asia Pacific Journal of Social Sciences	1
Nurse Media Journal of Nursing	1
Prehospital and Disaster Medicine	1
School mental Health	1
Journal of business continuity and emergency planning	1
Palgrave Macmillan, Singapore.	1
Handbook of Disaster Research (Springer International Publishing)	1
Springer International Publishing	1
Latin America and Caribbean- Blogs	1
International Journal of Disaster Risk Reduction	1
UNDRR- United Nations Office for Disaster Risk Reduction	1
Web Page - (Disaster Education in Japan: Preparing for Natural Disasters to Protect Kids' Lives), Kids Web Japan, Ministry of Foreign Affairs of Japan	1
What is Disaster? Alternative Approaches to Disaster Analysis. Odisha State Open University E-pathshala	1
The Sociological Review- Journal Article	1
Social Science and Medicine - Journal	1
World Development	1
Histoire Sociale- Journal Article	1
Annals of the International Communication Association – A Journal Article	1
Prepared for the global assessment report on disaster risk reduction- A journal article	1
The concept of community: Reading with Interpretations- A Book	1
Note Series- Urban Environmental Management	1
IUCN International Union for Conservation of Nature) - Book	1
Book, United Nations Economic and Social Affairs.	1
Research Journal of Recent Sciences, ISSN	1
Journal article European Review of Public Law	1
People Definition & Meaning - Merriam-Webster	1
21st Conference of the Parties to the United Nations Framework Convention on Climate Change.	1

continued on following page

Table 3. Continued

Publication Name	*Number of Articles*
Climate Vulnerability and Resilience in the Global South: Human Adaptations for Sustainable Future	*1*
SKOPE, ESRC-Economic and Social Research Council	*1*
Journal of Physical Education, New Zealand	*1*
M/C journal (Media/Culture)	*1*
Current Psychiatry Reports	*1*
Reuters U.S Web Page	*1*
HOKUGA- A journal article	*1*
Critical Stages / Scenes Critiques- The IATC webjournal/Revue web de l'AICT	*1*
Asian Studies	*1*
Ph.D. Thesis, The Dark Realism of Miyazawa Kenji, Portland State University	*1*
The Asian Conference on Cultural Studies	*1*
Risk Hazard and Crisis in Public Policy	*1*

(Google Scholar Search Database, Zotero Reference Management Software, 1973-2022)

Table 4. Frequency of natural hazards and disasters in 2022

Kinds	**Frequency**
Drought	22
Earthquake	31
Extreme temperature	12
Flood	**176**
Landslide	17
Mass movement (dry)	0
Storm	**108**
Volcanic activity	5
Wildfire	15

(Centre for Research on the Epidemiology of Disasters. (2022). Disasters in Numbers: Climate in Action. Https://Cred.Be/Sites/Default/Files/2022_EMDAT_report.Pdf - (Google et al.)

like the environment, space in which they interact, and experiences formed in the places, is the basis for Community (Clark, 1973, p. 397).

A community is an amalgamation of different sections of people's actions, interests, drives, purposes, and value systems they uphold. It has its territory and boundaries, considering individual interest and collaborative betterment; it seeks to know the relationship of humans with self and others, recognizing the dimensions of freedom with inherent values and traditions that threaten us and the threats we face. The Political group in the community plays an essential role in deciding the quality and behavior of the population. (Greer, 2017).

The boundaries signify awareness of the reality, ethnicity, feelings, and thoughts of people, culture symbolically and witness other cultures language and hence gain the need of sharing, Community where all the members come together, solve each other's problems, care for each other and thus strengthen community spirit and also share specific common characteristics like language, ethnicity, race(Jewkes & Murcott, 1996, pp. 557–561). Communication among the community is a constituent that has changed, especially meditation communication through mobile phones and the internet, which has brought a new arena to social relationships and has added to existing knowledge and the convenience of reaching each other despite the distance. (Katz et al., 2004, pp. 315–326).

At this point, the discussion needs to move on to the aspects of community: People, Culture, Knowledge, and Skills.

People

The concept of People is about human persons who are together due to their culture, ethnicity, linguistics, feelings of family, and beliefs and are organized in a group affected by social, political, and economic aspects (People et al. - Merriam-Webster, no date). The concept of People in the constitution discusses that people are state components. The Federal Constitutional Court (FCC), concerning the decision to add to the Treaty of Lisbon, explores the concept that all people are equal and deserve equal rights and opportunities from the point of view of democracy. People belong to a group in which their nationality and ethos are common, and people can be viewed as part of the representation at different levels, local, regional, political, and national. A group of people collectively associated and have formed into meaning through several historical developments(Gross, 2013, pp. 11–25).

Culture

The concept of culture relates to beauty, elegance, arts, aesthetics, tradition, and rituals, as well as exploring the people's style of living. The Latin meaning of the word culture means care, and the French meaning comes to dig the ground, cultivating the essence of differentiating and appreciating unique perspectives, developing sensibility and refinedness, and hence essential for human growth and development. Exploring culture anthropologically includes the behavior learned from observing society, people around, and direct families, the area people live in, and their religion.

Moreover, within this unique culture, people are part of the culture and distinct from others regarding achievements, interests, thoughts, feelings, etc. Every culture is also known for its people's accomplishments, relics like famous texts, language, literature, cloth, food, home, style and steps of making, and different practices while living it also is somewhere in our consciousness and effort (Berger, 2000).

Culture is a world with strategies, tools, symbols, stories, traditions, faith, practices, survival strategies, and techniques that continue from generations and hold the ethos of the particular group that has survived many challenges of life and disasters. Usually, culture from a two-dimensional perspective looks like material and non-material. Material culture is a tangible creation of the group of people in the community, like buildings, clothes, and crafts. Non-material cultures include the values, behavior, language they speak, the rules and norms of the community, symbols they understand, family structures, belief systems, political and socio-economic structures, folk, and knowledge.

Skill

The concept of skill in the professional world expects technical aptitude and the basic knowledge of how to perform your work. Various skills are grouped into soft and hard skills essential in different aspects of life, and core skills are discussed. Skill includes self, values, attitude, integrity, emotional, social, observing beauty, and bringing changes. Skill is focused on in different vocational education and training and assessed differently. The organization focuses on developing aptitude and attitude to prepare them for the future (Payne, 2004, pp. 1–7). The skill also includes the bodily movements technique, competency, well-made decisions, necessary actions, correcting errors, repeating, practicing to attain expertise, learning mechanisms, studying environmental scenarios, and developing creativity and tactics (Ovens & Smith, 2006, pp. 1–11).

The discussion focused on how communities determine factors that help manage a safe environment through the resources they use daily to meet basic needs. How will the depletion of natural resources affect the future of humans and other species, and how will it affect natural changes through differences in air quality, climate challenges, and other direct consequences? Therefore, as a community, choices for a particularly sustainable lifestyle will start with individuals using it to conserve resources efficiently for the future (Srinivas, 2020). The National Disaster Management Authority's policy discussion report on disaster risk reduction and recovery measures for people displaced by coastal and river erosion shows the role of communities and their people, culture, knowledge, and skills to make them prepare, understand, and learn to combat disasters.

People, Environment and Sustainability

The environment and people are intertwined with the responsibility to survive, and they can strategize to prevent any disaster. People need to practice a sustainable living style and take the initiative for conservation and care for the environment(Mandip, 2012, p. 1). Together, people learn and acknowledge the environmental values, stories, and rituals and reflect on them daily. People's awareness of the vast resources available and their environmental concerns must take firm actions and change their behavior (Tilbury & Wortman, 2004, pp. 4–5). People's participation in a disaster requires involvement and awareness of their self-capability.

The roadmap for preventing disaster and disaster preparedness gives importance to people's participation, which is a must, along with all other strategies. The policy also focuses on women's involvement in disaster management, significantly reducing any loss due to disaster. People have first-hand experience of disasters, so the role of the people in disaster management is crucial in saving lives and resources (Das, 2012, p. 41). A combination of different skilled people in a particular community or group of a community can come together to generate awareness programs through the infrastructure available there. They can create a capacity-building program through drama, music, radio, dance, mime, different art or literary competitions, picnics, and community lunches that can happen in the community recreation center. They are building infrastructure that is environment-friendly and disaster-safe so that untimely losses of lives can be avoided, so they are learning and becoming experts in techniques for constructing the infrastructure.

This emphasizes our preparedness to provide Education by experts and stakeholders with expertise in implementing disaster reduction methods through proper training and learning. To form a core group for different kinds of disasters that would include the experts and local people interested, show the skills to be trained appropriately, and make them capable of playing the role of leader for spreading awareness among the communities('Government of India - UNDP Disaster Risk Management Programme 2002-2009: Evaluation and Review of Lessons Learnt', no date, pp. 6–21).

Every individual in the community must be provided awareness regarding the various issues related to the local resources like the forest, the biodiversity, multiple species, quality of air, water, and quantity of rain, production of crops, and various training programs on their skills and the individualized community practices and local strategies people have come through their divergent thinking for co-existing together with nature sustainably. Constant evaluation of situation-based strategy provided by members of particular communities guides us to understand whether our environment is getting the care it should receive from us as the species of mother Earth by seeing the signals, patterns of nature, our health, and food changes. Hence, the local community is like a guardian to protect the traditions, nature, people, and the earth, and therefore understand environmental challenges better (Srinivas, 2020). The move suggests measures, such as creating internal norms to preserve forests and conserve the environment. Certain healthy practices to keep the forest healthy and productive and specific small-scale business training helped them become economically independent (United Nations, 2008, p. 130,151-155,169,221-224,281).

Interaction Between Culture, Knowledge, and Skills Concerning People and Environment

Involving communities in assessing the vulnerability of nature and understanding the natural environment, causes of climatic change, cultural practices, and food practices all together determined the vast role of the community as a group. Each individual contributes to a safe environment and senses any disaster risk (Hoyos et al., 2020). A local leader should lead community participation in conserving natural resources for a secure environment with local stakeholder participating in saving and managing resources. S/he should consider regional inequalities and their diverse interest, focusing on the survival and development of the community as a whole while focusing on the community's role in saving nature. This approach must also consider the perspective of a local group presented here with the resources and how they perceive the outside world.

Community Culture

They value specific actions that are not allowed. It is vital in conserving resources, diversity of flora and fauna, and preserving a safe environment. An example is that in Rajasthan, Khejri trees are not supposed to be cut if they are in an ordinary place dedicated to grazing and holy things. Certain traditions do not allow the killing of certain specific animals. Particular worship is done for trees, sky, earth, rain, moon, and stars to unit value with the available resources, and specific tribal communities like in Odisha State, India Juang and Saora follow the timing and observe seasons for collecting nontimber products from the forests. In every community, there are groups of individuals with expertise in specific skills like bee hiving, knowledge about reptiles, knowledge about tools and techniques, knowledge about medicines, and sacred areas, and few of them are experts in the different types of species. They are experts in food-based products like agriculture, vegetable, and sea-based products.

The community should rely on the indigenous knowledge and skills the people have and utilize them for survival and keeping the environment safe. Women in every community are the agents of in-depth understanding of the strategies for economic survival through different indigenous products and mobilizing the natural resources for more significant efforts for biodiversity conservation.

Moreover, the cultural style of living all contributes to a pattern of lifestyle and skills that cooperate for a common goal of a safe environment like if they can see some signs of disasters through actions of animals, birds and trees etc. (Agrawal and Gibson, 1999, pp. 629–649). Our population includes different kinds of communities, traditional practices, languages, resources, and other tools they use to save their resources and population during and after the disaster. Protecting the more vulnerable groups and empowering them through their indigenous language and background would help them prepare for disaster.

The ethnicity and language of indigenous communities of different races and understanding the risk of perceiving any disaster situation would allow people to combat it more collaboratively than independently (Ito, 2021, pp. 43–88). Different cultural and ethnic practices, like some, emphasize harmony and existence with the natural environment. Some give signals through folklore; folktales like Native Americans share folktales of animals in low terrain and show windstorms. In some places, animals also give signals through their different sounds for any disaster to come. In certain cultures, taking medical care is not accepted, and there are some related myths or practices, so the community must be aware of these cultural practices accordingly. It should prepare each target group's health and safety care and know how to find a middle ground between these situations. Understanding the cultural values and beliefs for dealing with disaster losses is essential to providing proper care and support (Rahmani et al., 2022, p. 482).

Integrating Challenges and Strategies for Disaster Management

The tragedy is also ascertained through vulnerable groups in times of tragedy and after the disaster, creating discrimination among people and disturbing supplies. Women and children are more vulnerable to disaster than men. The community has to play its role in preparedness, awareness, and perception of any disaster; they need to be culturally sound for a particular ethnicity so that the community disaster organization can communicate and care for them(Rahmani et al., 2022, p. 482).

People experienced disasters multiple times, used knowledge, and developed new innovative ideas and techniques to manage disasters effectively and minimize causalities. Strategies used by people include natural directions like wind, the temperature of the water, the Salinity of water, clouds' color and shape, observing specific behavior of birds, and the way the rainbow comes. For example, wind from the southeast direction is the signal for the storm, and from the northeast direction is the signal for not-so-severe cyclones. People and the meteorological department consult about the early signals to understand the warning signals easily. Other small and immediate strategies include holding on to solid trees, catching hold of floatable items, and using various food preservation techniques for emergencies during disasters. A few indigenous food preservation techniques in Bangladesh include packing rice and thick dark syrup, a by-product of sugar called Molasses in polythene, and keeping them in an earthen pot. That pot is kept after digging the ground; people instinctively think of self-protection. Moreover, they concern themselves with what dress they are wearing so that it will be convenient for them to swim if required.

People collectively join together to help each other with traditional roots and herbs as medicines after the disaster, as it takes a few days for the relief camp necessities to arrive. Therefore, they survive on edible plants, roots, and stems. As they share, they form bonds and become resilient to such disasters by sharing strengths and cooperating. People often also use the wind direction to plant specific trees and mangroves, which are very strong and can withstand severe storms and cyclones, like Persimmon trees, Keora, and Sundari Mangrove species. Sometimes, a few floods are good and evil, as a few floods bring out alluvial soil. The strategies used during a flood are making a raft using a Banana trunk, being alert, and using bamboo poles to increase the height of the sleeping bed(Haque, 2019, pp. 1–8); traditional methods include keeping heavy weight things like sandbags, a pile of bricks, frozen cement to stop the flood, L- shaped pipe used for extending silencer pipes of vehicles to bring above the water level for safety from floods. New modes of transportation like boats, rafts, and vans made up of empty barrels are used during floods. Together, people can prevent water leakages and protect the embankments. People can face and manage disasters if they have a sound knowledge of and experience with climate.

Traditionally, people used to protect their houses by preserving the house's base with a paste of mud, jute fiber, and husk. To save themselves from drought, environmental and food crop management ensures food security by building ponds, canals, and water reservoirs, preserving the rain. Water helps them to manage disasters effectively. People from different generations have observed the environment, animals' behavior, clouds, wind, and rain patterns to safeguard and manage disasters (Haque, 2019, pp. 8–12). They are empowering people, especially women and children, to have sound knowledge to understand the disaster in every dimension and address any emergencies due to disaster. Community-based organizations providing them with infrastructure, capacity-building programs, proper resource management of the environment, and green growth are the report's priorities for disaster risk reduction.

Cultural indications through symbols of the sea, low tide, birds, and animals have been passed through the stories from generations to study the environment and observe the symbols and hence help the community to be alert and face the disaster effectively, in a few cases if the community is relocated due to severity of the disaster, in that case, the relocation must consider their livelihood and survival strategies from the disaster. Integrating the community's experiences with scientific knowledge would help to find survival strategies and reduce the risk of disaster(Kulatunga, 2010, pp. 304–311).

It is easier for skilled individuals to understand disaster risks and effectively find strategies to manage them. A physician needs specific skills to respond to a disaster, like being an expert in treating people post-disaster, living in minimum primary conditions, and working in a group. They must have received unique training experiences in disaster management and be able to make the right decisions in critical situations (Noguchi et al., 2016, pp. 397–398). Individuals' social and emotional skills come to the test during and after the disaster, as disasters are destructive, tragic, painful, and involve suffering and loss of humans and resources. Enhancing emotional skills and social skills would bring hope and confidence to acknowledge the issues and find remedies for them by providing support care, finding self-efficacy, self-sustenance, and proper mental health to reduce stress, listening and seeing others experience and supporting each other and hence develop emotional and social skills to face challenges due to the environment (Powell & Bui, 2016, pp. 107–116).

Knowledge is another aspect required in an emergency to tackle any disaster; community health volunteers with sufficient knowledge can show their skills to manage disaster and its aftereffects like first aid care, early signals of disaster, the ability to tackle any rescue mission, logistics skills, basic communication skills, team skills, and the ability to sort out the order in terms of the most affected patient to be treated the first fundamental concept of knowledge about different disaster preparedness and challenges including strategies also be able to face emergencies (Kamal et al., 2012, pp. 371–378). Knowledge about a community song that motivates them to understand nature, people, and challenges would help them stay strong and united in difficult times. Active readiness, immediate decisions, and proper measures will help recover quickly from a disaster.

Environment plays a crucial role in managing disasters and reducing risk. For example, some barriers are available naturally, like dunes, coral reefs, and mangroves; they have this complex root system that stays together to guard against strong waves. Therefore, protecting the environment, preserving the ecosystem, and continuing reforestation is essential.

Lastly, the disaster law must be strengthened to help the environment and people. Channeling the resources and knowledge of expertise, with proper implementation techniques on the field, would help to manage environmental concerns and disaster risks properly (Disaster et al. in India: Status Report 2020 | Prevention Web, January 2021).To protect the environment, the proper choice of crops and drought resistance can reduce food losses in the agricultural sector, and the appropriate selection of animal breeds is beneficial to managing disasters (Pathirage et al., 2014, pp. 4–12). Reconstruction after the disaster requires skill and knowledge, adequate information, alertness, enough experience, competencies, effective decision-making, proper coordination, and rehabilitation measures need to be taken to make the reconstruction project successful(Pathirage et al., 2014, pp. 13–14), Technology knowledge, social knowledge about environment, resources available, skill available, understanding environmental factors and ways of protecting environments natural barrier, practical measures to maintain the ecosystem and vegetation. Proper legal matters can be implemented to prevent losses and tragedy during and after the disaster. Managing the finances of the government effectively protects the people and the country. Proper management and implementation of operations in the field require expertise and adequate training to deal with disasters(Pathirage et al., 2014, p. 20). From the above discussion, it is evident that the role and aspects of the community are depicted in the form of a figure.

The second phase of the chapter discusses the philosophical exploration of disasters and the contribution of philosopher Miyazawa Kenji.

Discussion of Philosophers Regarding Disaster and Management

Philosophers have shown little concern about disaster and exploring its different dimensions; even if a few definitions are available, they are not defined wholeheartedly. Some philosophers have discussed the philosophical issues of disaster, but the definitions have not been given priority. The latest commitment of philosophical

Figure 1. Role of community and its part in the management of disaster
(The relationship-based diagram is developed by the authors based on the review analysis done from 1973- 2022 during the first phase of the chapter).

development regarding disaster began when Plato discussed the 1755 Lisbon earthquake. It also ignited and initiated a discussion between Rousseau and Voltaire in which they questioned Leibniz's optimistic worldview, while Rousseau answered, pointing out the dimension of "God and Evil." This reflects the imbalances in society, which naturally create disasters. Already available definitions of disasters emphasize the measure of harm involved. If the harm is not much, then it is included in non-disastrous events; instead, if more harm is involved, then it is called a disaster, but this meaning of disaster is ambiguous as it is tough to measure and define how much harm is considered dangerous and hence categorized into disaster. According to Voice, disasters shatter moral ethics and effectual nationality, but some say disaster is more than harm. It is beyond physical harm or any destruction. Although disaster is considered a violent activity, it also defines what it means to be a non-violent disaster. Naturally, we can all relate to disaster with crisis, and violence can be explained in terms of disaster. The question arises of sometimes signifying disaster with a sudden accident, but this may not be true. For example, in the case of Zimbabwe, slowly losing their dignity and nationality with time is also a crisis and, hence, disastrous.

According to Naomi Jack, the term disaster explains disaster in terms of actions and behaviors we need to know about the disaster. An event can only become a disaster if there is a danger signal. Immediate emergency assistance is required; a disaster is a chain of events where there is loss of lives and disruption of normalcy in society due to natural accidents or activities of humans. Disaster is considered unpredictable, unwanted, shock, and tales of heroism, success, and failure. However, disaster seems to be seen in how it naturally occurs here. Medical ethics have contributed to Triage as an idea relevant to disaster. Triage means deciding who is more in need of treatment during disaster and post-disaster. In medical ethics, disaster is serving the highest benefit for the highest number, as given by Gerald. R. Winslow. Kenneth Kipnis suggests disaster in medical ethics regarding calamity or destruction, which maximizes patients compared to region clinical capacity. Several arguments relate to what will constitute disaster, whether a war, spreading contagious disease, or restricting individual freedom(Sandin, 2018, pp. 14–20). Philosophers have argued that moral ethics may not always be followed during a disaster. Here, focus is given to the best possible outcomes through our actions, and that outcome must be accepted by many, but there are chances that sometimes our decisions may not bring the best outcome. Leonard Kahn suggests that an ethical proposition should have a clause related to disaster. Moral ethics considers two different thinking, one in terms of consequences and another in terms of deontology. Heidi has explained that sometimes if you cross the mark of set rules.,there is no morality, and also, if there is a situation where crossing the mark of set rules is necessary for moral ethics, for example, love and war, torturing someone dangerous to save millions of people. There is another discussion given by Sorell about the Hobbesian perspective of disaster, which provides the difference between disasters occurring naturally and emergencies that occur due to political disorders. Sorell has also mentioned how this distinction may not be appropriate as there are different kinds of emergencies and how Hobbes has not used the word emergency but instead used the word war. So, from the above discussion, it is evident that philosophers may not talk about disasters directly. However, they indirectly discuss emergencies, hazards, crises, critical situations, consequences, uncertainty, risk, and moral ethics (Sandin, 2018, pp. 20–24).

Resilience in Disaster Management Based on Available Literature

Resilience was first visible in ecology in 1970 to assess the ability of a system to internalize changes and persist. Resilience is a multidisciplinary term in different disciplines (Shimizu, 2012, p. 41). Scientific Knowledge about cultural and environmental resources makes the community resilient towards disaster management as culture provides specific identity, values, and belief systems (Kulatunga, 2010, p. 310). Resilience is the collaboration of the effort to stand firm against all odds and stay balanced; it is a testament to persistence, adaptability to different variables, and achieving sustainable survival and livelihood with minimal loss; it is an expression of coming back after facing challenges economically, socially, resource-wise, inadequate facilities and lack of information. Resilience is the phenomenon of returning healthy and facing destruction, evolving together, and having the capability to face uncertainties and accept changes. It is also the attitude of predicting, planning, and managing the disaster, ensuring efficiency and effectiveness of technology, construction, society, and economy (Huq et al., 2021, pp. 431–432).

Resilient communities globally need to deal with disasters and manage better livelihoods after disasters, and knowledge management plays a key role. Recovering from a disaster is sometimes slow and requires a lot of monetary support and coordination. However, from a positive perspective, these times will unite the community and discuss society's learning and functioning. These small efforts to share, care, and overcome the difficult times together will make them resilient. Together with resilient people, the community needs to take proper care to build resilient infrastructure, buildings, toilets, dams, supply of water, borewells, roads, seawalls, and raised platforms for protection from disaster.

Educating the community makes them resilient, avoids cultural risk in the future, and brings a safe culture that understands and is aware of the risks and issues during disasters. Through education, training, techniques, facts, technology, and media, organizations like schools can provide awareness to help the community. Children understand basic knowledge about disasters, and organizations can learn from the communities, their ethnicity, and practices to manage disasters. This can help the volunteers, policymakers, and community stakeholders benefit the citizens (Pathirage et al., 2014, pp. 3–10).

Technology can play a crucial role in communities' recovery and provide immediate relief packages through sensing and satellite images, provide real-time data, and make decisions about allocating resources and support to disaster-prone areas. It saves time, helps the community become resilient, protects it, and provides experiences and information for future disasters (Pathirage et al., 2014, p. 16).

Disaster resilience includes accepting, adjusting, surviving, and alleviating, which supports and helps to decrease the effect of disaster. Due to non-homogeneity among communities, there is a lack of clarity in identifying, interpreting, and applying resilience and also assessing resilience. DRIFT (Disaster Resilience Integrated System for Transformation)

The policy should also be developed and implemented to make sustainable community and physical resilience (Shimizu, 2012, pp. 42–46). Iiate village of Fukushima prefecture was a community of resilient villages and are practicing social resilience based on a cultural concept which has naturally been passed through generations in the area says ''Madei no Chikara''means that power is in your hands, which means to put all your hard work, mind and heart and time by collaborating with families, people, communities, relatives and societies. Villagers have made a sustainable plan with people, culture, and nature and are networked for food, agriculture, education, hospitals, and global exchange of culture programs. Similarly, Tono city in Iwate Prefecture initiated a project that included the people living in the community, organizations and industries, people, and the environment with a philosophy of ''Sato,'' which means village or settlement. The faith Sato brings forward is to build in one's house a sustainable space, way of life, culture, nature, traditions, and history (Shimizu, 2012, pp. 55–56).

Miyazawa Kenji's Philosophy in Unfolding the Role of the Community During Disasters and a Safe Environment Through His Works

Several disaster prevention strategies and evacuation training are being conducted in Japan, such as In Motoshiro Elementary School. Evacuation training is practiced to prevent fires and earthquakes; similarly, in Machida, Tsurukawa Daini Elementary School in Tokyo conducts disaster prevention every year, where students learn how to protect their lives during disasters. They were aware of the points to remember during evacuation from dangerous situations. Disaster-simulated training is also provided for children and parents; for example, during earthquakes, public transport will not be available; therefore, parents have to come and pick up the students from school.

From many disaster prevention simulated experiences available in Japan, Honjo Life Safety Learning Center, where families can tour the whole center and participate in first-hand experiences, for example, situations like difficulty in opening car doors during a flood, families also can experience the feeling of the typhoon; they have a participatory maze experience that is filled with smoke for helping them learn how to evacuate during a fire, these experiences help families to discuss remedies and prepare accordingly for any dangerous situations. Similarly, in Shinagawa, first-hand fun learning experiences on preventing disasters are available (Disaster Education in Japan: Preparing for Natural Disasters to Protect Kids' Lives, no date).

Resilience can be operationalized in policy and disaster management (Shimizu, 2012, pp. 40–42). The study talks about Miyazawa's view of nature; his focus is that nature must not be disturbed due to modernity and its essence of pureness; the immense contribution of nature must be respected and mobilized correctly for a safe environment (Siercks, 2022, pp. 1–18). Through his stories, Kenji Miyazawa has always provided a path to examine more complex issues of people and nature. In this story, he focuses on farming issues faced by farmers and how they can understand nature and science to improve their lives, so he talks about cooperation between science and nature to make people live better and more sustainably. He has focused throughout the story on how the drought has taken so many lives and how the hero has shown leadership and dedication to protect the land and the lives of farmers. He also emphasized the challenges due to drought, such as food security being the primary concern, and how the main character takes help from the local communities and motivates them to solve problems like economics and injustice society faces. He was also part of the local community to improve their lives, was a qualified agriculturalist, and worked as a farmer to support the local people. He believed in the Education of farmers in soil, nature, and fertilizer farmers must have scientific thinking. He asked the farmers to be creative and happy, artists to be satisfied, and perform theatre together as a community to stay happy and healthy in difficult times and situations. He also transferred his knowledge and skills about productive farming to the farmer. The study focuses on the active bond humans and nature share. Nature does care for our basic needs, and we must also take care of nature and understand the details and signs it provides when not in care. Nature needs to be observed, alerted, and understood critically using science.

Miyazawa also states how Disasters like Famine resulted in the unavailability of basic food needs, affecting certain vulnerable groups, especially children and young girls, who face the adversity of disaster more painfully by being pawns and labor work. The community should move together during disasters rather than exploit the more vulnerable group. Miyazawa has talked about how the tragedy brings out other societal and fundamental human rights issues if not appropriately dealt like the more vulnerable group being affected through forced labor, exploiting labor laws, insecurities, and loan sharks with no livelihood earnings.

His story "The Life of Budori" also focuses on how our hardships, struggles, and the experiences we have gained due to the local and experienced impact in the forefront help us to polish our skills, gain new knowledge to reduce the risk of disaster, and save many lives and be a forefront leader in the role in the community (Ito, 2021, pp. 1–97). The author focuses on the greenness of space in the environment through his poems and mentions how the peaceful coexistence of humans, species, and nature is essential to sustain the future.

In the story, Indra's Net Miyazawa mentions how we all are interlinked through the different dependencies on each other's presence on the earth. However, if any link is broken, it will affect everyone. Coexisting culture in the community is essential for the locals to stay peaceful and sustainable in a safe environment. He talks about how our connecting links are so vulnerable to natural phenomena that we must think good for nature, resources in the heart, animals, and plants to not get extinct from the earth. His poem Strong in the Rain provides hope, resilience, and motivation for people to accept and use their experiences for a better world. Miyazawa talks about wind and light and shows that we are part of the bigger space that nature expects.

As a community, we should strive for everyone's happiness and peace for survival. Harmony, Compassion, and Empathy for all species would help us keep the environment safe and sustainable. The study also reflects on how Miyazawa's stories reflect that our pain and suffering can get the better of us by observing others' suffering and working as a group to overcome this suffering of others and, hence, saving others.

Miyazawa studied nature through various self-trips to know the depth of its presence and understand it more scientifically (Pulvers, 2013). The study discusses the literary works of Kenji Miyazawa, who provided a vision for a resilient future for natural phenomena, community, people, and resources. The story of "Matasaburo" of the Wind," by Kenji Miyazawa, shares that individual roles can collectively collaborate towards more significant progress and development. Particular focus in the story is given to the wind and what kind of work the wind involves in the environment.

Here, Miyazawa focuses on the illegal use of minerals, destroying the natural ecosystem and exploiting the resources. Miyazawa also discusses how important it is to know our environment, nature, the river, the sun, moon, light, air, and soil; knowledge in detail can guide you about the signs of disasters and read through the messages they give us on how to utilize these mediums or resources of nature and manage them sustainably securing the future (Larson, 2021, pp. 1–8). The relationship between humans and nature is through communicating with each other and the changing phenomenon of walking toward true happiness. Everything is interconnected and simultaneously changes to be in a better place. Through the story of Kenju's Forest Playground, the author views the bond between Kenju and The Cedar tree as pure, so it changed accordingly and, in the future, the land to be filled with Cedar Trees and hence the cooperative communication with nature brings happiness for both humans and nature. So, the community needs to form a true path of bonding with nature and prepare individuals for the change it puts through the heart (ŠKOF, 2013, pp. 89–104).

The discussion in the article focuses on forming a bond through shared understanding, facing everyday challenges, showing resilience, and surviving the disaster. It helps us bind and motivates us to help each other by caring for our pain, like the loss of lives, basic needs, insecurities, and poverty. However, here comes the role of the community to move ahead towards progress, happiness, and growth. So, in these difficult times, Miyazawa's poem Strong in the Rain helps the community to look upon each other in the adversities and cross the path towards happiness and prosperity (Kubota,2011, March 21). The study focuses on how the community can collaborate for performance through art and theatre using available resources. Miyazawa's poem Haru-no Shura was performed to motivate the people to unite for a common purpose (Niino, 2012).

FINDINGS

The findings focus on determining people's roles, considering culture, knowledge, and skill in preventing and managing the disaster, and finding out the strategies and challenges of a community for managing disaster leads to

1. Community is the driving moral force that instills moral values, finds space for humanity and well-being, and protects humankind and integrity. The essence of the community is traditional with a flexible nature accommodating modernity and hence grows together with people, culture, knowledge, and the skills learned.
2. Sustainable lifestyle, awareness of available resources and self-capability, tapping the skills, learning and using signals based on tales passed from generations, building organizational training centers, and preparing men, women, youths, and children for disaster risk reduction. The indigenous knowledge, skills, and cultural practices for a sustainable lifestyle are what people need to have and utilize for survival and to keep the environment safe.
3. Focusing on disasters like droughts, preserving water management, and finding different ways to use water for agricultural needs are essential to avoiding poverty and food insufficiency. Building signal codes for any calamity will save human lives and train communities and locals to save human lives by providing them

with the proper types of equipment. Specific operational guidelines can be provided to locals, and integrating them with their traditional methods would help them prepare for any disaster. All the stakeholders should perform their respective responsibilities to reduce the risk of disaster and climate change.

4. Challenges after disaster lead to the accumulation of several wastes that may be hazardous and destructive waste, such as soil and municipal waste, which are dangerous to humanity; species dependent on water and the underwater environment get disturbed. After disaster, recycling discarded matter, sewage treatment, proper drainage system, proper water management, rehabilitation of soil, and waste management need to be taken to protect the environment and the health and hygiene of human beings.
5. Disaster Knowledge factors emphasize decreasing losses due to disasters and providing fast and practical assistance to victims for speedy recovery. Knowledge related to accurate and valid data about disaster information based on the database available and experiences faced would ensure disaster management with effectiveness.

The discussion on Philosophical explorations related to disasters, resilience, and the contribution of Philosopher Miyazawa Kenji leads to the following findings.

1. The slightest interest of philosophers in disaster limits the broad outset of the meaning. However, few philosophers have debated the philosophical issues of disaster that impact the balance within the society, questioned Leibniz's optimistic view of the world, and measured disaster in terms of harm, hurt, pain, crisis, and violence. Few philosophers have explored the meaning of disaster in events, actions, and behavior. Disaster is also synonymic with shock, unpredictability, stories of heroism, emergencies, hazards, risk and uncertainty, preparedness, and survival.
2. Medical history brings out the Theory of Triage, where serving those in more need during and post-disaster also speaks that the best actions are expected for the best outcomes may not always keep the moral ethics. Disaster and moral ethics are considered inseparable.
3. According to Ecology, resilience is a multidisciplinary term that means instilling change internally, staying persistent, and acquiring scientific knowledge about available resources based on traditions, beliefs, and values. Resilience is synonymic with standing united against difficult situations, adaptability, bouncing back in odd times, keeping calm, maintaining good health despite challenges, acceptance, enhancing capability in times of uncertainties, the attitude of planning, and taking responsibility.

4. A resilient community plans foresight, works, and collaborates with the organization at the state, national, and international levels to encourage the development of resilience for disaster and lessen the risk and losses physically, politically, and economically. Hence, the term disaster resilience must be part of community engagement and be ready with whatever resources are available.
5. Miyazawa Kenji created a unified way of living where every being is interconnected. If one is disturbed, another also gets affected, specifically between humans and nature. Nature must not be disturbed but rather respected for its significant contribution to our lives. His stories and poems guide the commoner to accept reality and understand the science behind nature to protect your community and motherland.
6. Miyazawa Kenji, through his tales, has emphasized the significance of forefront leadership, polished skill due to first-hand experiences, united community, accepting challenges, and finding solutions to help the society overcome the challenges put forward by these disasters.
7. Miyazawa Kenji emphasized Productive farming by having scientific and sound knowledge about the region, environment, soil, and its strengths and weaknesses to combat food security and economic crises during and post-disaster.
8. Miyazawa Kenji shows that sharing and observing pain and suffering caused by disaster can help self and community to overcome the pain, and hence, we save each other.
9. Miyazawa Kenji, through his stories and poems, makes us walk in the vision of a resilient future that will bring forward significant development of the community and its construct.
10. Miyazawa Kenji also warns of not harming nature, not doing any illegal activities, and exploiting resources; if this is not the case, then the dangerous repercussions we as a community have to face in terms of disaster.

DISCUSSION

The findings based on the internal elements that are parts of the community are the first-hand experience of the disaster. Therefore, they play an essential role in protecting the environment and developing strategies using their culture, knowledge, and skills. The review analysis finds that resilience has always been part of the community and how slowly it has become part of stories, poems, and folklores for helping the youth understand survival skills and acknowledging the fact that disaster is risky; therefore, appropriate prevention strategies at the local level with the help of construct of community is necessary. This disaster prevention strategy works as first aid is necessary before emergencies occur.

The existing literature discusses various topics related to disaster management, such as community, resilience, skills, culture, people, prevention strategies for disaster management, and sustainable lifestyle and livelihood. Miyazawa Kenji also witnessed the experience of disaster; therefore, through his teachings, poems, and stories, he talked about the galaxy, cosmos, and universe and was motivated to learn about the scientific facts and evidence of nature in your region for better survival strategies.

This study is limited to four elements: people, culture, knowledge, and skills, which contribute to disaster prevention and management and their interaction with the environment, resilience, and contribution of Miyazawa Kenji for sustainable life, as well as disaster prevention and management.

CONCLUSION

The fundamental philosophies that conclude the various ethos and philosophy each community carries such as Madei no Chikara- the power in our brain, heart, families, communities, relatives, and societies collaborate with organizations to provide security and safety. Sato- building a resilient environment first in our homes, resilient way of space, nature, environment, culture, and traditions. Similarly, Miyazawa Kenji's stories depict the reality of those times, how nature interacts with us, and provide signals to alert us to any danger approaching. He also reminds us that if people do not protect the environment and do not care for it, then this imbalance is harmful, and therefore, moving towards sustainability is essential. For this, resilience plays a key role. Future research can focus on different philosophical perspectives of disaster and how local organization and innovation can make a community self-sufficient and explore realistically sustainable living styles.

REFERENCES

小池直子. (2012). *Beyond the Practicality of Time and Materialism of Life*. Available at: http://hokuga.hgu.jp/dspace/bitstream/123456789/1907/1/KOIKE.pdf

21st Conference of the Parties to the United Nations Framework Convention on Climate Change. (2015). *Understanding Human Rights and Climate Change*. www.ohchr.org/Documents/Issues/ClimateChange/COP21.pdf

Agrawal, A., & Gibson, C. C. (1999). Enchantment and disenchantment: The role of community in natural resource conservation. *World Development*, *27*(4), 629–649. doi:10.1016/S0305-750X(98)00161-2

Berger, A. A. (2000). The meanings of culture: Culture: Its many meanings. *M/C Journal, 3*(2). Available at: https://journal.mediaculture.org.au/mcjournal/article/view/1833?__hstc=87814215.82af9c9a98fa600b1bb630f9cde2cb5f.1540944000065.1540944000066.1540944000067.1&__hssc=87814215.1.1540944000068&__hsfp=2025384311

Centre for Research on the Epidemiology of Disasters. (2023). *Disaster Year in Review: Cred Crunch.* https://reliefweb.int/report/world/cred-crunch-newsletter-issue-no-70-april-2023-disasters-year-review-2022

Clark, D. B. (1973). The Concept of Community: A Re-Examination. *The Sociological Review*, *21*(3), 397–416. doi:10.1111/j.1467-954X.1973.tb00230.x

Das, P. (2012). Disaster management in India: Policy review and institutional structure. *Asiapac Journal of Socical Sciences*, *4*, 37–52.

Disaster Education in Japan: Preparing for Natural Disasters to Protect Kids' Lives. (n.d.). *Web Japan.* Available at: https://web-japan.org/kidsweb/ja/cool/20/202011_disaster-prevention-education_ja.html

Disaster risk reduction in India: Status report 2020. (n.d.). Available at: https://www.preventionweb.net/publication/disaster-risk-reduction-india-status-report-2020-0

Government of India. (n.d.). *UNDP Disaster Risk Management Programme 2002-2009: Evaluation and Review of Lessons Learnt.* Author.

Greer, S. (2017). *The Concept of Community: Readings with Interpretations.* Routledge. doi:10.4324/9781315131443

Gross, T. (2013). "We the People"-What is the Legal Meaning of "People"? *Revue europeenne de droit public= European review of public law= Eur. Zeitschrift des offtl.Rechts= Rivista europea di diritto pubblico, 25*(4), 1791–1822.

Haque, M. (2019). Indigenous knowledge and practices in disaster management: Experiences of the coastal people of Bangladesh. Disaster Risk Reduction: Community Resilience and Responses, 59–72.

Huq, M. E. (2021). Resilience for Disaster Management: Opportunities and Challenges. In Climate Vulnerability and Resilience in the Global South. Cham: Springer International Publishing (Climate Change Management). doi:10.1007/978-3-030-77259-8_22

Ito, E. (2021). *The Dark Realism of Miyazawa Kenji: Social Activism in the Science-Fiction Children's Story, The Life of Gusukō Budori, and its Precursor Work.* Dissertations and Theses. doi:10.15760/etd.7725

Jewkes, R., & Murcott, A. (1996). Meanings of community. *Social Science & Medicine*, *43*(4), 555–563. doi:10.1016/0277-9536(95)00439-4 PMID:8844956

Kamal, A., Songwathana, P., & Sia, W. S. (2012). Knowledge and skills of emergency care during a disaster for community health volunteers: A literature review. *Nurse Media Journal of Nursing*, *2*(2), 371–381.

Katz, J. E., Rice, R. E., Acord, S., Dasgupta, K., & David, K. (2004). 'Personal Mediated Communication and the Concept of Community In Theory and Practice. *Annals of the International Communication Association*, *28*(1), 315–371. doi:10.1080/23808985.2004.11679039

Kulatunga, U. (2010). Impact of culture towards disaster risk reduction. *International Journal of Strategic Property Management*, *14*(4), 304–313. doi:10.3846/ijspm.2010.23

Larson, M. (n.d.). *Reading Kenji Miyazawa after 3.11: Region, Utopia, and Modernity*. Available at: https://researchmap.jp/19760415/published_papers/33470174/attachment_file.pdf

Mandip, G. (2012). Green HRM: People management commitment to environmental sustainability. *Research Journal of Recent Sciences, ISSN*, *2277*, 2502.

Mishra. (n.d.). *What is Disaster? Alternative Approaches to Disaster Analysis*. Odisha State Open University E-pathshala. https://osou.ac.in/eresources/Disaster.pdf

National Disaster Management Authority. (2022). *National Policy on Mitigation and Rehabilitation Measures for people displaced by coastal and river erosion*. https://ndma.gov.in/sites/default/files/PDF/National_Policy_on_MRM_CRE.pdf

NDMD (National Disaster Management Division). (2004). *'Disaster Management in India'- A Status Report*. Ministry of Home Affairs, Government of India.

Niino, M. (2012, June). Caught between Solidarity and Hesitation: The 3/11 Earthquake and Theatre in Japan. *Critical Stages/Scènes*, (6), https://www.critical-stages.org/6/caught-between-solidarity-and-hesitation-the-311-earthquake-and-theatre-in-japan/

Noguchi, N., Inoue, S., Shimanoe, C., Shibayama, K., Matsunaga, H., Tanaka, S., Ishibashi, A., & Shinchi, K. (2016). What kinds of skills are necessary for physicians involved in International disaster response? *Prehospital and Disaster Medicine*, *31*(4), 397–406. doi:10.1017/S1049023X16000418 PMID:27221114

Ovens, A., & Smith, W. (2006). Skill: Making sense of a complex concept. *Journal of Physical Education New Zealand*, *39*(1), 72.

Pathirage, C. (2014). *Knowledge factors and associated challenges for successful disaster knowledge sharing*. Prepared for the global assessment report on disaster risk reduction.

Payne, J. (2004). *The changing meaning of skill*. Available at: https://ora.ox.ac.uk/objects/uuid:32bdde50-3585-4af4-99c6-9ae28344e68d

People Definition & Meaning. (n.d.). Available at: https://www.merriam-webster.com/dictionary/people

Perry, R. W. (2018). Defining Disaster: An Evolving Concept. In Handbook of Disaster Research. Cham: Springer International Publishing. doi:10.1007/978-3-319-63254-4_1

Powell, T. M., & Bui, T. (2016). Supporting Social and Emotional Skills After a Disaster: Findings from a Mixed Methods Study. *School Mental Health*, *8*(1), 106–119. doi:10.1007/s12310-016-9180-5

Pulvers, R. (2013). Miyazawa Kenji's Prophetic Green Vision: Japan's Great Writer/Poet on The 80th Anniversary of His Death. *Asia-Pacific Journal-Japan Focus*, *11*(44).

Rahmani, M., Muzwagi, A., & Pumariega, A. J. (2022). Cultural Factors in Disaster Response Among Diverse Children and Youth Around the World. *Current Psychiatry Reports*, *24*(10), 481–491. doi:10.1007/s11920-022-01356-x PMID:35953637

Sandin, P. (2018). Conceptualizations of disasters in philosophy. Disasters: Core concepts and ethical theories, 13–26. doi:10.1007/978-3-319-92722-0_2

Shimizu, M. (2012). Resilience in Disaster Management and Public Policy: A Case Study of the Tohoku Disaster. *Risk, Hazards & Crisis in Public Policy*, *3*(4), 40–59. doi:10.1002/rhc3.17

Siercks, E. J. (2022). *Scaling the Nation: Local Literary Production and National Literature in Postwar Japan, 1946-1955*. University of California. Available at https://search.proquest.com/openview/5fe6ccd8e023e9c581ebdbb98631a7bd/1?pq-origsite=gscholar&cbl=18750&diss=y

Škof, N.M. (2013). Miyazawa Kenji: Interpretation of his Literature in Present Japan. *Asian Studies (Quezon City, Philippines)*, *1*(2), 91–106.

Srinivas, H. (2023, May 20). *Environmental Management and Community Participation: Enhancing Local Programmes*. GDRC Research Output: Concept Note Series E-169. Kobe, Japan: Global Development Research Center. https://www.gdrc.org/uem/iso14001/ems-comparti.html

Tilbury, D., & Wortman, D. (2004). *Engaging people in sustainability*. IUCN. Available at: https://books.google.com/books?hl=en&lr=&id=FypMpjqPv0C&oi=fnd&pg=PR5&dq=people+and+environment+and++sustainability&ots=v3gPYSnqtu&sig=5DeX9edfvyAli63lCfhlUCQl-Es

United Nations and United Nations. (Ed.). (2008). *Achieving sustainable development and promoting development cooperation: dialogues at the Economic and Social Council*. United Nations.

Walsh, J. C., & High, S. (1999). Rethinking the concept of community. *Histoire sociale/Social History*. Available at: https://hssh3.journals.yorku.ca/index.php/hssh/article/view/4623

ADDITIONAL READING

Hiratsuka, A., & Wakae, K. (2019). Study on Sustainable Rainwater Resource Utilization—Towards Deepening of Homo Environments. *Journal of Water Resource and Protection*, *11*(5), 491–528. doi:10.4236/jwarp.2019.115029

Melissa, T., Ashantha, G., Kaveh, D., Alireza, A., & Marion, L. (2019). Engaging residents from different ethnic and language backgrounds in disaster preparedness. *International Journal of Disaster Risk Reduction*, *39*, 101245. Advance online publication. doi:10.1016/j.ijdrr.2019.101245

Moritoki, Š. (2013). Miyazawa Kenji: Interpretation of his Literature in Present Japan. *Asian Studies (Quezon City, Philippines)*, *1*(2), 89–104. doi:10.4312/as.2013.1.2.91-106

APPENDIX

List of Reviewed Studies

Seven key terms are listed from the reviewed studies for review analysis: disaster, community, People, Sustainable Development, Culture, Skill, Miyazawa Kenji, and Resilience. The Sub-topics under this are community, disaster management, culture, disaster risk reduction, knowledge, skill, response, emergency, post-disaster, philosophy, Indigenous People, Conservation, Resources, and Disaster Resilience. Specific sub-topics also overlap with the topics. The included studies vary from 1973 to 2022.

Table 5. List of reviewed studies

Sl. No,	Topic	Sub-Topic	Author	Year	Published as a Journal/Book/ Government Data/Dissertations/ News Articles etc and Publisher Name
1.	Disaster	People, Community, Knowledge, Skills, Culture, Management	Asian Disaster Preparedness Centre	2002-2009	GOI-UNDP, Disaster Risk Management Programme 2002-2009, Evaluation and Review of Lessons Learned
		Disaster Management	Disaster Management in India – A Status Report	2004	GOI, Ministry of Home Affairs, NDMD.
		Culture,Disaster Risk Reduction	Udayangani Kulatunga	2010	International Journal of Strategic Property Management
		Management	Pulak Das	2012	Asia Pacific Journal of Social Sciences
		Knowledge, skill, emergency care	Kamal et al.,	2012	Nurse Media Journal of Nursing
		Skill, Response	Noguchi *et al.*,	2016	Prehospital and Disaster Medicine
		Skill, Post-disaster	Tara.M.Powell, Tuyen Bui	2016	School mental Health
		Skill,Disaster Management	Joseph Albanese, James Patura	2018	Journal of business continuity and emergency planning
		Indigenous Knowledge, Disaster Risk Reduction	Mahfuzul Haque	2018	Palgrave Macmillan, Singapore.
		Approaches to disaster	Ronald.W. Perry	2018	Handbook of Disaster Research (Springer International Publishing)
		Philosophy,	Per Sandin	2018	Springer International Publishing

continued on following page

Table 5. Continued

Sl. No,	Topic	Sub-Topic	Author	Year	Published as a Journal/Book/ Government Data/Dissertations/ News Articles etc and Publisher Name
		Indigenous People, Disaster Management	Hoyos et al.	2020	Latin America and Caribbean-Blogs
		Culture, Reconstruction	Siriwardhana et al.,	2021	International Journal of Disaster Risk Reduction
		Disaster Risk Reduction	Perwaiz, Aslam; Parviainen, Janne; Somboon, Pannawadee et al.	2021	UNDRR
		Prevention of disasters in Japan	Web Page	-	Web Page - (Disaster Education in Japan: Preparing for Natural Disasters to Protect Kids' Lives), Kids Web Japan, Ministry of Foreign Affairs of Japan
		Management of disaster	Mishra et al.		What is Disaster? Alternative Approaches to Disaster Analysis. Odisha State Open University E-pathshala
2.	Community	Community	David. B. Clark	1973	The Sociological Review- Journal Article
		Community	Rachel Jewkes, Anne Murcott	1996	Social Science and Medicine - Journal
		Community, Conservation, Environment, Resources	Arun Agrawal Clark. C. Gibson	1999	World Development
		Community	John C. Walsh, Steven High	1999	Histoire Sociale- Journal Article
		Community	Katz et al.	2004	Annals of the International Communication Association – A Journal Article
		People, community, knowledge, skills and Japan	Pathirage *et al.*	2014	Prepared for the global assessment report on disaster risk reduction- A journal article
		Community	Scott Greer	2017	The concept of community: Reading with Interpretations- A Book
		Community, Environment, people	Hari Srinivas	2020	Note Series- Urban Environmental Management
3.	People, Sustainable Development	People, Environment, Sustainability	Daniella Tilbury, David Wortman	2004	IUCN International Union for Conservation of Nature) - Book

continued on following page

Table 5. Continued

Sl. No,	Topic	Sub-Topic	Author	Year	Published as a Journal/Book/ Government Data/Dissertations/ News Articles etc and Publisher Name
		Sustainable Development	United Nations	2008	Book, United Nations Economic and Social Affairs.
		People, Environment, Sustainability, Management	Gill Mandip	2012	Research Journal of Recent Sciences, ISSN
		People	Thomas Gross	2013	Journal article European Review of Public Law
		People	Web Page	-	People Definition & Meaning - Merriam-Webster
		Environment Climate Change	Government Data	2015	21st Conference of the Parties to the United Nations Framework Convention on Climate Change.
		Climate change Environment	Huq *et al.*	2021	Climate Vulnerability and Resilience in the Global South: Human Adaptations for Sustainable Future
4.	Skill	Skill	Jonathan Payne	2004	SKOPE, ESRC-Economic and Social Research Council
		Skill	Alan Ovens, Wayne Smith	2006	Journal of Physical Education, New Zealand
5.	Culture	Culture	Arthur Asa Berger	2000	M/C journal (Media/Culture)
		Culture,	Rahmani et al.,	2022	Current Psychiatry Reports
6.	Miyazawa Kenji	Miyazawa Kenji	Yoko Kubota	2011	Reuters U.S Web Page
		Miyazawa Kenji	Naoko Koike	(2012).	HOKUGA- A journal article
		Miyazawa Kenji	Morihiro Niino	2012	Critical Stages / Scenes Critiques- The IATC webjournal/Revue web de l'AICT
		Miyazawa Kenji	Nagisa.M. Škof	2013	Asian Studies
		Miyazawa Kenji	Elsiemae Ann Ito	2020	Ph.D. Thesis, The Dark Realism of Miyazawa Kenji, Portland State University
		Miyazawa Kenji	Michael Larson	2021	The Asian Conference on Cultural Studies
7.	Resilience	Culture, Disaster Risk Reduction, Resilience	Udayangani Kulatunga	2010	International Journal of Strategic Property Management
		Resilience, Disaster Management	Mika Shimizu	2012	Risk Hazard and Crisis in Public Policy
		Miyazawa Kenji, disaster, resilience	Eric James Siercks	2022	Book, University of California, Lo Angels

Chapter 6
Examining the Impact of Landslide Hazards in the PeriUrban Areas by Global and Local Modelling Techniques

Arif Çagdaş Aydinoglu
Gebze Technical University, Turkey

Eda Ustaoglu
Abdullah Gul University, Turkey

Suleyman Sisman
https://orcid.org/0000-0002-0924-1092
Gebze Technical University, Turkey

Gehver Aydogdu
Ministry of Environment, Urbanization, and Climate Change, Turkey

ABSTRACT

Landslides can be considered one of the most severe natural hazards globally, and their management has a key role to inman safety. Landslide susceptibility maps can help the sustainable management of peri-urban areas such as determining the target areas of projects to develop landslide resistant areas, forest planning, infrastructure planning, and land use zoning. This study aims to analyse the relationship between landslide occurrence and its determinants in a peri-urban region, namely Taşlıdere Basin located in Güneysu District of Rize, Turkey, through providing both global and local regression models. To consider spatial non-stationarity, geographically weighted regression was used as a local model. It was found that the local model outperformed the global model in the estimation of landslide susceptibility determinants. The spatial analysis results indicated that almost all variables had heterogeneous variation over the study area. Therefore, this study provides a methodology for understanding the local dynamics of landslide occurrence.

DOI: 10.4018/979-8-3693-2721-0.ch006

INTRODUCTION

Landslides could be one of the effects of climate change, increasing their frequency and posing increasing risk to human society (Lin et al., 2020; Zhang et al., 2021; WHO, 2023). This is especially true in places with a high population density and expanding land use development (Fuchs et al., 2017) where landslides not only damaged infrastructure and settlements but also resulted in fatalities and serious injuries (Leroy et al., 2022). The status and impact of landslides are anticipated to worsen in Turkey as a result of climate change with a rise in total precipitation and increase in the prevalence and intensity of rainfall events. This was confirmed for the peri-urban area namely Taşlıdere Basin located in the Güneysu District of Rize, Turkey, probably one of the most landslide-prone regions in all of country. This situation calls for better landslide management, especially in developing nations such as Turkey where the hazard is worsening due to a number of factors, including inadequate infrastructure, increased vulnerability, low degree of resilience, and the absence of powerful, efficient and sustainable mitigation measures. In this context, this chapter focuses on the geographic relationship between landslide occurrence and its affecting factors in Taşlıdere Basin.

The use of global regression analysis in the assessment of landslide hazards is common in the literature. The issue with the global regression models is that they presume input variable independence and a constant relationship between variables across space. In fact, the relationship between landslide susceptibility and response variables may change in different locations. Therefore, the application of global models in certain situations may lead to biased parameter estimations, misleading significance tests, and causing false conclusions and inefficient intervention (Li et al., 2017). Local spatial modeling approaches such as Geographically Weighted Regression (GWR) can be utilized to analyse the geographic relationships, providing an improved understanding of spatial processes. The subject method has been utilized in various environmental studies for producing appropriate estimations of the spatial interactions (Nilsson, 2014; Landry et al., 2021). This chapter aims at contributing to the literature through adopting both global (Logistic Regression, LR) and local (GWR) regression models to search the relationship between landslide susceptibility and geophysical, environmental, and locational fifteen variables in a highly landslide-prone region of Turkey namely Taşlıdere Basin.

LITERATURE REVIEW

Factors in the natural environment that affect the likelihood of landslides include physical factors, lithology, geomorphological factors, faults, soil conditions, land cover/use and rainfall. The likelihood that a landslide will occur in a specific location during a specified time frame is known as landslide hazard (Varnes, 1984). Landslide hazard is related to two aspects: First, Landslide susceptibility is a measure of how likely it is that a particular set of influencing terrain conditions (e.g. lithology, slope, land cover etc.) would result in a landslide in a given location (Crozier and Glade, 2005). Second, the timing of landslides is related to the incidence of triggering events (e.g. rainfall, earthquakes etc.). In the literature, a wide range of approaches and procedures for landslide hazard assessment have been created and used including both qualitative and quantitative techniques. The first group is centred on geomorphological assessment, in which the landslide hazard is directly estimated using the skills and experiences of professionals; a direct correlation between previous landslides and causative terrain factors is identified and used to construct landslide hazard (Huabin, 2005; Mandal and Maiti, 2015). The second group is based on quantitative methods on the relationship between causal factors and landslides (Tsangaratos et al., 2017). Statistical methods are commonly used for the investigation of the functional links between landslide controlling elements and the probability of landslides. Multivariate and logistic regression (Tsangaratos and Ilia, 2016), bivariate statistical models (Kornejady et al., 2018), fuzzy logic (Aghda et al., 2018), multicriteria analysis (Karpouza et al., 2021), and artificial neural network (Solanki et al., 2022) are some examples of statistical methods.

There are limited studies conducted for the Taşlıdere Basin in Turkey aiming at assessing the landslides in the subject location. Yüksek et al. (2013) is an example of a macro level study which focused on both qualitative and quantitative evaluation of big floods in the Eastern Black Sea Basin of Turkey, also covering the Taşlıdere Basin. A more recent study by Süme et al. (2021) focused on landslide mitigation methods for Kireçhane and Kırklartepe locations within the Taşlıdere Basin with a particular focus on a detailed drainage project in the area. Differently, Aydınoğlu and Altürk (2021) developed landslide susceptibility maps for the Taşlıdere Basin using statistical and Machine Learning (ML) techniques. A final example is Şen (2019) who assessed the flood risk of the water basins in Rize including the Taşlıdere Basin using climatic and hydrological models. However, these studies were generally applied to either assessing the flood risk in the study area or the landslide was evaluated by developing a susceptibility map through focusing on selected factors and applying global models or introducing mitigation methods to lower the risk of landslide susceptibility. If the relationship between landslide susceptibility and determinant factors are set constant as it is the case in the global models, there may be bias in the models as it does not consider spatial non-stationarity.

MATERIALS AND METHODS

The implemented methodological steps and the relationships between used approaches are summarized in Figure 1.

Study Area

The Güneysu district, about 107 km^2, is situated in the southeast part of Rize province of Turkey. Taşlıdere Basin located in this district was selected as a study area covering about 330.92 km^2 (Figure 2). When the frequency range and the deMultipth of the landslide events in the study area were examined, it was observed that shallow landslides occurred mostly (Aydınoğlu & Altürk, 2021). Landslides have frequently occurred due to geological, topographic, and climatic characteristics, and unsuitable land use. Can et al. (2005) reported that 121 mudflows occurred in 2002 due to short-term heavy rainfall and these flows have caused 17 loss of human lives and the destruction in 4 properties. Besides, Dağ and Bulut (2012) indicated that in Çayeli (Rize) and its surroundings, particularly after heavy precipitation, shallow landslides occurred and caused huge economic damage and loss of life (Nefeslioglu, Gokceoglu, & Sonmez, 2008; (Nefeslioglu & Gokceoglu, 2011).

The study area is rainy in all seasons, with cool summers and warm winters as examples of typical Black Sea climate. The study area located in the basin has an average annual rainfall of 2233.5 mm which is the highest precipitation in Turkey according to meteorological measurements, an average annual high temperature of 18 °C, and the average temperature of 14.1 °C. According to Erinç's (1965) Precipitation Activity Index, the climate type is very humid and the vegetation type is a very humid forest zone in the study area. This area has dense and lush natural vegetation. The area has broad-leaved coastal forests at an altitude between the coastal

Figure 1. Methodological process steps

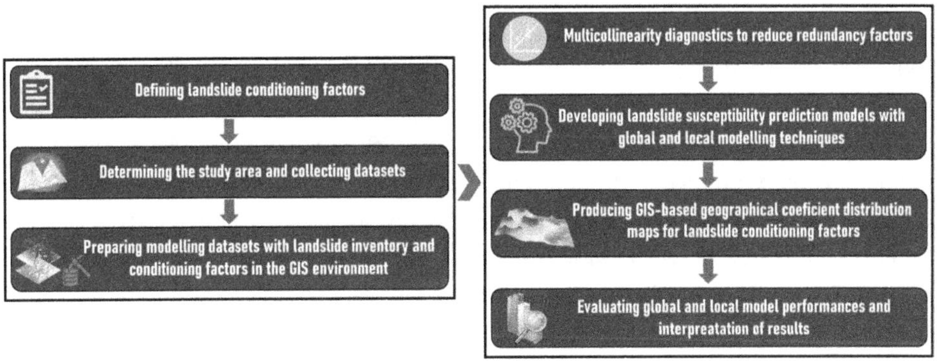

Figure 2. The study area (Taşlıdere Basin from Rize/Turkey)
General Directorate of Combating with Desertification and Erosion (GDCDE), 2016

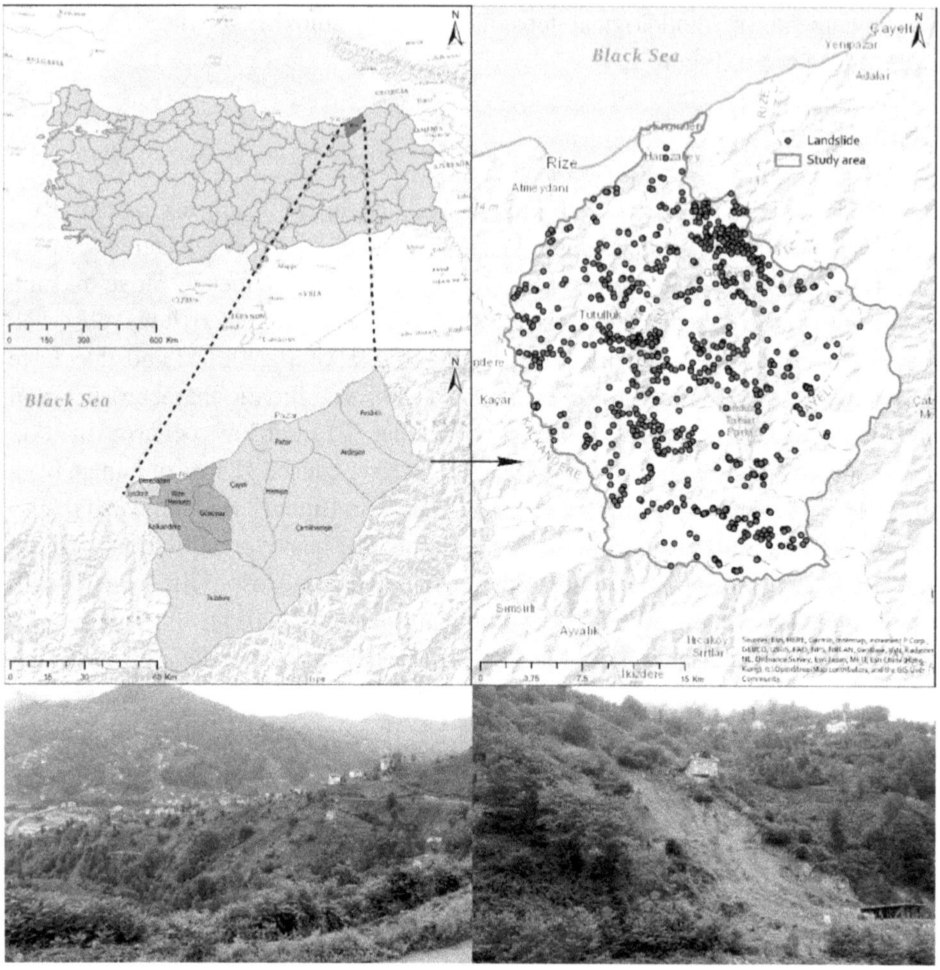

level and 750 m from the coastal level. This part is characterised by the formation of dense and lush forests and the formation of a rich forest called Colchic Flora. The 800-1400 m altitude step is determined as a mixed forest zone. Coniferous species dominate as the altitude increases especially higher than 1600 m. Forests leave their place to alpine meadows at locations higher than 2000m. While 57.15% of the study area is covered with fertile or degraded forest textures, the remaining 36.96% is an agricultural area is mainly utilised as tea cultivation area (GDCDE, 2016).

The study area is within the Eastern Pontides tectonic units (Ketin, 1966; Yılmaz, Güner, & Şaroğlu, 1998). The region has large-scale faulting in NE–SW direction. This fault structure was formed by the effect of compression in the N–S direction (Şengör & Kidd, 1979; Barka, 1984; Şaroğlu & Yılmaz, 1986; Koçyiğit, 1985; Yılmaz et al., 1997; Yilmaz et al., 2000). The pointed tectonic unit where the basin is located was affected by the Austric, Anatolian, and Attic tectonic units. In the NE-SW trending fault system, a thrust is observed along the tectonic boundary between the Upper Cretaceous-aged Caglayan formation and the Eocene-aged Kabakoy formation (Yılmaz et al., 1998). The North Anatolian Fault Zone (KAF in Turkish) is another active fault system that is 175 km away from the study area. This fault zone, a right-sided strike-slip fault, is approximately 1200 km long and has a width varying between 100 m and 10 km (Şaroğlu, Emre, & Boray, f1987; Şengör & Kidd, 1979).

In the Eastern Black Sea region, chemical degradation processes are observed at lower elevations while physical degradation is observed at higher elevations. Periglacial-glacial morphological shapes are common at altitudes above 2000 m in the region (Erinç, 1944; Atalay, 1982). Nevertheless, the Taşlıdere basin is mainly located in the region where chemical degradation is observed because chemical degradation is more intense in areas below 2000 m altitude. The study area is in the E–W trending East Black Sea mountain system and in a region where high relief features are viewed. The topographic distance within the basin changes in a short time and the topographic slope increases towards the south depending on the elevation. Increasing slope levels correspond to the central and northern parts of the basin where morphology such as hill and ridge is prevalent. The drainage structure of the basin has been affected by the changing slope values in a short distance and these high-energy drainage systems created young valleys with "V" shaped transverse profiles. In general, the distribution of lithological units affected the morphology in the study area where igneous rocks are common. The drainage system, which has a dendritic appearance in the basin, shows a semi-parallel feature with lithology in some parts in the south (GDCDE, 2016).

Preparation of the Dataset

Landslide Inventory Data

The landslide inventory dataset includes about the location, activity, type, and physical characteristics of landslide events that have occurred in the Taşlıdere Basin. As part of the projects undertaken by the General Directorate of Combating with Desertification and Erosion (GDCDE, 2016), the shallow landslide inventory less than 5m depth of sliding surface was determined by using color infrared stereo aerial photographs at a scale of 1: 15,000 and high-resolution satellite images. The dataset include 878 landslide events with point geometry is shown in the study area. Landslide inventories are utilized to acquire LS, hazard, and risk maps.

Landslide Conditioning Factors

LS studies previously conducted were examined to determine landslide conditioning factors given that there is no a common standard approach for the determination of the factors. The factors were determined considering the general characteristics of the study area and the availability of the datasets was also important. In line with the aims of this chapter, fifteen landslide factors including elevation, aspect, slope, lithology, land cover, Topographical Roughness Index (TRI), Sediment Transport Index (STI), Stream Power Index (SPI), Topographical Wetness Index (TWI), plan and profile curvatures, proximity to drainage and road network, road and drainage network density were selected as explanatory variables (Arabameri, Pradhan, & Rezaei, 2019; Faraji Sabokbar, Shadman Roodposhti, & Tazik, 2014; Hong, Pradhan, Sameen, Chen, & Xu, 2017; Chowdhuri et al., 2020; Sahin, Colkesen, Acmali, Akgun, & Aydinoglu, 2020; Yalcin, Reis, Aydinoglu, & Yomralioglu, 2011; Yilmaz, 2009; Li, Liu, Han, & Dou, 2020; Chen et al., 2019). All conditioning factors were brought and edited within the GIS. For these factors defined in Table 1, GIS analyses were performed by using related data sources to produce geographic datasets for conditioning factors in ArcGIS Pro software. Datasets defining factors were generated in a raster data format with a 10x10m gespatial resolution (Figure 3).

The continuous datasets such as elevation, TRI, plan and profile curvature, STI, proximity to the drainage network, drainage network density, SPI, TWI and road network density were used with data values in their original format. The categorical datasets such as land cover, aspect and lithology variables were transformed into dummies representing the presence (1) or absence (0) of categories. In the literature, one of the most preferred approaches is to use variables reflecting categorical and continuous data structure together as dummy and original values. In this chapter, lithology was transformed into six dummy variables. To prevent local redundancy problem of dummy variables, land cover and aspect variables were reclassified before being transformed into dummy variables. In this context, the land cover factor was grouped into four categories: 1-Agriculture, 2-Forest (fertile forest, poor forest, forest soil), 3-Settlement, and 4-Stony. However, the stony variable that has only two landslide case points was not included in the models. The aspect factor was reclassified in four categories: North, South, East, and West.Considering categorical variables, defining one of their sub-categories as a reference category is a necessary process step for modeling. Therefore, the "Agriculture" group for land cover, "Kaçkar granitoid" group for lithology and "North" group for aspect were determined as reference categories because these categories include a large part of the study area.

The geographical database includes both dependent variable (landslide inventory) and explanatory variables (conditioning factors). The relationships between explanatory variables were evaluated by using Pearson Correlation Analysis and

Table 1. Descriptions and sources of conditioning factors determined in the study

Factors	Definition	Data Source
Elevation (Figure 3a)	Elevation values representing height difference from the sea surface for each location/ pixel	DEM produced from 10m contours (Provided by General Directorate of Mapping)
Slope (Figure 3b)	The maximum change in elevation from neighbour raster cells	Slope analysis derived from DEM
Aspect (Figure 3c)	The downslope direction measured clock-wise in degrees from the north	Aspect analysis derived from DEM
Land Cover (Figure 3d)	Land cover classes defining forest, agricultural, watery, stony, and residential areas	Produced from forest management maps (Provided by the General Directorate of Forestry)
Lithology (Figure 3e)	Classes expressing general geological formation	Digitized from 1:100,000 scaled Geology Map (Provided by the General Directorate of Mineral Research and Exploration)
Topographical Roughness Index (TRI) (Figure 3f)	The elevation difference between neighboring cells in the focal areas	Calculating the height value and the minimum height difference and its ratio to the maximum and minimum height difference by focal statistics derived from DEM
Plan Curvature (Figure 3g)	The change in slope orientation along with a certain topographic height	Curvature analysis derived from DEM
Profile Curvature (Figure 3h)	The change in the hillside downward slope	Curvature analysis derived from DEM
Sediment Transport Index (STI) (Figure 3i)	The potential of sediment transport in the sediment network	Calculated by $LS = (m+1)\left(\frac{\alpha}{22.13}\right)^m \left(\frac{\sin\beta}{0.896}\right)^n$ equation with flow accumulation (α) and slope information (β) derived from DEM (m=0.4, n=1.3)
Proximity to drainage network (Figure 3j)	The shortest euclidean distance from any drainage network for each location	Distance analysis for vector layers of stream network, derived from DEM
Drainage network density (Figure 3k)	Stream density for each location	Line density analysis from vector layers of stream network
Stream Power Index (SPI) (Figure 3l)	Erosion power of flowing water	Calculated by $SPI = \alpha \tan(\beta)$ equation with flow accumulation (α) and slope information (β) derived from DEM

continued on following page

Table 1. Continued

Factors	Definition	Data Source
Topographical Wetness Index (TWI) (Figure 3m)	The effect of topography on hydrological actions, the geospatial distribution of soil moisture, and surface saturation	Calculated by $TWI = \ln\left(\frac{\alpha}{\tan\beta + C}\right)$ equation with flow accumulation (α) (and slope information (β) derived from DEM (C=0.001)
Proximity to Road Network (Figure 3n)	The shortest euclidean distance from any road for each location	Distance analysis from vector layers of the road network (Provided by the General Directorate of Forestry)
Road network density (Figure 3o)	Road density for each location	Line density analysis from vector layers of the road network

Source and year: This table was created by authors, 2024

Variance Inflation Factor (VIF) index to obtain more accurate models. VIF used in LS studies for multicollinearity diagnosis determines the levels of intercorrelation between explanatory variables. It is calculated as shown in Eq. 1 (Allison, 2012; Kalantar et al., 2020):

$$VIF_i = \frac{1}{1 - R_i^2} \tag{1}$$

where indicates the correlation value between explanatory factors in the dataset. To avoid multicollinearity, the VIF should be 5 or at least less than 10. In the literature, another technique used commonly to avoid multicollinearity is Pearson's Correlation Analysis. It is a statistical method measuring the relationship between two quantitative variables. According to this method, correlation coefficients are obtained as given in Eq. 2 (Kalantar et al., 2020):

$$r_{xy} = \sum_{i=1}^{n} \frac{X_i - \bar{X}}{\sum_{k=1}^{n} (X_i - \bar{X})^2} \times \frac{Y_i - \bar{Y}}{\sum_{k=1}^{n} (Y_i - \bar{Y})^2} \tag{2}$$

where r_{xy} indicates correlation coefficient and it is defined with a value between [-1, 1]. If this value is -1 or close to 1, it is understood that there are strong relationships between the explanatory variables. X_i and Y_i indicate the related value of X and Y for i_{th} factor. $\bar{X}$ and $\bar{Y}$ are the mean value for X and Y explanatory variables. Explanatory factors that are highly correlated remark linear dependence

Figure 3. Produced maps with GIS analysis techniques for landslide conditioning factors

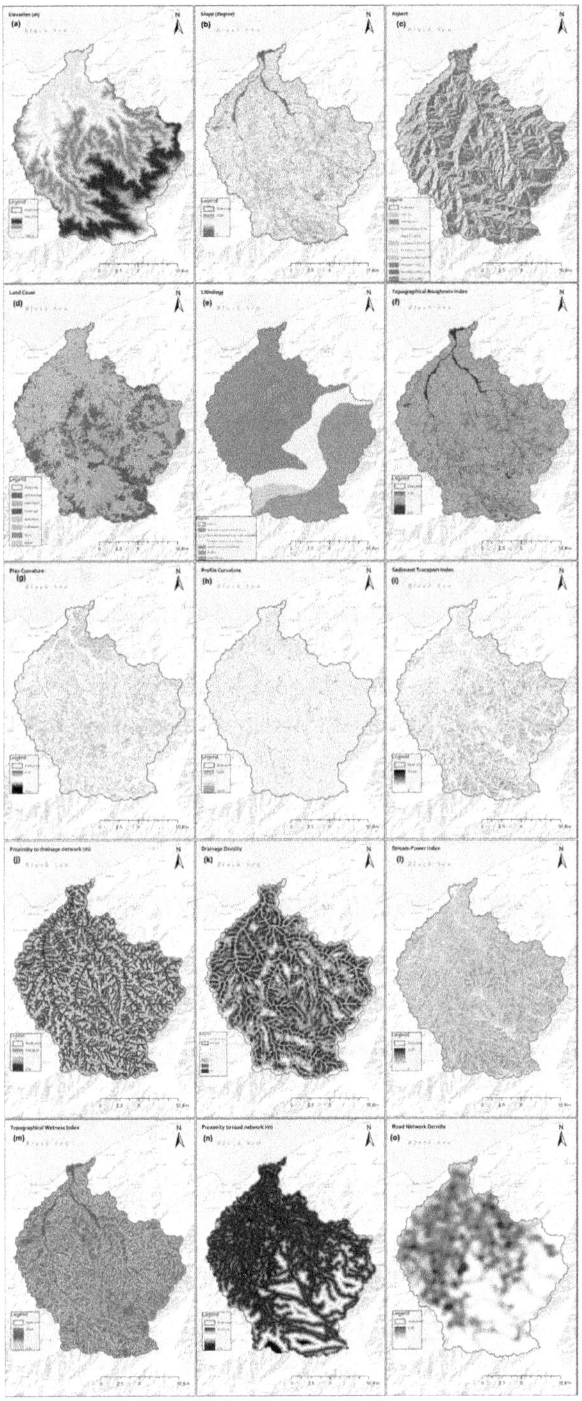

where they almost have the same impact on the dependent variable. Hence, to obtain more significant results, one of the explanatory variables can be excluded from the modeling process. To avoid multicollinearity in the literature, the independent variables with a correlation coefficient of more than 0.70 are not included in the model (Kalantar, Ueda, Lay, Al-Najjar, & Halin, 2019; Kalantar et al., 2020).

MODELING TECHNIQUES

Logistic Regression (LR)

Logistic Regression (LR) is a global regression technique to investigate the relationships between a categorical dependent variable and independent variables (Yalcin et al., 2011). LR is used commonly to evaluate the presence of landslide events by using the values of explanatory variables (Chalkias, Kalogirou, & Ferentinou, 2014). LR is the statement of probability values of the dependent variable according to the explanatory variables, which define the cause-effect relationship between the dependent and explanatory variables (Althuwaynee, Pradhan, Park, & Lee, 2014). In the LR model, a link function is added to the traditional linear model and the dependent variable is categorical, while the explanatory variables may be continuous or categorical data which do not necessarily have a normal distribution (Zhao, Wang, Jiang, Liu, & Wei, 2019). In LS studies, the global LR model is one the most popular techniques to uncover the association between the presence or absence of a landslide by using various factors. The global LR model is defined by Eq. 3 (Erener & Düzgün, 2010; Kavzoglu, Sahin, & Colkesen, 2014):

$$Logit(p) = \ln(\frac{p_i}{1-p_i}) = \beta_0 + \beta_1 \times X_1 + \beta_2 \times X_2 + ... + \beta_n \times X_n \quad (3)$$

From Eq. 3, the probability values are calculated for landslide occurrence and given in Eq. 4.

$$p(Y=1) = p_i = \frac{1}{1+e^{-(\beta_0+\beta_1 \times X_1+\beta_2 \times X_2+...+\beta_n \times X_n)}} \quad (4)$$

Where indicates the probability of landslide occurrence in the susceptibility index is 1 $p = (Y = 1)$, $p_i / (1 - p_i)$ is the frequency ratio or odds ratio, β_0 is the model intercept (constant), $X_1, X_2, ..., X_n$ are the explanatory variables and

$\beta_1, \beta_2, ..., \beta_n$ are model coefficients for the explanatory variables, also these coefficients measure the impact of any explanatory variable in the global LR model to explain variations in the independent variable.

Geographically Weighted Regression (GWR)

The estimated coefficients by using global regression models (such as LR) don't change across space and consider geographical stationarity in the modeling relationships between variables. This situation is disadvantageous for global models (Fotheringham, Charlton, & Brunsdon, 2001). Especially LR models struggle to meet the requirements of LS studies (Chalkias et al., 2020). Differently, local models based on statistical techniques may be more suitable to understand geographically changing relationships being more accurate to model real-world problems.

The GWR approach, a local modelling technique, is a type of kernel-weighted model which is a geographic adaptation of global regression approaches. The GWR model allows to vary geographically in modeling relationships between the dependent variable and explanatory variables (Brunsdon, Fotheringham, & Charlton, 1996). While global models generate only one model equation for all the study area, the GWR generates an equation for each landslide points separately (Yu, Wang, Niu, & Hu, 2016). In other words, in the GWR, model coefficients are predicted based on the geographical coordinates of the data and these coefficients may show different values and signs at different data points (Arabameri et al., 2019). The GWR is calculated as given in Eq. 5 (Fotheringham & Oshan, 2016):

$$y_i = \beta_0(u_i, v_i) + \beta_1(u_i, v_i) \times X_1 + \beta_2(u_i, v_i) \times X_2 + ... + \beta_n(u_i, v_i) \times X_{in} + \varepsilon_i \quad (5)$$

where indicates the landslide occurrence for location i, (u_i, v_i) indicate coordinates for i in the study area, $\beta_0(u_i, v_i)$ is the model constant term for i, n indicates the number of sample and $X_{i_1}, ..., X_{i_n}$ are the explanatory variables used in the GWR model, $\beta_1(u_i, v_i), ..., \beta_n(u_i, v_i)$ are local regression coefficients for related independent variables at i and ε_i is an error term at i.

In the local GWR model, a distance-based weighting function is utilized for providing more influence on the regression coefficient $\beta_n(u_i, v_i)$ of data points located closer than data points located further apart. The regression coefficients are predicted from Eq. 6 (Fotheringham, Brunsdon, & Charlton, 2003; Chalkias et al., 2020; Feuillet et al., 2014):

$$\beta_0(u_i, v_i) = (X^T W(u_i, v_i) X)^{-1} X^T W(u_i, v_i) y \tag{6}$$

where $\hat{\beta}(u_i, v_i)$ indicates the vector of model coefficient estimations which are location-specific, is a matrix of the explanatory variables, $W_{(u_i,v_i)}$ are the geospatial weights matrix with a size of $(n \times n)$ and y is the data vector for the landslide occurrence. $W_{(u_i,v_i)}$ is a diagonal matrix and it is computed for each point i at which regression coefficients are estimated. While the matrix in Eq. 7 includes diagonal elements representing the geographic weighting of the n observed data for point i, the non-diagonal elements of the matrix have the "0" value.

$$W_i = \begin{pmatrix} w_{i1} & 0 & 0 & \cdots & 0 \\ 0 & w_{i2} & 0 & \cdots & 0 \\ 0 & 0 & w_{i3} & \cdots & 0 \\ \vdots & \vdots & \vdots & \ddots & 0 \\ 0 & 0 & 0 & 0 & w_{in} \end{pmatrix} \tag{7}$$

In the GWR calibration approach, there are two different weighting approaches called Fixed and Adaptive spatial kernel to assign a weight to each data around a point i. In the Fixed type, the kernel extent is defined with the distance to the regression point. The bandwidth parameter is same for all locations. Weights are influenced by the density and dispersion of data points in the space for this kernel. The kernel extent is defined with the number of nearest neighbor's locations of interest in the Adaptive type. The kernel may change for any location in space. This kernel adapts itself to local variations and density (Oshan, Li, Kang, Wolf, & Fotheringham, 2019). Fixed and adaptive kernels may have two forms: Gaussian or Bi-square (Chalkias et al., 2020; Du, Wang, & Guo, 2014; Hong et al., 2018). In this scope of study, an adaptive Gaussian was used as kernel function for the GWR model. This function is defined in Eq. 8:

$$W_{ij} = \exp(-d_{ij}^2 / \theta_{i(k)}^2) \tag{8}$$

where W_{ij} indicate the weight of observation for location j to predict the local coefficient at location i, d_{ij} indicate the distance between points i and j and $\theta_{i(k)}$ is an adaptive bandwidth size that is determined by using k_{th} Nearest Neighbor (KNN) euclidean distance. Here the determination of the optimum bandwidth may significantly affect the performance of the GWR model. Therefore, to obtain the optimum bandwidth and to select the most suitable procedure, the purpose was to achieve the smallest value of corrected Akaike Information Criterion (AICc) in each trial (Akaike, 1973).

RESULTS

Landslide Occurrence Statistics

Understanding the physical conditions specific to the study area and the processes triggering landslides is of great importance in assessing the possibility of landslides occurrance. Variables were calculated for each landslide inventory in the GIS environment to present the occurrence percentages of landslides in the sub-categories of each factor. Results are presented in Table 2.

The elevation of the watershed increases from a minimum of 0 m at the sea surface towards a maximum of 2,442 m at the south and the average height is 845 m. Slope affects the gravity-dependent motion of water and other materials. Deep landslides are observed in the medium-slope areas between 10° and 30°, and severe landslides are seen on the steeper slopes.

Approximately 46% of the study area has north, 30% south, 14% west, and 10% east aspect values. The roughness class of the watershed was 34.67% in the medium and 62.47% in the high classes. Roads are an effective factor in landslides as they can put extra load on the slope and cut the hillside continuity. The road network density values in the study area were determined as minimum 0, maximum 9.67, and average 3.53. The distance of the watershed to the road is 205.17 m on average and 2,150 m at the maximum. Streams affect slope stability by abrading the slope from the heel and saturating the material with water at the stream level. The drainage network density of the study area expresses the rate at which basins are split by rivers. The distance to the drainage network in the study area is 807.31 m at maximum, 0 at minimum, and an average of 132.21 m, and the standard deviation is 102.61 m.

MULTICOLLINEARITY ANALYSIS

In this chapter, the GWR approach was followed with a procedure including two steps to obtain more significant models and to reduce redundancy factors. In the first step, Pearson's Correlation Analysis was implemented to measure the relationship between landslide conditioning factors including explanatory variables. The analysis results are presented in Figure 4. Highly correlated ($r > 0.70$) factors were detected

Table 2. The distribution of the landslides according to each group of conditioning factors

Factor	Class	Landslide (%)	Factor	Class	Landslide (%)
Land cover	Forest	49,66	**Aspect**	North	48,41
	Agriculture	49,54		South	28,25
	Settlement	0,57		East	9,80
	Stony	0,23		West	13,55
Elevation (m)	0-250	9,00	**Stream Power Index (SPI)**	0-1	29,84
	250-500	30,98		1-2	37,70
	500-750	21,41		2-3	22,55
	750-1000	14,92		3-4	5,35
	1000-1250	10,14		>4	4,56
	1250-1500	5,13	**Sediment Transport Index (STI)**	1	4,78
	1500-1750	4,44		2	2,96
	1750-2000	3,08		3	14,24
	2000-2250	0,91		4	58,43
	>2250	0,00		5	19,59
Plan Curvature (Plan_Curv)	Concave	33,03	**Profile Curvature (Profile_Cu)**	Concave	35,08
	Flat	27,56		Flat	28,36
	Convex	39,41		Convex	36,56
Proximity to Drainage Network (DN_Proximity)	0-100	44,31	**Drainage Density (DN_Density)**	0,0013-1	2,16
	100-200	32,35		1-2	15,83
	200-300	17,43		2-3	33,14
	300-400	4,44		3-4	32,80
	>400	1,48		>4	16,06
Proximity to Road Network (RD_Proximity)	0-100	60,78	**Road Network Density (RD_Density)**	0-0,13	2,51
	100-200	22,69		0,13-2	11,85
	200-300	5,13		2-4	23,58
	300-400	3,88		4-6	34,28
	>400	7,53		>6	27,79
Geology	Bzdp	53,30	**Slope**	0-10	1,37
	Bzlps	4,21		10-20	14,01
	Rhdlp	9,91		20-30	32,80
	Anblp	15,38		30-40	39,41
	Alluv	0,80		>40	12,42
	Kaçkar	16,40	**Topographical Roughness Index (TRI)**	Low	0,11
				Middle	31,78
				High	68,11

Note: Bzdp: Basalt andesite dacite lava and pyroclasts; **Bzlps:** Basalt andesite lava and sub-pyroclastic sand pebbles; **Rhdlp:** Rhyodacite lava and pyroclasts; **Anblp:** Andesite basalt lava and pyroclasts; **Alluv:** Alluvium; **Kaçkar:** Kaçkar Granitoid.

Source and year: This table was created by authors, 2024

Figure 4. Pearson's correlation analysis between explanatory independent variables

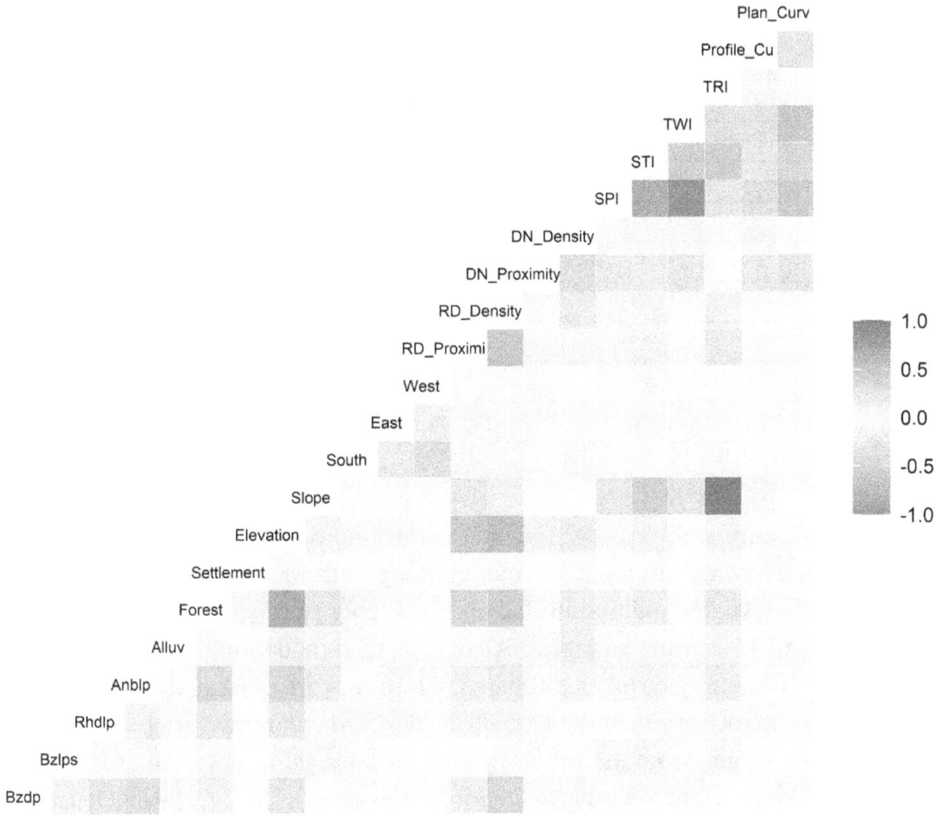

between Slope - TRI factors and SPI - TWI factors. TRI and SPI factors were eliminated to avoid the multicollinearity problem.

In the second step, VIF values were calculated between landslide conditioning factors for multicollinearity diagnosis. The VIF values were found to be less than 10 for all factors. There is no multicollinearity problem between explanatory variables as it is given in Table 3.

Global Modeling With the LR

The LR is one of the most prefferred techniques for LS because it does not include complex mathematical terms. Various factors on landslide occurrence are evaluated with the sign of the estimated coefficients with the LR model. The positive coefficients of the parameters represent landslides with a higher probability of occurring, while the negative coefficients of the parameters represent landslides with a lower probability of occurring (Sahin et al., 2020). The LR model was developed by using the "glm"

Table 3. Results of multicollinearity diagnostic between explanatory variables

Factors	Multicollinearity (VIF)	Factors	Multicollinearity (VIF)
Bzdp	3,2266	East	1,1697
Bzlps	1,2150	West	1,2372
Rhdlp	1,5075	RD_Proximity	1,7358
Anblp	3,2186	RD_Density	2,6257
Alluv	1,2786	DN_Proximity	1,3630
Forest	1,9764	DN_Density	1,3299
Settlement	1,0233	STI	1,9186
Elevation	3,5355	TWI	2,2575
Slope	1,9381	Profile_Curv	1,1943
South	1,3280	Plan_Curv	1,7393

function of R software (glm, 2022) which is free and open-source. Table 4 shows estimated coefficients in global LR model along with the statistical significance level of the landslide conditioning factors, including twenty-one factors with 10 continuous and 11 dummy variables. Also, statistical modeling results for each of the factor coefficients as well as estimated parameters are presented.

According to the results of the global LR model, the dummy variables “Bzdp”, “Bzlps” and “Anblp” from the lithology category, the dummy variables “South”, “East” and “West” from the aspect category, “elevation”, “slope”, “RD_Density”, “DN_Density”, “STI” and “Plan_Curv” from other categories were positively associated with landslides. The dummy variables “Rhdlp” and “Alluv” from the lithology category, “Forest” and “Settlement” dummy variables from the land cover category, as well as “RD_Proximity”, “DN_Proximity”, “TWI” and “Profile_Curv” variables were negatively associated with the landslides.

Statistical significance level (i.e. p-values) results of the global LR model indicated that there are not many factors that contribute significantly to the model. Finally, eight conditioning factors (except intercept) were found significant showing different statistical significance levels. These factors were “Bzdp”, “South”, “Elevation”, “Slope”, “RD_Proximity”, “RD_Density”, “DN_Proximity” and “STI”. These significant factors regarding their signs and significance levels are in line with the findings of LS studies; and therefore, were used as inputs in the local GWR model.

Table 4. The global LR model parameter estimates and the statistical significance levels of landslide conditioning factors

Explanatory Variables	Dummy Variables	Parameter Estimate	Sdt. Error	z Value	Pr(>|z|)
Intercept		-2,003598	0,802550	-2,497000	0,012541*
Lithology	*Bzdp*	0,783396	0,215950	3,628000	0,000286***
	Bzlps	0,290363	0,321853	0,902000	0,366971
	Rhdlp	-0,238282	0,221629	-1,075000	0,282312
	Anblp	0,184961	0,277274	0,667000	0,504729
	Alluv	-0,757403	0,802550	-2,497000	0,012541
	KackarGr	Reference category			
Land cover	Agriculture	Reference category			
	Forest	-0,261482	0,169196	-1,545000	0,122241
	Settlement	-1,027771	0,625662	-1,643000	0,100446
Aspect	North	Reference category			
	South	0,463735	0,142960	3,244000	0,001179**
	East	0,334082	0,215010	1,554000	0,120233
	West	0,155465	0,178198	0,872000	0,382975
Elevation		0,000770	0,000231	3,336000	0,000849***
Slope		0,0207 24	0,009154	2,264000	0,023580*
RD_Proximity		-0,001372	0,000454	-3,024000	0,002495**
RD_Density		0,159676	0,045322	3,523000	0,000426***
DN_Proximity		-0,001918	0,001049	-1,829000	0,067428.
DN_Density		0,071201	0,056405	1,262000	0,206833
STI		0,008163	0,003647	2,238000	0,025189*
TWI		-0,041019	0,062580	-0,655000	0,512173
Profile_Curv		-0,046869	0,087551	-0,535000	0,592424
Plan_Curv		0,134462	0,098335	1,367000	0,171501

Note: Significance: 0 '***' 0.001 '**' 0.01 '*' 0.05 '.' 0.1

Local Modeling With the GWR

The GWR technique was implemented to investigate the local-scale effects of the landslide factors including six continuous and two dummy variables that were determined to be significant in the LR model. The logistic (binary) form was preferred for the GWR model. It was implemented with the free GWR 4.0 software package developed by Nakaya (Nakaya, 2022). To obtain a more accurate local model, the

Table 5. Trial procedure results to obtain optimal bandwidth value

Trial no	Bandwidth			AICc
	Min	Max	Optimum	
1	10	1270	58	1538,903
2	20	1270	52	1538,195
3	30	1270	56	1538,837
4	40	1270	52	1538,195
5	50	1270	52	1538,195
6	60	1270	60	1539,192
7	70	1270	70	1540,831
8	80	1270	80	1544,458

Table 6 Summary statistics for the estimated parameters of the GWR model

Explanatory Variables	Dummy Variables	Min	Max	Mean	Std. Dev.
Intercept		-5,202845	3,318731	-1,55163	1,589405
Lithology	Bzdp	-1,248274	5,276740	0,910693	0,973185
Aspect	South	-0,930052	1,574582	0,376455	0,455750
Elevation		-0,003157	0,003581	0,000764	0,001217
Slope		-0,044276	0,081828	0,018994	0,027406
RD_Proximity		-0,009926	0,010164	-0,002021	0,003486
RD_Density		-0,306010	0,686558	0,175282	0,187892
DN_Proximity		-0,012998	0,003405	-0,002385	0,003062
STI		-0,011615	0,040537	0,006165	0,009318

optimum bandwidth value with 52 neighbors was determined by performing a trial procedure. The results of this trial procedure are presented in Table 5.

Following the GWR analysis, the landslide conditioning factors are summarized by various indicator statistics in Table 6.

Comparison of GWR and LR Model

Global LR and Local GWR models were evaluated by using the corrected AIC criterion (AICc) and overall pseudo R^2 values, which are frequently used in the literature also including the deviance. The value of AICc equals to 1641,284 for LR and 1538,195 for GWR. Deviance and overall pseudo R^2 for the LR were 1598,50

Table 7. Comparison of the LR and GWR models for diagnosis

Goodness-of-Fit Criteria	LR	GWR	Difference
AICc	1641,284	1538,195	103,089
Deviance	1598,50	1321,971	276,529
Overall pseudo R^2	0,15	0,42	0,27

and 0,15% respectively, while for the GWR model, deviance is equal to 1321.971 and overall pseudo R^2 is 0,42% (Table 7). The modeling results clearly indicated that the GWR model achieved more accuracy than the LR model in terms of three performance criteria.

Local Geographical Distribution of Landslide Conditioning Factors

After the local GWR model was developed with eight variables, coefficient distributions were mapped in the GIS environment to investigate the geographic variation of each conditioning variable in the study area. By using this GWR model, a set of results was computed for each landslide inventory in the dataset, including the local coefficients related to each landslide conditioning factor and pseudo R^2 values. These computed local parameter coefficients measure the level of the geographic relationships between the explanatory variables and the landslide occurrence at each landslide location. For the local explanatory parameter estimates and pseudo R^2 values were performed with GIS-based Inverse Distance Weighted (IDW) analysis, and the maps of the coefficient geographic distributions of their variability are presented in Figure 5.

DISCUSSION

Various statistics and machine learning approaches have been utilized to predict landslide occurrence. Although many criteria causing disaster hazards can be determined, the effect of each conditioning factor on landslide hazard has been generally evaluated with global models. In the study area, among 25 conditioning factors, the Land Cover (Agriculture), Aspect (North), and Lithology (Kaçkar granitoid) were determined as reference categories, TRI and SPI factors were eliminated to avoid the multicollinearity. As a result of the global LR model, except intercept, lithology (Bzdep), aspect (South), elevation, slope, proximity to the road network and drainage network, road network density, and STI were determined to

Figure 5. The geographic variation of local explanatory parameter estimates

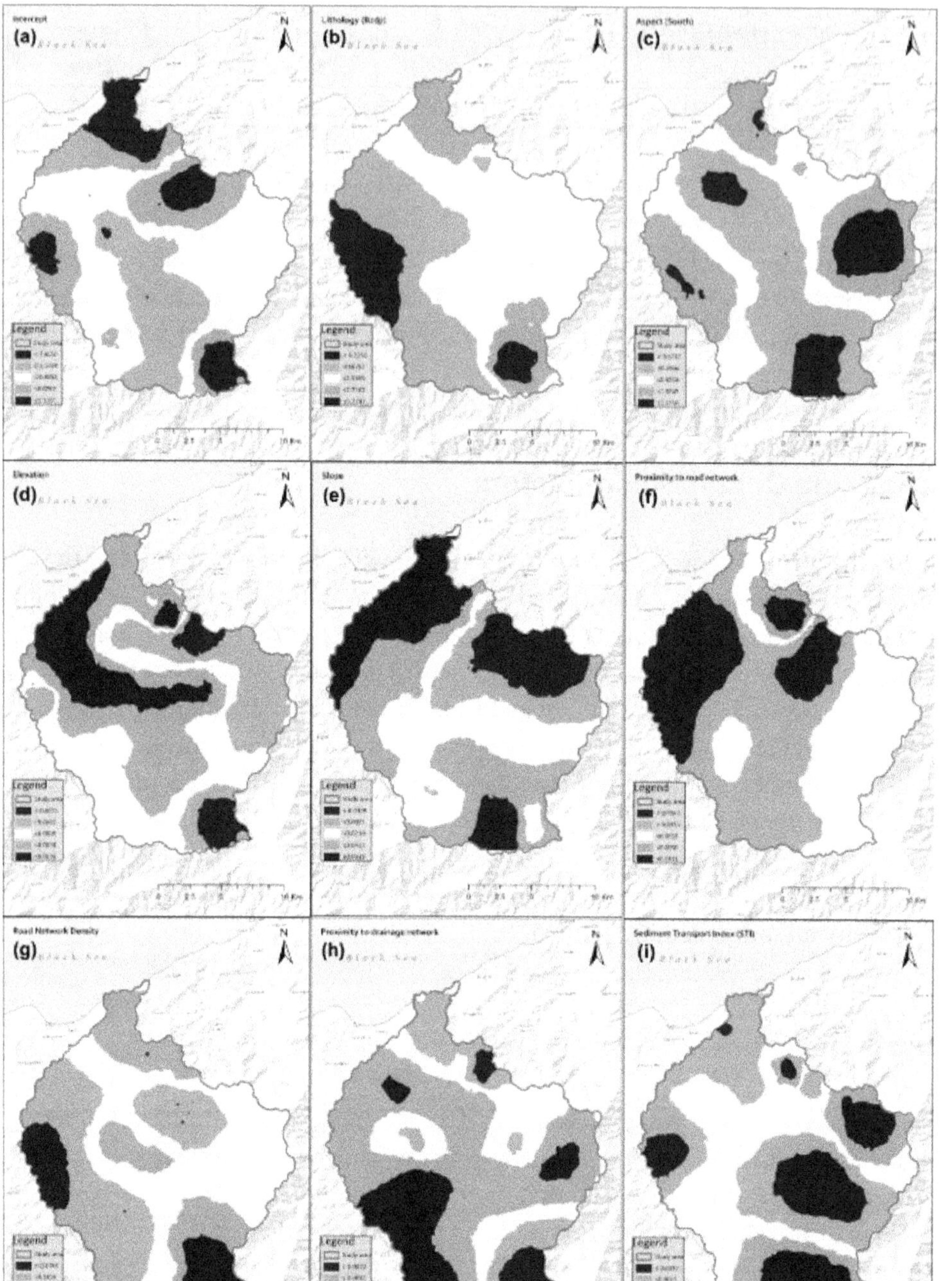

have statistical significance level among the remaining 20 factors. The local GWR model achieved more accuracy than the global LR model was developed to analyze local variations of these eight conditioning factors.

The lithology (Bzdep) factor is positively related to the landslides in almost entire of the study area except for a region in the western part. It is one of the most important factors with a high coefficient value. Its effect increases towards the north and southeastern of the study area especially certain regions, while it has a middle-level effect towards the inland and a low-level effect in the eastern of the study area (Figure 5b). The aspect (South) factor is positively related to LS in most parts of the study area. It has a strong relationship with landslide occurrence in some regions of the north, eastern and western of the study area. It is negatively related to the LS from south to north in the middle of the study area. It reaches the highest negative effect in some parts of the south and northwestern of the study area (Figure 5c).

The elevation factor is associated with LS in most parts of the study area. Especially degree of this relationship increases towards the west and the north of the study area. Differently, the elevation variable is associated with LS negatively in the middle, the southeast, and the northeast of the study area (Figure 5d). The slope factor was similar to the elevation variable. In most parts of the study area, it is associated positively with high values of LS like elevation. While the relationship between slope and landslide occurrence is constantly higher in the northern and the western part of the study area, the degree of this relationship decreases towards the east part of the study area (Figure 5e).

The proximity to the road network is negatively related to the landslides in most parts of the study area and positively in the specific region northeast of the study area. It has the largest effect in the middle and the south of the study area (Figure 5f). The road network density is positively related to the landslides in most parts of the study area, like the proximity to the road network. However, it has a negative effect in a small region located in the west of the study area (Figure 5g). The proximity to the drainage network indicates a heterogeneous relationship with LS in almost all of the study area. It has both positive and negative relationships with landslides in certain regions of the study area. However, the power of the positive relationship increases constantly towards southwestern and western parts of the study area while the power of the negative relationship increases especially towards the southeastern part (Figure 5h). The STI factor is positively related to the landslides in most parts of the study area. In certain regions of the eastern and southern parts of the study area, STI has a strong positive relationship with LS while it has a strong negative relationship in certain regions of the eastern and western of the study area (Figure 5i).

Figure 6. The geographical variation of the local pseudo-R^2 value of the GWR model

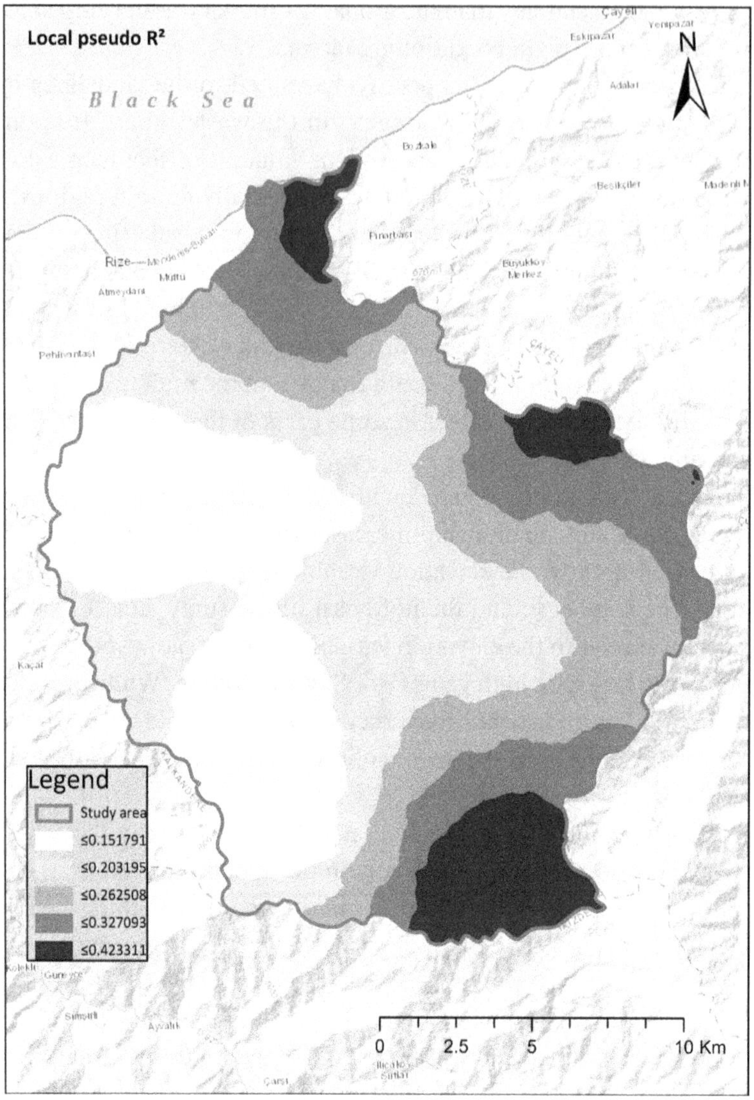

Besides geographic distribution maps, an overall local pseudo-R^2 value distribution map was also produced for the study area (Figure 6). The pseudo-R^2 varies from 0,10 - 0, 42 locally. The model achieves higher values of pseudo-R^2 by increasing from the west to the east and on the overall study area from the north to the south.

CONCLUSION

Landslides are one of the most hazardous natural disasters and a major barrier to sustainable urban development in Taşlıdere Basin, Rize, which is one of the specific areas where landslides are most common in Turkey. Especially due to heavy rainfall that is one of the main characteristics of the region as well as the destruction of natural vegetation by human-driven processes, the destruction of forests, dysfunctional drainage networks, and unplanned new road and infrastructure constructions, landslides are major hazard factors for peri-urban areas of the Güneysu District where Taşlıdere Basin is located. Another reason for the landslide hazards in the subject location is due to the dispersed settlements and illegal constructions in these areas that may influence the soil structure and increase the intensity of landslides. Therefore, examining the factors influencing the landslide occurance appears as a vital aspect of land use planning in hazard prone locations such as the Taşlıdere Basin.

Based on the point data obtained from the GDCDE and providing a comparison of LR model and a local regression model, this chapter investigated the geospatial pattern of landslide casualty factors and quantified the impact of geophysical, environmental, and locational fifteen variables on landslide casualty events. The objective of the local model is to map and identify the non-stationarity of the link between casualty factors and landslides. So, this modelling assists in the detection of the subsequent geographical variability in these relationships and enables the determination of the crucial casualty factors that have to be accounted for a landslide occurance analysis. The analysis of the current study indicated that the use of a local model like GWR can improve a common global model such as LR for assessing landslide occurance. In particular, the performance and reliability of the model increased when the geographical heterogeneity of the relationship between landslide occurance and casualty factors is integrated into the landslide hazards evaluation as it is affirmed by the goodness-of-fit criteria estimated for the models.

According to the results of the global LR model, lithology, aspect, elevation, slope, RD_Density, DN_Density, STI and Plan_Curv are positively associated with the landslides. The dummy variables "Rhdlp" and "Alluv" from the lithology category, "Forest" and "Settlement" dummy variables from the land cover category, as well as "RD_Proximity", "DN_Proximity", "TWI" and "Profile_Curv" variables are negatively associated with the landslides. In the GWR model, the spatial non-stationarity was considered in the estimation of lithology, aspect, elevation, slope, RD_Proximity, RD_Density, DN_Proximity and STI variables. The geographic variation of the estimated coefficients of these variables are presented in Figure 5. From the Figure, the intercepts range from -3.46 to 3.31, the lithology parameter estimates range from 0.22 to 5.27, the aspect estimates range from 0.17 to 1.57, the elevation estimates range from 0.001 to 0.003, the slope estimates range from

-0.019 to 0.081, the proximity to road network estimates range from -0.004 to 0.01, the road network density estimates range from -0.076 to 0.686, the proximity to drinage network estimates range from -0.007 to 0.003, and the sediment transport index estimates range from -0.003 to 0.041.

This chapter provides a methodology for understanding the local dynamics effective in the occurrence of landslides. The quantification of driving factors of LS could guide governments and policymakers in their decision-making for the future natural hazard prevention in Taşlıdere Basin. Landslide and flood control projects should be developed for the re-establishment of dysfunctional drainage networks in areas identified as highly susceptible and proposed to be protected. In these areas, the effectiveness of forests is of great importance in the prevention of flooding and infiltration of precipitation by gradually leaking precipitation into the soil. Therefore, the protection of forests and the rehabilitation of degraded forest areas with tree species of high interception rate and pile root, especially transpiration and interception rates should be encouraged. The majority of land use that is highly susceptible to landslides is the tea gardens. It is recommended for the tea gardens to construct open and close drainage systems as well as structures that have drainage capability. It is also recommended that livestock farming in the study area be encouraged through providing financial support and funding. The road and infrastructure constructions are intensive in the area which causes environmental damages. Therefore, the construction work should be performed under control with the required measures be taken. For low susceptible areas, the planning and policy actions can be directed to new urban use development whereas high susceptible areas can be strengthened and re-used for urban recreation purposes. In addition, the produced maps in this chapter could be utilized directly in the urban master plans and policies to identify the suitable locations where the growth of peri-urban areas could be directed.

REFERENCES

Aghda, S. M. F., Bagheri, V., & Razifard, M. (2018). Landslide susceptibility mapping using fuzzy logic system and its influences on mainlines in Lashgarak Region, Tehran, Iran. *Geotechnical and Geological Engineering*, *36*, 915–937. doi:10.1007/s10706-017-0365-y

Akaike, H. (1973). Information theory and an extension of the maximum likelihood principle. In B.N. Petrov, and P. Ciaki (Eds.), *2nd International Symposium on Information Theory* (pp. 267– 81). Budapest: Akademiai Kiad6.

Allison, P. (2012). Logistic regression using SAS: Theory and application (2nd ed.). SAS Institute.

Althuwaynee, O. F., Pradhan, B., Park, H.-J., & Lee, J. H. (2014). A novel ensemble decision tree-based CHi-squared Automatic Interaction Detection (CHAID) and multivariate logistic regression models in landslide susceptibility mapping. *Landslides*, *11*(6), 1063–1078. doi:10.1007/s10346-014-0466-0

Arabameri, A., Pradhan, B., & Rezaei, K. (2019). Gully erosion zonation mapping using integrated geographically weighted regression with certainty factor and random forest models in GIS. *Journal of Environmental Management, 232*(September), 928–942. doi:10.1016/j.jenvman.2018.11.110

Atalay, I. (1982). Türkiye Jeomorfolojisine Giriş (Genişletilmiş 2. Baskı), *Ege Üniversitesi Edebiyat Fakültesi Yayınları, 9*(1982)

Aydınoğlu, A. Ç., & Altürk, G. (2021). Heyelan Duyarlılık Haritalarının İstatistik ve Makine Öğrenmesi Teknikleri Kullanılarak Üretilmesi: Taşlıdere Havzası Örneği (Rize). *Coğrafya Dergisi. The Journal of Geography*, *0*(43), 159–176. doi:10.26650/JGEOG2021-814561

Barka, A. (1984). Erzincan havzasının bazı nanotektonik özellikleri, 1. Ulusal Deprem Sempozyumu, 223-247.

Brunsdon, C., Fotheringham, A. S., & Charlton, M. E. (1996). Geographically Weighted Regression: A Method for Exploring Spatial Nonstationarity. *Geographical Analysis*, *28*(4), 281–298. doi:10.1111/j.1538-4632.1996.tb00936.x

Can, T., Duman, T. Y., Nefeslioglu, H. A., Durmaz, S., Gokceoglu, C., & Sonmez, H. (2005). Earthflows in a small catchment from Eastern Black Sea Region Turkey: conditional Environmental factors and susceptibility assessments. In *Proceedings of the International Symposium on Latest Natural Disasters: New Challenges for Engineering Geology, Geotechnics and Civil Protection* (pp. 82-83). Academic Press.

Chalkias, C., Kalogirou, S., & Ferentinou, M. (2014). Landslide susceptibility, Peloponnese Peninsula in South Greece. *Journal of Maps*, *10*(2), 211–222. doi:10.1080/17445647.2014.884022

Chalkias, C., Polykretis, C., Karymbalis, E., Soldati, M., Ghinoi, A., & Ferentinou, M. (2020). Exploring spatial non-stationarity in the relationships between landslide susceptibility and conditioning factors: A local modeling approach using geographically weighted regression. *Bulletin of Engineering Geology and the Environment*, *79*(6), 2799–2814. doi:10.1007/s10064-020-01733-x

Chen, W., Zhao, X., Shahabi, H., Shirzadi, A., Khosravi, K., Chai, H., Zhang, S., Zhang, L., Ma, J., Chen, Y., Wang, X., Bin Ahmad, B., & Li, R. (2019). Spatial prediction of landslide susceptibility by combining evidential belief function, logistic regression and logistic model tree. *Geocarto International*, *34*(11), 1177–1201. doi:10.1080/10106049.2019.1588393

Chowdhuri, I., Pal, S. C., Arabameri, A., Ngo, P. T. T., Chakrabortty, R., Malik, S., Das, B., & Roy, P. (2020). Ensemble approach to develop landslide susceptibility map in landslide dominated Sikkim Himalayan region, India. *Environmental Earth Sciences*, *79*(20), 1–28. doi:10.1007/s12665-020-09227-5

Crozier, M. J., & Glade, T. (2005). Landslide hazard and risk: issues, concepts and approach. In T. Glade, M. Anderson, & M. J. Crozier (Eds.), *Landslide hazard and risk*. Wiley. doi:10.1002/9780470012659.ch1

Dağ, S., & Bulut, F. (2012). Coğrafi Bilgi Sistemleri Tabanlı Heyelan Duyarlılık Haritalarının Hazırlanmasına Bir Örnek: Çayeli (Rize, KD Türkiye). *Jeoloji Muhendisligi Dergisi*, *36*(1), 35–62.

Du, S., Wang, Q., & Guo, L. (2014). Spatially varying relationships between land-cover change and driving factors at multiple sampling scales. *Journal of Environmental Management*, *137*, 101–110. doi:10.1016/j.jenvman.2014.01.037 PMID:24603033

Erener, A., & Düzgün, H. S. B. (2010). Improvement of statistical landslide susceptibility mapping by using spatial and global regression methods in the case of More and Romsdal (Norway). *Landslides*, *7*(1), 55–68. doi:10.1007/s10346-009-0188-x

Erinç, S. (1944). *Glazialmorphologische Untersuchungen im Nordostanatolischen Randgebirge*. PhD Dissertation Series, Geographical Institute of the University of Istanbul.

Erinç, S. (1965). *Yağış müessiriyeti üzerine bir deneme ve yeni bir indis. İstanbul Üniversitesi. Edebiyat Fakültesi*. Coğrafya Enstitüsü Yayınları.

Faraji Sabokbar, H., Shadman Roodposhti, M., & Tazik, E. (2014). Landslide susceptibility mapping using geographically-weighted principal component analysis. *Geomorphology*, *226*, 15–24. doi:10.1016/j.geomorph.2014.07.026

Feuillet, T., Coquin, J., Mercier, D., Cossart, E., Decaulne, A., Jónsson, H. P., & Sæmundssonthorsteinn, T. (2014). Focusing on the spatial non-stationarity of landslide predisposing factors in northern Iceland: Do paraglacial factors vary over space? *Progress in Physical Geography*, *38*(3), 354–377. doi:10.1177/0309133314528944

Fotheringham, A. S., Brunsdon, C., & Charlton, M. (2003). *Geographically weighted regression: the analysis of spatially varying relationships* (1st ed.). John Wiley & Sons.

Fotheringham, A. S., Charlton, M. E., & Brunsdon, C. (2001). Spatial Variations in School Performance: A Local Analysis Using Geographically Weighted Regression. *Geographical and Environmental Modelling*, *5*(1), 43–66. doi:10.1080/13615930120032617

Fotheringham, A. S., & Oshan, T. M. (2016). Geographically weighted regression and multicollinearity: Dispelling the myth. *Journal of Geographical Systems*, *18*(4), 303–329. doi:10.1007/s10109-016-0239-5

Fuchs, S., Röthlisberger, V., Thaler, T., Zischg, A., & Keiler, M. (2017). Natural hazard management from a coevolutionary perspective: Exposure and policy response in the European Alps. *Annals of the Association of American Geographers*, *107*(2), 382–392. doi:10.1080/24694452.2016.1235494 PMID:28267154

GDCDE. (2016). *Rize Province Güneysu Micro-catchment Landslide Susceptibility and Hazard Map Production Project.* General Directorate of Combating with Desertification and Erosion, Ankara, Turkey. Retrived From https://webdosya.csb.gov.tr/db/cem/icerikler/taslidire-heyelan-20211108102301.pdf

glm. (2022). *glm: Fitting Generalized Linear Models.* Retrieved from https://www.rdocumentation.org/packages/stats/versions/3.6.2/topics/glm

Hong, H., Pradhan, B., Sameen, M. I., Chen, W., & Xu, C. (2017). Spatial prediction of rotational landslide using geographically weighted regression, logistic regression, and support vector machine models in Xing Guo area (China). *Geomatics, Natural Hazards & Risk*, *8*(2), 1997–2022. doi:10.1080/19475705.2017.1403974

Hong, H., Pradhan, B., Sameen, M. I., Kalantar, B., Zhu, A., & Chen, W. (2018). Improving the accuracy of landslide susceptibility model using a novel region-partitioning approach. *Landslides*, *15*(4), 753–772. doi:10.1007/s10346-017-0906-8

Huabin, W., Gangjun, L., Weiya, X., & Gonghui, W. (2005). GIS-based landslide hazard assessment: An overview. *Progress in Physical Geography*, *29*(4), 548–567. doi:10.1191/0309133305pp462ra

Kalantar, B., Ueda, N., Lay, U. S., Al-Najjar, H. A. H., & Halin, A. A. (2019). Conditioning Factors Determination for Landslide Susceptibility Mapping Using Support Vector Machine Learning. *International Geoscience and Remote Sensing Symposium (IGARSS)*, 9626–9629. 10.1109/IGARSS.2019.8898340

Kalantar, B., Ueda, N., Saeidi, V., Ahmadi, K., Halin, A. A., & Shabani, F. (2020). Landslide Susceptibility Mapping: Machine and Ensemble Learning Based on Remote Sensing Big Data. *Remote Sensing (Basel)*, *12*(11), 1737. doi:10.3390/rs12111737

Karpouza, M., Chousianitis, K., Bathrellos, G. D., Skilodimou, H. D., Kaviris, G., & Antonarakou, A. (2021). Hazard zonation mapping of earthquake-induced secondary effects using spatial multi-criteria analysis. *Natural Hazards*, *109*(1), 637–669. doi:10.1007/s11069-021-04852-0

Kavzoglu, T., Sahin, E. K., & Colkesen, I. (2014). Landslide susceptibility mapping using GIS-based multi-criteria decision analysis, support vector machines, and logistic regression. *Landslides*, *11*(3), 425–439. doi:10.1007/s10346-013-0391-7

Ketin, İ. (1966). Tectonic units of Anatolia (Asia minor). *Bulletin of the Mineral Research and Exploration*, *66*(66), 20–37. https://dergipark.org.tr/en/pub/bulletinofmre/issue/3892/51961

Koçyiğit, A. (1985). The Karayazı faul. *Bulletion of Geological Society of Turkey*, *28*(1), 67–72.

Kornejady, A., Ownegh, M., Rahmati, O., & Bahremand, A. (2018). Landslide susceptibility assessment using three bivariate models considering the new topo-hydrological factor: HAND. *Geocarto International*, *33*(11), 1155–1185. doi:10.1080/10106049.2017.1334832

Landry, S. M., Koeser, A. K., Kane, B., Hilbert, D. R., McLean, D. C., Andreu, M., & Staudhammer, C. L. (2021). Urban forest response to Hurricane Irma: The role of landscape characteristics and sociodemographic context. *Urban Forestry & Urban Greening*, *61*, 127093. doi:10.1016/j.ufug.2021.127093

Leroy, S. A. G., Gracheva, R., & Medvedev, A. (2022). Natural hazards and disasters around the Caspian Sea. *Natural Hazards*, *114*(3), 2435–2478. doi:10.1007/s11069-022-05522-5

Li, C., Li, F., Wu, Z., & Cheng, J. (2017). Exploring spatially varying and scale-dependent relationships between soil contamination and landscape patterns using geographically weighted regression. *Applied Geography (Sevenoaks, England)*, *82*, 101–114. doi:10.1016/j.apgeog.2017.03.007

Li, Y., Liu, X., Han, Z., & Dou, J. (2020). Spatial proximity-based geographically weighted regression model for landslide susceptibility assessment: A case study of Qingchuan area, China. *Applied Sciences (Basel, Switzerland)*, *10*(3), 1107. Advance online publication. doi:10.3390/app10031107

Lin, Q., Wang, Y., Glade, T., Zhang, J., & Zhang, Y. (2020). Assessing the spatiotemporal impact of climate change on event rainfall characteristics influencing landslide occurances based on multiple GCM projections in China. *Climatic Change*, *162*(2), 761–779. doi:10.1007/s10584-020-02750-1

Mandal, S., & Maiti, R. (2015). *Semi-quantitative approaches for landslide assessment and prediction*. Springer. doi:10.1007/978-981-287-146-6

Nakaya, T. (2022). *GWR 4.0, Geographically Weighted Modelling Software*. Retrieved from https://gwr.maynoothuniversity.ie/gwr4-software/

Nefeslioglu, H. A., & Gokceoglu, C. (2011). Probabilistic risk assessment in medium scale for rainfall-induced earthflows: Catakli catchment area (Cayeli, Rize, Turkey). *Mathematical Problems in Engineering*, *2011*, 1–21. Advance online publication. doi:10.1155/2011/280431

Nefeslioglu, H. A., Gokceoglu, C., & Sonmez, H. (2008). An assessment on the use of logistic regression and artificial neural networks with different sampling strategies for the preparation of landslide susceptibility maps. *Engineering Geology*, *97*(3–4), 171–191. doi:10.1016/j.enggeo.2008.01.004

Nilsson, P. (2014). Natural amenities in urban space-A geographically weighted regression approach. *Landscape and Urban Planning*, *121*, 45–54. doi:10.1016/j.landurbplan.2013.08.017

Oshan, T., Li, Z., Kang, W., Wolf, L., & Fotheringham, A. (2019). MGWR: A Python Implementation of Multiscale Geographically Weighted Regression for Investigating Process Spatial Heterogeneity and Scale. *ISPRS International Journal of Geo-Information*, *8*(6), 269. doi:10.3390/ijgi8060269

Sahin, E. K., Colkesen, I., Acmali, S. S., Akgun, A., & Aydinoglu, A. C. (2020). Developing comprehensive geocomputation tools for landslide susceptibility mapping: LSM tool pack. *Computers & Geosciences*, *144*(July), 104592. doi:10.1016/j.cageo.2020.104592

Şaroğlu, F., Emre, Ö., & Boray, A. (1987). Türkiye'nin Diri Fayları Ve Depremsellikleri (Active Faults and Their Seismicity of Turkey) General Directorate of Mineral Research and Exploration, Ankara-Turkey (1987). Report No: 8174.

Şaroğlu, F., & Yılmaz, Y. (1986). Geological Evolution and Basin Models During Neotectonic Episode in the Eastern Anatolia. *Bulletin of the Mineral Research and Exploration*, *107*(107), 70–93. https://dergipark.org.tr/en/pub/bulletinofmre/issue/3926/52248

Şen, O. (2019). *İklim ve hidrolojik modeller ile Rize'de bulunan su havzalarının taşkın risk tayini*. Unpublished PhD Thesis, Istanbul: Istanbul Technical University, Institute of Science and Technology.

Şengör, A. M. C., & Kidd, W. S. F. (1979). Post-collisional tectonics of the Turkish-Iranian plateau and a comparison with Tibet. *Tectonophysics*, *55*(3–4), 361–376. doi:10.1016/0040-1951(79)90184-7

Solanki, A., Gupta, V., & Joshi, M. (2022). Application of machine learning algorithms in landslide susceptibility mapping, Kali Valley, Kumaun Himalaya, India. *Geocarto International*, *37*(27), 16846–16871. Advance online publication. doi:10.1080/10106049.2022.2120546

Süme, V., Yüksek, T., & Kaya, A. (2021). Rize Taşlıdere Havzasında heyelan kontrolü: Örnek olay incelemesi, Kireçhane ve Kırklartepe Yöreleri. *Journal of Anatolian Environmental and Animal Sciences*, *6*(4), 526–531. doi:10.35229/jaes.994874

Tsangaratos, P., & Ilia, I. (2016). Comparison of a logistic regression and Naive Bayes classifier in landslide susceptibility assssments: The influence of models complexity and training dataset size. *Catena*, *145*, 164–179. doi:10.1016/j.catena.2016.06.004

Tsangaratos, P., Ilia, I., Hong, H., Chen, W., & Xu, C. (2017). Applying information theory and GIS-based quantitative methods to produce landslide susceptibility maps in Nancheng County, China. *Landslides*, *14*(3), 1091–1111. doi:10.1007/s10346-016-0769-4

Varnes, D. J. (1984). Landslide hazard zonation, a review of principles and practice. IAEG Commission on Landslides. Paris: UNESCO.

WHO. (2023). *Landslides*. Retrived from: https://www.who.int/health-topics/landslides#tab=tab_1

Yalcin, A., Reis, S., Aydinoglu, A. C., & Yomralioglu, T. (2011). A GIS-based comparative study of frequency ratio, analytical hierarchy process, bivariate statistics and logistics regression methods for landslide susceptibility mapping in Trabzon, NE Turkey. *Catena*, *85*(3), 274–287. doi:10.1016/j.catena.2011.01.014

Yilmaz, A., Adamia, S., Chabukiani, A., Chkhotua, T., Erdoğan, K., Tuzcu, S., & Karabiyikoglu, M. (2000). Structural Correlation of the Southern Transcaucasus (Georgia)-Eastern Pontides (Turkey). Geological Society, London, Special Publications, 173(1), 171–182. doi:10.1144/GSL.SP.2000.173.01.08

Yilmaz, I. (2009). Landslide susceptibility mapping using frequency ratio, logistic regression, artificial neural networks and their comparison: A case study from Kat landslides (Tokat-Turkey). *Computers & Geosciences, 35*(6), 1125–1138. doi:10.1016/j.cageo.2008.08.007

Yılmaz, Y., Tüysüz, O., Yiğitbaş, E., Genç, Ş. C., & Şengör, A. M. C. (1997). Geology and tectonic evolution of the Pontides. *Regional and petroleum geology of the Black Sea and surrounding region, 68*, 183-226.

Yılmaz, Y., Güner, Y., & Şaroğlu, F. (1998). Geology of the quaternary volcanic centres of the east Anatolia. *Journal of Volcanology and Geothermal Research, 85*(1–4), 173–210. doi:10.1016/S0377-0273(98)00055-9

Yüksek, Ö., Kankal, M., & Üçüncü, O. (2013). Assessment of big floods in the Eastern Black Sea Basin of Turkey. *Environmental Monitoring and Assessment, 185*(1), 797–814. doi:10.1007/s10661-012-2592-2 PMID:22411030

Zhang, Y., Wang, Y., Chen, Y., Xu, Y., Zhang, G., Lin, Q., & Luo, R. (2021). Projection of changes in flash flood occurance under climate change at tourist attractions. *Journal of Hydrology (Amsterdam), 595*, 126039. doi:10.1016/j.jhydrol.2021.126039

Zhao, Y., Wang, R., Jiang, Y., Liu, H., & Wei, Z. (2019). GIS-based logistic regression for rainfall-induced landslide susceptibility mapping under different grid sizes in Yueqing, Southeastern China. *Engineering Geology, 259*(December), 105147. doi:10.1016/j.enggeo.2019.105147

Chapter 7
Empirical Insights Into the Impact of COVID-19 on Employment Patterns in South Africa's Accommodation Sector

Reshma Sucheran
Durban University of Technology, South Africa

ABSTRACT

The tourism sector in South Africa plays a crucial role in contributing significantly to economic growth and employment. However, the COVID-19 pandemic has had an unparalleled impact on the global tourism industry, exposing vulnerabilities and uncertainties. This study addresses a gap in the literature by examining the effects of the pandemic on employment within the accommodation sector in South Africa. Online surveys were conducted with managers using a quantitative research methodology. The findings reveal a significant decline in employment, salary reductions, and widespread redundancies. Notably, the study finds that more experienced establishments demonstrated resilience against job losses, and chain-owned establishments exhibited greater resilience in mitigating salary reductions. The study's implications extend to informing strategic measures for recovery, resilience, and long-term viability within the accommodation sector.

DOI: 10.4018/979-8-3693-2721-0.ch007

INTRODUCTION

The tourism sector is pivotal in South Africa's economic framework, catalysing economic growth and presenting substantial employment opportunities across diverse demographic groups (South African Tourism, 2021). Recognised for its labour-intensive nature and critical role as a significant employer, the tourism industry encompasses various domains such as hotels, restaurants, transportation, and related sectors (Khan et al., 2020). Nevertheless, historical data highlights the susceptibility of the tourism business to significant occurrences and disasters, resulting in substantial reductions in salaries and employment levels (Gössling et al., 2020; Khan et al., 2020; Prideaux et al., 2020). While the hospitality industry has faced challenges from previous diseases, the scale of the COVID-19 pandemic has been unparalleled, exerting a profound impact on the global tourism business (Nhamo et al., 2020). As Baum et al. (2020) highlighted, the fluidity of tourism businesses and their workforce highlights their vulnerability and instability during external crises. The pandemic has manifested in widespread layoffs, furloughs, and business closures, prompting concerns among hospitality workers regarding job security (Cheer, 2020; Baum et al., 2020; Ann & Blum, 2020; Mayer, 2020).

According to the World Travel & Tourism Council (WTTC), the tourism sector witnessed a staggering 97 million job losses in 2020 (WTTC, 2020). In South Africa, the ramifications of the COVID-19 pandemic on the tourism sector have been substantial, as evidenced by a 71% decline in foreign arrivals in 2020 compared to the previous year (Statistics South Africa, 2021). Many tourism enterprises in the country experienced a complete cessation of global tourism, leading to prolonged revenue depletion while fixed expenses persisted (Maylie, 2020). Given its role in generating approximately 740,000 direct employment opportunities and over 1.5 million indirect jobs, with small businesses forming a significant portion of the sector, the tourism industry in South Africa faced an uncertain future, with numerous businesses at risk of not enduring and others striving to adapt to the challenging circumstances.

Despite the extensive body of literature examining various aspects of the impact of the COVID-19 pandemic on the tourism industry in South Africa, including economic impacts, environmental implications, crisis management, response and recovery measures, and protected areas (Sucheran, 2021, 2022; van der Merwe et al., 2021; Anyanwu & Salami, 2021; Arndt et al., 2020; Lewis et al., 2021; Booyens et al., 2022; Rogerson & Rogerson, 2020; Lekgau & Tichaawa, 2021; Rogerson, 2021; Spenceley et al., 2021; Giddy & Rogerson, 2021), a discernible gap remains in the exploration of the effects on employment within the accommodation industry. This gap is significant for several reasons, warranting the need for the present research study. Firstly, South Africa's accommodation industry exhibits varied structures, from large hotel chains to small, locally owned establishments, resulting in different vulnerabilities and adaptive capacities. Secondly, investigating specific challenges and opportunities within this sector is vital for comprehending the overall impact on employment. And thirdly, the industry's adaptability and effective recovery strategies are crucial for the resilience of the broader tourism sector. Recognising the unique challenges accommodation businesses face is crucial for guiding policy recommendations and interventions to support recovery and ensure sustainability, ultimately contributing to the resilience of the broader tourism sector. Therefore, researching the impacts of the COVID-19 pandemic on employment in accommodation establishments in South Africa is imperative to fill this notable gap in understanding and informing strategic measures for the industry's recovery and long-term viability.

LITERATURE REVIEW

The COVID-19 pandemic has distinguished itself from previous epidemics in its global reach and the prolonged recovery time required. The far-reaching impact of the pandemic is evident in various sectors, with a particular focus on the hospitality and tourism industry. Nicola et al. (2020) emphasise the considerable influence of the pandemic on businesses, leading to the displacement of jobs in tourism and hospitality due to international and local travel restrictions, lockdowns, and heightened infection concerns (WTTC, 2020). This unprecedented disruption quickly prompted the industry to lay off personnel, halt operations, and experience significant financial losses, causing disturbances in the human resources ecosystem within the accommodation sector (Jasmine, 2019; Jones & Comfort, 2020; Al-Fadly, 2020; Merkle, 2022; Soehardi et al., 2020; Ozili & Arun, 2020). Furthermore, the pandemic revealed the vulnerability of employment contracts and the instability of workers, particularly those from marginalised social demographics (Tom et al., 2020).

Hotels faced the need to alter schedules, work duties, early retirements, layoffs, or unpaid vacations, contributing to an extraordinary rise in unemployment comparable to the Great Depression (Moraa & Kipngetich, 2021). Moreover, the psychological anguish and lack of social security intensified job losses, with employees in the tourism and hospitality sector particularly vulnerable due to poor wages and education (Baum et al., 2016; Williamson et al., 2017; Robinson et al., 2019). The uncertainty of jobs, exacerbated by industry seasonality and reliance on cheap labour, presented challenges for these workers, making it difficult to pursue education and alternative employment opportunities (Robinson, 2019). Older individuals faced additional hurdles in securing new jobs due to age and academic qualifications (Almedia & Santos, 2020). The pandemic accentuated existing issues in the hospitality business, including inadequate wages, strained labour-management relations, extended work schedules, and limited job growth, amplifying employment concerns during the crisis (Kurmann et al., 2020).

Organisational resilience, the ability to adjust and restart regular operations following a disaster, became imperative for entities facing challenges during the pandemic (Wren-Lewis, 2020). Minimising losses, resolving hospitality staff absenteeism, and adapting to new circumstances were critical for recovery. However, the potential movement of skilled labour to other industries during restructuring threatened economic recovery. The pandemic triggered significant changes in work practices, with remote work, virtual meetings, and flexible hours challenging the traditional norms in the tourism and hospitality industry. While companies took steps to satisfy workers and continue operations, some hotel managers opted for strategic crisis management training during the pandemic. This led to reduced room division staff and expansion in food and beverage operations, emphasising staff safety and multitasking training to adapt, lead, and communicate effectively (Yacoub & ElHajjar, 2020).

Some hotels committed to sustainable employment through staff engagement, online training, and open communication. Post-pandemic, Lai and Wong (2020) advised staff training in outbreak communication, standard precautions, infection control, and cleanliness to guarantee long-term sustainability. However, the reality was that several hotels had to resort to staff cuts, salary reductions, and withholding perks despite the epidemic exposing vulnerabilities in hospitality organisations and their workers globally. Governments globally also responded to the outbreak with various initiatives. UAE officials extended overseas workers' stays until 2020 to support tourist workers who lost employment due to the pandemic (Haak-Saheem, 2020). The condition of hospitality businesses and their workforce proved highly unstable and susceptible to external crises, indicating their vulnerability on a global scale. Tom et al. (2020) concluded that the pandemic has laid bare the systemic issues of inequality and the fragility of employment within the hospitality sector, necessitating a re-evaluation of labour practices and a push towards more equitable and sustainable employment models post-pandemic.

In South Africa, where the tourism business is vital to the economy, the COVID-19 epidemic severely restricted travel and reduced tourism. Most tourism enterprises opted for decreased pay over furloughs or layoffs to minimise economic damage (Department of Tourism, 2021). The South African Tourist Industry Recovery Plan aims to restore pre-crisis employment levels by 2024, employing various initiatives and enablers to mitigate the impacts of the COVID-19 pandemic, which cost 125,000 jobs (National Department of Tourism, 2021). The recovery plan acknowledges the challenges faced by the sector and outlines strategies to rejuvenate tourism-related employment and economic activity. The experience of South Africa further illustrates the broader economic consequences of the pandemic and the strategic initiatives required for recovery, and the literature calls for re-evaluating labour practices and establishing more equitable and sustainable employment models in the post-pandemic era.

METHODOLOGY

The research was conducted in South Africa during June and July 2020, corresponding to implementing Level 3 lockdowns following the preceding Level 5 and Level 4 lockdowns from March 26th to April 20th and May 1st to May 31st of 2020, respectively. The primary focus of the study was to investigate the perspectives of accommodation establishment managers regarding the influence of the COVID-19 outbreak on employment during this specific timeframe. The research data utilised in this study is part of a more extensive dataset that explored the economic ramifications of the pandemic on all lodging establishments in South Africa.

A quantitative research methodology was employed to collect empirical data, utilising online surveys administered via Google Forms. Accommodation establishment managers across South Africa were targeted, representing various categories such as hotels, guesthouses, bed and breakfasts, backpackers and hostels, caravan and campsites, country houses, game lodges, nature lodges, and self-catering establishments. A comprehensive list of accommodation facilities and their corresponding contact information was obtained from South African Tourism and the Tourism Grading Council of South Africa. The study employed a saturation sampling approach, focusing on approximately 3,000 establishments, with the participants contacted through electronic mail. The online surveys were conducted over four weeks spanning June and July 2020, aiming to achieve representativeness by targeting a minimum sample of 360 establishments. Ultimately, a total of 482 establishments participated in the survey. An online survey was deemed appropriate given the prevailing travel restrictions and social distancing measures, allowing data collection from a broad geographic region while prioritising participant safety. The anonymity and confidentiality of all participants were guaranteed throughout the research process.

The self-administered questionnaire utilised in the survey was designed by drawing inspiration from established methodologies (AlBattat & Som, 2014, 2019; Hung et al., 2018). However, the survey implemented in this study is novel and employed in its current format for the first time. The questionnaire comprised three distinct sections. The initial section focused on the attributes of hotels, exploring aspects such as star classification, ownership model, establishment size (quantity of guestrooms), workforce size, and the nature of amenities offered. The second section examined the economic dimensions of accommodation establishments, specifically centring on the financial condition during the lockdown period, the effects on staff, and present and future economic apprehensions confronting hotels. Data analysis for this study was performed utilising the Statistical Package for Social Sciences (SPSS) Version 26. The analysis involved a detailed examination and summarisation of the data employing descriptive statistics such as frequencies, percentages, and mean values. Incorporating chi-square analysis in this study's data analysis helped assess associations and dependencies between crucial variables. Given the multifaceted nature of the pandemic's impact on the accommodation sector, it was crucial to explore relationships among various factors, such as accommodation characteristics, that may influence employment dynamics. The study results were also enriched with verbatim quotes that incorporate constructive qualitative feedback obtained from the respondents.

RESULTS AND DISCUSSION

The research utilised a questionnaire as its research instrument, consisting of 55 items measured at either a nominal or an ordinal level. The questionnaire was structured into five sections, each addressing themes, including biographical data, concerns, economic measures, staffing, and education. The Cronbach's alpha scores for all the questionnaire items were assessed, and the reliability scores for each section exceeded the recommended Cronbach's alpha value. This suggests a satisfactory level of consistent scoring for the respective sections of the research.

Profile of Accommodation Establishments in the Study

The findings in Table 2 shed light on the diverse background of accommodation establishments under examination, offering an understanding of the sector's

Table 1. Cronbach's alpha

	Number of Items	Cronbach's Alpha
Concerns	4	0.806
Economic Measures	8	0.751
Staffing	18	0.916
Education	4	0.726

Author's own compilation, 2022

Table 2. Characteristics of accommodation establishments

Characteristics		N	%
Star Grading	One-star	1	0.2
	Two-star	10	2.1
	Three-star	175	36.7
	Four-star	191	40.0
	Five-star	39	8.2
	Ungraded	61	12.8
Type of accommodation establishment	Hotel/Resort	49	10.3
	Motel/Inn	2	0.4
	Lodge (bush/country)	51	10.7
	Guesthouse	181	37.9
	Self-catering	82	17.2
	Game Lodge	10	2.1
	Country House	9	1.9
	Bed and Breakfast	77	16.1
	Caravan and Campsite	7	1.5
	Backpackers and Hostel	9	1.9
Type of ownership	Independently owned, self-managed	464	97.3
	Chain-owned, managed by chain	11	2.3
	Unsure	2	0.4
Size (number of rooms)	Less than 10 rooms	261	54.7
	11-30 rooms	162	34.0
	31-60 rooms	30	6.3
	61-100 rooms	15	3.1
	More than 100 rooms	9	1.9
Years in operation	Less than five years	61	12.8
	5 to 10 years	88	18.4
	11 to 15 years	130	27.3
	16 to 20 years	84	17.6
	21 to 25 years	63	13.2
	More than 25 years	51	10.7
Number of employees before COVID-19	Less than ten employees	338	69.8
	11 to 30 employees	100	21.0
	31 to 60 employees	27	5.7
	61-100 employees	7	1.5
	101-130 employees	4	0.8
	131-150 employees	1	0.2
	More than 150 employees	5	1.0

Author's own compilation, 2022

multifaceted characteristics. The distribution of star-graded establishments demonstrates a prevalence of 4-star and 5-star establishments, comprising 40% and 36.7% of the sample, respectively, with a more minor yet significant 12.8% unrated. This distribution highlights the dataset's concentration of higher-graded accommodation, indicating a notable presence of upscale establishments. The dominance of guesthouses as the primary type of lodging (37.9%) highlights the sector's diversity, reflecting the distinctive South African hospitality landscape where smaller, personalised accommodations, often independently owned and self-managed, play a pivotal role. The variability in accommodation facility size is evident in the study, with a majority (54.7%) comprising fewer than ten rooms, indicating a dominance of boutique or niche accommodations. Conversely, a significant proportion (34.0%) with 11 to 30 rooms suggests a diverse range of lodging sizes, bearing implications for adaptive capacities in the face of external shocks such as the COVID-19 pandemic. The data on the duration of operation reveals that 68.8% of accommodation establishments have operated for 11 years or more, indicating a subset with stability and resilience. Workforce size dynamics before the pandemic highlight that a substantial majority (69.8%) of establishments had fewer than ten employees, in line with the common characteristic of smaller accommodations with lean staffing structures. However, 21.0% with a staff size ranging from 11 to 30 employees suggests a subset of larger establishments. Understanding the composition and distinctive characteristics of the sector lays the basis for a more in-depth analysis of the employment implications stemming from the COVID-19 pandemic on accommodation establishments in South Africa.

Table 3. Employment impacts

	Lost Jobs		Staff on Reduced Salaries		Staff Made Redundant	
	N	%	N	%	N	%
None	29	6.1	77	16.1	50	10.5
Few (1% to 20%)	52	10.9	40	8.4	55	11.5
Some (21% to 50%)	148	31.0	52	10.9	146	30.6
Many (51% to 80%)	54	11.3	9	1.9	47	9.9
Most (81% to 100%)	169	35.4	293	61.4	134	28.1
Unsure	25	5.2	6	1.3	45	9.4
Total	**477**	**100**	**477**	**100**	**477**	**100**

Author's own compilation, 2022

Direct Employment Impacts

The survey data shown in Table 3 reveals a substantial 35.4% of managers confirming the termination of staff employment (81% to 100%) directly attributable to the pandemic, indicating a significant reduction in the workforce and a widespread experience of job loss. The causative factors for these included limitations on all travel, diminished demand for accommodation services, and the temporary or permanent closure of accommodation-providing entities. Furthermore, 61.4% of respondents affirmed that most of their staff (81-100%) encountered reduced pay during the pandemic. This compelled many accommodation businesses to implement cost-reduction strategies, including salary cuts, to navigate the economic downturn caused by the pandemic. While strategic for business viability, these cost-cutting measures may substantially impact affected employees' financial stability and overall job satisfaction. The survey findings also disclosed that 28.1% of managers reported the redundancy of 81% to 100% of their staff. Redundancies typically arise when businesses struggle to maintain their workforce, necessitating personnel reductions or operational cessation. The consequences of these employment reductions are far-reaching, contributing to enduring economic instability and elevated rates of joblessness within the local context, thereby affecting workers and their households. Overall, the data confirms that the industry underwent significant transformations marked by a considerable decline in employment, salary reductions, and widespread redundancies. While some accommodation establishments managed to retain a fraction of their workforce through wage reductions, others were compelled to make challenging decisions to curtail expenses, resulting in employee layoffs and redundancies. These findings resonate with existing research, such as that by Tom et al. (2020), confirming the difficult decisions hospitality businesses face due to the economic pressures of the pandemic. Additionally, studies like that of Hyasat (2022), Lagos et al. (2021), and Merkle (2022) corroborate similar challenges, including work stoppages and salary reductions.

In addition to the quantitative data presented above, direct responses from participants vividly communicate the profound employment challenges that businesses and their local communities face. One respondent expressed the widespread impacts: *"This has an enormous impact on the community surrounding the establishment. As all 54 staff members come from the local community, this means that the whole community is suffering as the business is not able to pay full wages/salaries."* This observation highlights the interconnectedness of accommodation businesses and local communities, emphasising that the repercussions of employment challenges extend beyond the organisational boundaries. Another participant further encapsulates the harsh reality of the situation: *"I already had to retrench all my staff indefinitely"*, reflecting the immediate and severe impact of the pandemic on employment stability within the sector. The response from another participant: *"Zero income means zero ability to pay staff. More unemployment, more poverty"* indicates the cascading effect of economic distress on both employees and the broader society and emphasises the potential exacerbation of social issues such as unemployment and poverty due to the economic standstill brought about by the pandemic.

Another verbatim response reveals the perspective of small accommodation businesses: *"Small single self-catering units like mine are at most risk of bankrupting. I had staff that I have had to let go, I feel terrible my one lady worked for me for 11 years. She is in dire straits same as me"*. This quote offers an insight into the vulnerability of smaller establishments, shedding light on the emotional toll experienced by business owners compelled to make difficult decisions regarding their long-term and loyal employees. The dire financial predicament accommodation businesses face is further captured in the verbatim response: *"Almost all of our retirement savings have been exhausted. We can last at best about another 6 weeks"*. This admission outlines the imminent threat to the survival of accommodation businesses, emphasising the urgency for financial support and highlighting the limited capacity for resilience without external assistance. Another respondent provided insight into the practical implications of reduced occupancy, stating, *"Of my 23 employees, only two are working regularly and two others as and when the need arises. The rest are sitting at home. Occupancy rate does not warrant more to work"*. This response highlights the direct correlation between reduced demand, staffing decisions, and the resultant idle workforce due to decreased operations.

However, in the face of the challenging circumstances brought about by the COVID-19 pandemic, some South African accommodation sector businesses demonstrated resilience by employing diverse strategies to retain their staff. These businesses, recognizing the unprecedented nature of the crisis, took extraordinary measures to safeguard their employees' well-being. One respondent showed this commitment by stating, *"I am currently using my retirement fund to operate and pay employees full employment even though we have zero guests as of now"*, which highlights the exceptional dedication of specific business owners who went above and beyond to sustain employment despite the absence of revenue. Similarly, another participant articulated heightened concerns about the future: *"We are very concerned about the business and our staff when UIF TERS runs out. Fortunately, we have some very quickly diminishing savings which we use to pay fixed costs and to support our staff."* This again highlights the financial strain businesses face and the utilisation of dwindling resources to uphold operational and employee support responsibilities.

A critical discussion also emerged from the data as some businesses faced the difficult decision of balancing staff retention with financial viability. One respondent emphasized this dilemma, stating, *"Fortunately, the Guesthouse is not our only source of income, but we are paying staff from our own savings, and only having outside staff in, e.g., Gardner. Inside staff have been laid off on reduced pay, indefinitely, as they were not with us long enough to claim UIF."* This demonstrates the complex trade-offs businesses grappled with, striving to retain core staff while making tough decisions regarding others. Another respondent echoed the financial challenges, stressing the need for pragmatic actions, *"We cannot carry over R100 000 every 3 months just on salaries. A labour lawyer is currently assisting us with suspensions. I would rather re-employ when things come right. Our guests are all European guests. Until flights return safely, we are going to struggle."* Furthermore, a business owner's proactive approach was illustrated in the statement: *"I registered my business as a certified accommodation/self-isolation facility before lockdown. I met with staff to explain the possible financial impact if we don't do business during this time."* This stance, coupled with an innovative solution of allowing employees to self-isolate on the premises, reflects a strategic effort to mitigate financial risks and maintain employee well-being.

Key Business Concerns

Respondents were tasked with evaluating their levels of concern on a scale ranging from 1 to 5, where '1' denoted 'least concerned' and '5' indicated 'most concerned.'

Figure 1. Key business concerns during the COVID-19 pandemic

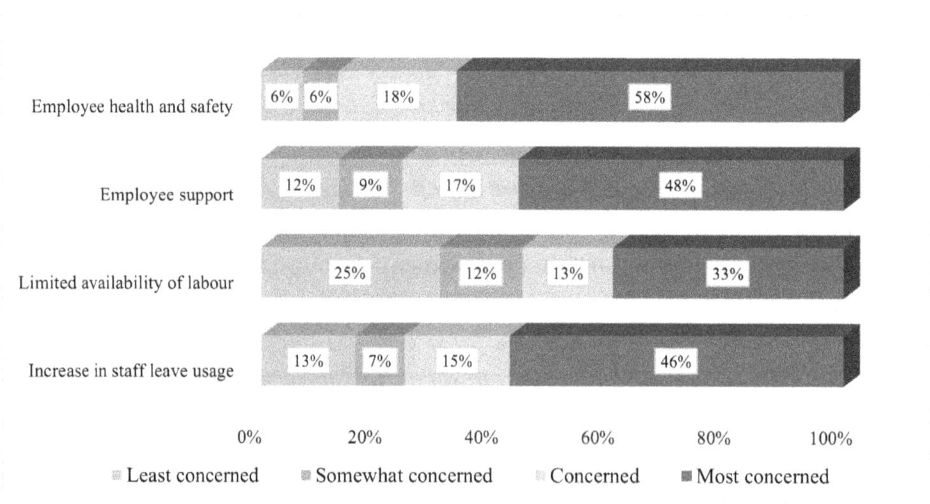

The research found that accommodation establishments during the COVID-19 pandemic exhibited notable concerns across various elements. Foremost among these concerns was employee health and safety, acquiring a mean rating of 4.17 on the 1 to 5 scale. This outcome highlights the paramount importance placed on ensuring the employee's well-being amidst the challenges posed by the pandemic. Employee support was the second most significant concern for accommodation establishments, supported by an average rating of 3.80. This aspect likely encompasses measures establishments undertake to bolster their employees during these difficult times, including providing necessary resources, effective communication channels, and emotional support. Šuligoj's (2022) research stresses the significance of clear instructions and education to alleviate employee fears and uncertainties. Training initiatives become crucial in equipping staff with the necessary skills to adapt to new operational procedures arising from pandemic-related restrictions. As emphasised by the author, understanding employee sentiment becomes imperative for navigating the hotel industry's ongoing and post-crisis work environment. Hyasat's (2022) findings reveal that the perceived neglectful organisational response can add to the stress and pressure experienced by hotel employees during the crisis.

Another noteworthy concern was observed in increased staff leave usage, with an average rating of 3.74. This suggests a surge in employees taking leave during the pandemic, potentially attributed to personal health concerns, childcare responsibilities, or related factors. The heightened leave usage may have presented operational challenges for establishments, necessitating a flexible approach to workforce management. Accommodation establishments expressed a relatively lower concern regarding the limited availability of labour, with an average rating of 3.16. The implication is that while some establishments experienced difficulty securing suitable replacement staff, it did not emerge as prominently concerning as employee health, support, and leave usage. Evaluating these concerns provides a comprehensive understanding of lodging establishments' priorities and challenges during the pandemic. The findings resonate with the global trends the World Travel and Tourism Council (WTTC, 2020) identified, which projected substantial job losses in the travel and tourism industry due to prolonged travel restrictions. This global perspective aligns with this localised study, emphasising the universality of the challenges the accommodation sector faced during the pandemic. Moreover, a study by Jones and Comfort (2020) examining the global implications of the COVID-19 crisis on tourism and sustainable development corroborates these findings as it highlights the interrelation of the pandemic with employment trends, particularly in sectors heavily reliant on tourism.

Table 4. Chi-square analysis: Characteristics of accommodation establishments by employment trends

	Star Grading	Type	Ownership	Size	Years in Operation
Number of employees before the start of COVID-19	.002*	.000*	.000*	.000*	.000*
Job losses	.023*	.670	.892	.103	.023*
Reduced salaries	.136	.394	.000*	.192	.193
Staff made redundant	.431	.610	.784	.243	.100

*Significant values ($p<0.05$) Chi-square analysis
Author's own compilation, 2022

Chi-Square Analysis

The study employed chi-square tests to assess the intricate relationships between the characteristics of establishments and employment dynamics before and during the COVID-19 pandemic.

NUMBER OF EMPLOYEES BEFORE THE ONSET OF THE COVID-19 PANDEMIC

The analysis uncovered a statistically significant association between hotels' star grading and staffing levels before the pandemic (χ^2 (30, n=477)=56.4, p=.002), indicating a non-independent relationship. Higher-star hotels, recognised for their extensive services and luxury offerings, maintained larger workforces, while lower-star counterparts operated more efficiently with streamlined staff and focused more on cost efficiency. Correspondingly, the study identified a highly significant association between the type of accommodation establishments and the number of employees before the pandemic where χ^2 (54, n=477)=215.4, p= <.001). The relationship is influenced by various factors inherent to each type of establishment. Generally, more complex establishments such as hotels and resorts tend to have more staff due to their extensive facilities, diverse services, and larger guest capacities. Guesthouses, B&Bs, and country houses may have smaller staff sizes, emphasising personalised service in more intimate settings. Motels, self-catering accommodations, and backpackers often operate with smaller staff structures, catering to travellers with different preferences. At the same time, game lodges, with unique offerings like safari experiences, may have specialized staff. On the other hand, camping facilities may have minimal permanent staff, relying more on seasonal or event-based hiring. The study draws attention to the importance of tailored approaches to employee management based on the inherent nature of each accommodation type, acknowledging that one-size-fits-all staffing solutions are impractical within this multifaceted sector. The study also reveals a highly significant association between ownership type (chain-owned or independently owned) and the number of employees before the pandemic (χ^2 (12, n=477)=74.6, p=<.001). Chain-owned establishments, often implement standardised operations and centralized management, leading to a systematic approach to staffing with predefined roles. Independently owned establishments, in contrast, exhibit greater flexibility in operations. Owners may play a more hands-on role, and individual preferences, business strategies, and the scale of the establishment influence staffing levels. Smaller, independently owned accommodations may maintain a more intimate setting with a smaller staff, while larger ones may adopt a staffing approach similar to chain-owned establishments.

The analysis further revealed a highly significant association between the size of accommodation establishments and the number of employees before the pandemic (χ^2 (24, n=477)=42.8, p=<.001). Larger establishments, denoted by a higher number of rooms, require a more substantial workforce, while smaller establishments focus on optimising efficiency with fewer staff. Generally, larger accommodation establishments with more rooms tend to have a larger staff complement to manage operations and provide services to guests. Such establishments often require more front-desk staff, housekeeping, maintenance personnel, and other support roles to ensure efficient day-to-day functioning. However, the nature of this relationship can be influenced by factors such as the establishment's business model, level of automation and technological integration, management practices, and the specific services offered. Therefore, the correlation between the size of a hotel and its staff size is dynamic and dependent on a range of operational and strategic considerations. Tom et al. (2020) noted that larger hoteliers demonstrated a greater capacity to retain their workforces during the COVID-19 crisis, emphasising the critical role of operational scale in preserving employment. The authors also noted that the closure of smaller businesses in the hospitality sector considerably affected employment, highlighting the vulnerability of small, medium and micro enterprises during the pandemic. A highly significant association between the number of years an accommodation establishment has been in operation and staffing levels was also revealed from the study, where χ^2 (30, n=477)=81.9, p=<.001). This correlation can be attributed to potential business growth, increased clientele, and increased services over time. This relationship is further influenced by factors such as technological advancements, economic conditions, market dynamics, and the specific type and size of the hotel. While some hotels may experience a growth in staff as they mature, others may adopt streamlined operations or leverage technology to optimise efficiency, leading to a reduced workforce. Established accommodation businesses with a more extended operational history may demonstrate more effective workforce management, operational expertise, and resilience.

Job Losses During the COVID-19 Pandemic

The study revealed a statistically significant association between job losses and the star grading of hotels where (χ^2 (30, n=477)=47.42, p=.023. Higher-star accommodation establishments faced distinctive challenges, potentially stemming from shifts in consumer behaviour and reduced demand for premium services. Notably, the economy segment exhibited the slightest vulnerability to job losses, followed by the midscale and upper-midscale segments. Ozdemir et al. (2021) also noted that while upscale hotels experienced relatively more significant effects than the economy, midscale, and upper-midscale segments, the upper upscale and luxury hotel segments were highly impacted by the COVID-19 pandemic. Similarly, a statistically significant association was identified between job losses and the duration of operation (χ^2 (30, n=477)=47.27, p=.023), indicating that more experienced establishments demonstrated resilience, addressing challenges more effectively and experiencing fewer job losses. This finding highlights the potential effect of operational experience in mitigating the adverse employment impacts of external crises.

Conversely, no statistically significant associations were observed between job losses and accommodation category (χ^2 (54, n=477)=48.29, p=.670), ownership type (χ^2 (12, n=477)=6.44, p=.892), or establishment size (χ^2 (24, n=477)=33.03, p=.103). This lack of significance implies a uniform impact across different types, ownership structures, and sizes of accommodation establishments during the pandemic, emphasising the overarching influence of economic and operational factors on job losses within the accommodation sector. The findings suggest that, regardless of the accommodation's type, ownership, or size, the sector faced similar challenges regarding staff reductions, highlighting the pervasive impact of broader contextual factors during the unprecedented challenges posed by the COVID-19 pandemic.

Reduced Salaries During the COVID-19 Pandemic

The research uncovered a statistically significant association between reduced salaries and the type of accommodation ownership where χ^2 (12, n=477)=45.38, p=<.001. By benefiting from substantial financial resources and corporate support, chain-owned establishments exhibited greater resilience in mitigating salary reductions than independently-owned establishments. However, Ozdemir et al. (2021) argue an alternative perspective, indicating that chain-managed hotels were most affected by the pandemic, followed by independent hotels. Their study concluded that the impact of the COVID-19 pandemic was most pronounced in chain- hotels compared to independent hotels. In contrast, no statistically significant associations were identified between reduced salaries and star grading (χ^2 (30, n=477)=38.5, p=.136), accommodation type (χ^2 (54, n=477)=56.16, p=.394), the size of accommodation establishment (χ^2 (24, n=477)=29.79, p=.192) and its duration of operation (χ^2 (30, n=477)=36.4, p=.193). The findings suggest that, regardless of the accommodation's star grading, type, size and duration of operation, the sector faced similar challenges in terms of salary reductions, pointing towards a shared vulnerability to economic pressures during the COVID-19 pandemic.

Staff Redundancy During the COVID-19 Pandemic

The study results revealed no statistically significant associations between staff redundancy and star grading (χ^2 (30, n=477)=30.68, p=.431), accommodation category (χ^2 (54, n=477)=50.48, p=.610), ownership type (χ^2 (12, n=477)=8.01, p=.784), accommodation size (χ^2 (24, n=477)=28.40, p=.243), and years in operation (χ^2 (30, n=477)=40.2, p=.100). This indicates that the challenges posed by the pandemic, such as reduced travel demand, restrictions, and economic uncertainties, affected accommodation establishments uniformly across these parameters.

Limitations and Future Research

The research findings on the impact of the COVID-19 pandemic on employment dynamics in the accommodation sector reveal several limitations and propose crucial opportunities for future research. The study predominantly adopts a managerial perspective, neglecting the experiences and viewpoints of frontline employees. Incorporating diverse stakeholder perspectives, including management and frontline employees, would enhance the overall understanding of the pandemic's impact on the workforce. The temporal scope of the study is also limited to the immediate and short-term impacts, suggesting a need for future research to extend the temporal scope, allowing for a better understanding of the evolving nature of employment dynamics and recovery trajectories over an extended period. Also, the reliance on survey data introduces the possibility of response bias, emphasising the need for future research to complement survey data with qualitative methods to cross-verify findings. Addressing these limitations and exploring suggested possibilities for future research will undoubtedly enhance the depth, accuracy, and applicability of knowledge regarding the impacts of the COVID-19 pandemic on employment in the accommodation sector.

CONCLUSION AND RECOMMENDATIONS

The research findings provide an in-depth understanding of the profound impact of the COVID-19 pandemic on employment within the accommodation sector in South Africa, focusing on job losses, reduced salaries, and redundancies. The survey data indicates a substantial workforce reduction, confirming extensive terminations, leading to most or all staff being laid off. This unprecedented job loss is directly attributed to the disruptive effects of the COVID-19 outbreak, including travel restrictions, reduced demand for accommodation facilities, and temporary or permanent closure of such facilities. The economic downturn induced by the pandemic led numerous enterprises to implement cost-cutting measures, with a significant portion of staff experiencing lower pay. Salary reductions became a crucial strategy for organisations to cope with the financial challenges precipitated by the pandemic, raising concerns about the potential impact on the financial stability of affected employees. The research further highlights the widespread occurrence of redundancies. Redundancies typically arise when establishments face difficulties in maintaining their workforce, leading to personnel reductions or the cessation of operations. The impact of such employment reductions extends beyond individual workers to affect households, contributing to economic instability and higher unemployment rates within local communities. The overall findings depict a complex and challenging landscape for the accommodation sector, characterised by significant declines in employment, reductions in salaries, and the implementation of layoff and redundancy measures.

The research findings provide valuable insights regarding the primary concerns of accommodation establishments during the COVID-19 pandemic. The data reveals a hierarchy of concerns, with establishments emphasising certain aspects over others. Among these concerns is the unequivocal priority placed on employee health and safety. This highlights the paramount importance that accommodation establishments attribute to ensuring the well-being of their workforce amidst the challenges posed by the pandemic. Employee support emerged as the second most significant concern, highlighting the industry's recognition of the need to bolster the resilience and morale of its workforce during the pandemic. The increase in staff leave usage is another substantial concern, showing that accommodation establishments encountered challenges associated with a surge in staff taking leave at the time.

The findings of this research further confirm a noteworthy correlation between job losses and hotel star grading, indicative of a vulnerability within higher-star-rated accommodations. This vulnerability appears linked to evolving consumer behaviors and the imposition of travel restrictions. Furthermore, the observed connection between job losses and the duration of operation suggests a potential resilience among well-established establishments in confronting the challenges posed by the pandemic. Notably, the lack of statistically significant associations between job losses and factors such as accommodation category, ownership type, and establishment size imply a consistent impact across diverse accommodation segments. Moreover, the non-significant correlations between reduced salaries and both star grading and accommodation type highlight a widespread influence on compensation, irrespective of the luxury status of the establishment.

The verbatim quotes from respondents are a powerful and authentic narrative and render a vivid and distressing picture of the extensive challenges brought about by the COVID-19 pandemic on employment in South Africa. These firsthand accounts clarify the immediate consequences businesses face and highlight the emotional strain experienced by business owners as they deal with the complexities of workforce management during these unprecedented times. From significant measures to sustain full employment to pragmatic decisions balancing financial constraints, the study exposes intricate dynamics of addressing employment challenges during the COVID-19 pandemic. Moreover, the narratives go beyond the confines of individual accommodation establishments, offering a glimpse into the broader societal implications of the employment crisis. The interconnectedness of businesses with their local communities becomes evident, indicating that the repercussions of reduced employment extend far beyond organisational boundaries. The urgency is profound as businesses exhaust their financial resources, face the imminent exhaustion of savings, and grapple with survival challenges. The critical discussion conveyed through these responses calls for immediate attention to the sector's unique vulnerabilities and the development of targeted strategies to bolster businesses, safeguard employment, and contribute to the broader economic recovery efforts. Moreover, the direct quotes emphasise the urgent need for strategic interventions and robust support mechanisms within the accommodation sector.

Several recommendations emerge from the findings of this study. The study recommends that accommodation establishments communicate transparently, involve employees in decision-making processes, and implement fair compensation adjustments to mitigate the negative impact on workforce morale. They should also explore new business models, service offerings, or partnerships to diversify revenue streams and enhance resilience. The study also recommends flexibility in staffing models, such as cross-training employees for multiple roles, which can enhance resilience in the face of uncertainties. Accommodation establishments should prioritise the holistic well-being of employees by implementing comprehensive support measures beyond traditional compensation structures. This includes mental health support, wellness programs, career counselling, and skill development programs that can assist affected employees in facing the challenges of job loss and career transitions. Recognising the impact of factors such as star grading, accommodation type, ownership type, establishment size, and operational duration on staffing levels, establishments should develop tailored strategies based on their specific characteristics. This involves understanding each type of establishment's unique challenges and adapting workforce management accordingly. Long-standing establishments should leverage their operational experience to build resilience and navigate challenges effectively. This may include investing in technology, adopting innovative solutions, and drawing on past experiences to inform strategic decisions in workforce management during crises. To enhance the adaptability and skill sets of the workforce, accommodation establishments should invest in continuous training and development programs. This prepares employees for evolving roles and responsibilities, contributing to their professional growth and the overall resilience of the establishment. As the sector evolves in the post-pandemic landscape, these insights can provide a foundation for informed decision-making and developing resilient strategies within the accommodation sector.

ACKNOWLEDGMENT

This work is based on research supported by the National Institute for the Humanities and Social Sciences (REF NO: BRI22/1215).

REFERENCES

Al-Fadly, A. (2020). Impact of COVID-19 on SMEs and employment. *Entrepreneurship and Sustainability Issues*, *8*(2), 629–648. doi:10.9770/jesi.2020.8.2(38)

Albattat, A. R., & Mat Som, A. P. (2019). *Disaster planning and preparedness in the hotel industry*. Emerald. doi:10.1108/9781787699373

AlBattat, A. R. S., & Som, A. P. M. (2014). Emergency planning and disaster recovery in Malaysian hospitality industry. *Procedia: Social and Behavioral Sciences*, *144*, 45–53. doi:10.1016/j.sbspro.2014.07.272

Almeida, F., & Santos, J. D. (2020). The effects of COVID-19 on job security and unemployment in Portugal. *The International Journal of Sociology and Social Policy*, *40*(9), 995–1003. doi:10.1108/IJSSP-07-2020-0291

Ann, S., & Blum, S. C. (2020). Motivating senior employees in the hospitality industry. *International Journal of Contemporary Hospitality Management*, *32*(1), 324–346. doi:10.1108/IJCHM-08-2018-0685

Anyanwu, J. C., & Salami, A. O. (2021). The impact of COVID-19 on African economies: An introduction. *African Development Review*, *33*(S1, Suppl 1), S1. doi:10.1111/1467-8268.12531 PMID:34149237

Anyanwu, J. C., & Salami, A. O. (2021). The impact of COVID-19 on African economies: An introduction. *African Development Review*, *331*(1). Advance online publication. doi:10.1111/1467-8268.12531 PMID:34149237

Arndt, C., Davies, R., Gabriel, S., Harris, L., Makrelov, K., Modise, B., . . . Anderson, L. (2020). Impact of Covid-19 on the South African economy. *Southern Africa-Towards Inclusive Economic Development Working Paper, 111*.

Baum, T., Kralj, A., Robinson, R. N., & Solnet, D. J. (2016). Tourism workforce research: A review, taxonomy and agenda. *Annals of Tourism Research*, *60*, 1–22. doi:10.1016/j.annals.2016.04.003

Baum, T., Mooney, S. K., Robinson, R. N., & Solnet, D. (2020). COVID-19's impact on the hospitality workforce–new crisis or amplification of the norm? *International Journal of Contemporary Hospitality Management*, *32*(9), 2813–2829. doi:10.1108/IJCHM-04-2020-0314

Booyens, I., Rogerson, C. M., Rogerson, J. M., & Baum, T. (2022). Covid-19 crisis management responses of small tourism firms in South Africa. *Tourism Review International*, *26*(1), 121–137. doi:10.3727/154427221X16245632411872

Cheer, J. M. (2020). Human flourishing, tourism transformation and COVID-19: A conceptual touchstone. *Tourism Geographies*, *22*(3), 1–11. doi:10.1080/14616688.2020.1765016

Giddy, J. K., & Rogerson, J. M. (2021). Nature-based tourism enterprise adaptive responses to COVID-19 in South Africa. *Geo Journal of Tourism and Geosites*, *36*(2spl), 36. doi:10.30892/gtg.362spl18-700

Haak-Saheem, W. (2020) Talent management in Covid -19 crisis – how Dubai manages and sustains its global talent pool. *Asian Business and Management, 19*(3), 298-301. https://doi.org/ doi:10.1057/s41291-020-

Hyasat, A. (2022). The impact of the covid-19 pandemic upon tourism & hospitality employees in Jordan. *Geo Journal of Tourism and Geosites*, *43*(2), 1099–1107. doi:10.30892/gtg.43331-925

JasmineC. A. (2019) Impacts of Covid-19 on company and efforts to support organization. doi:10.2139/ssrn.3590726

Jones, P., & Comfort, D. (2020). The COVID-19 crisis, tourism and sustainable development. *Athens Journal of Tourism*, *7*(2), 75–86. doi:10.30958/ajt.7-2-1

Khan, A., Bibi, S., Lyu, J., Latif, A., & Lorenzo, A. (2020). COVID-19 and sectoral employment trends: Assessing resilience in the US leisure and hospitality industry. *Current Issues in Tourism*, *24*(7), 952–969. doi:10.1080/13683500.2020.1850653

Kurmann, A., Lalé, E., & Ta, L. (2020). The Impact of COVID-19 on US Employment and Hours: Real-Time Estimates with Homebase Data. *EurekAlert*. www.eurekalert.org/pub_releases/2020-05/du-rds050520.php

Lagos, D. G., Poulaki, P., & Lambrou, P. (2021). COVID-19 and Its Impact on Tourism Industry. *Advances in Experimental Medicine and Biology*, *1318*, 815–824. doi:10.1007/978-3-030-63761-3_45 PMID:33973213

Lai, I. K. W., & Wong, J. W. C. (2020). Comparing crisis management practices in the hotel industry between initial and pandemic stages of COVID-19. *International Journal of Contemporary Hospitality Management*, *32*(10), 3135–3156. doi:10.1108/IJCHM-04-2020-0325

Lekgau, R. J., & Tichaawa, T. M. (2021). Adaptive strategies employed by the MICE sector in response to COVID-19. *Geo Journal of Tourism and Geosites*, *38*(4), 1203–1210. doi:10.30892/gtg.38427-761

Lekgau, R. J., & Tichaawa, T. M. (2021). Adaptive strategies employed by the MICE sector in response to COVID-19. *Geo Journal of Tourism and Geosites*, *38*(4), 1203–1210. doi:10.30892/gtg.38427-761

Lewis, F., Browne, M., & Houdet, J. (2021). COVID-19–What is the impact for tourism as a vehicle of SMME development and incentivizing environmental management in South Africa. *African Journal of Hospitality, Tourism and Leisure*, *10*(3), 987–998. doi:10.46222/ajhtl.19770720-144

Mayer, K. (2020). *Financial health is workers' biggest wellness concern*. https://hrexecutive.com/financial-health-is-employees-biggest-covid-19-wellness-concern/

Maylie, D. (2020). *South Africa's tourism industry next stop recovery*. Available at https://www.ifc.org/wps/wcm/connect/region__ext_content/ifc_external_corporate_site/subsaharan+africa/news/south+africas+tourism+industry+next+stop+recovery

Merkle, T., Tajeddini, K., Mensah Herv, D., Amoako-Atta, E., Hossain, M., Illés, C. B., & Dunay, A. (2022). Impact of COVID-19 Pandemic on Hotel Employees in the Greater Accra Region of Ghana. *Sustainability (Basel)*, *14*(5), 2509. doi:10.3390/su14052509

Moraa, R. O., & Kipngetich, R. (2021). Impact of COVID-19 on employee retention in the hospitality industry. *The Strategic Journal of Business and Change Management*, *8*(1), 81–90. doi:10.61426/sjbcm.v8i1.1895

Nhamo, G., Dube, K., & Chikodzi, D. (2020). Impacts and implications of COVID-19 on the global hotel industry and Airbnb. In *Counting the Cost of COVID-19 on the global tourism industry* (pp. 183–204). Springer. doi:10.1007/978-3-030-56231-1_8

Nicola, M., Alsafi, Z., Sohrabi, C., Kerwan, A., Al-Jabir, A., Iosifidis, C., Agha, M., & Agha, R. (2020). The socio-economic implications of the coronavirus pandemic (COVID-19): A review. *International Journal of Surgery*, *78*, 185–193. doi:10.1016/j.ijsu.2020.04.018 PMID:32305533

OziliP. K.ArunT. (2020). Spillover of COVID-19: impact on the global economy, available at SSRN: https://ssrn.com/abstract=3562570 or doi:10.2139/ssrn.3562570

Robinson, R. N., Martins, A., Solnet, D., & Baum, T. (2019). Sustaining precarity: Critically examining tourism and employment. *Journal of Sustainable Tourism*, *27*(7), 1008–1025. doi:10.1080/09669582.2018.1538230

Rogerson, C. M., & Rogerson, J. M. (2020). COVID-19 tourism impacts in South Africa: Government and industry responses. *Geo Journal of Tourism and Geosites*, *31*(3), 1083–1091. doi:10.30892/gtg.31321-544

Rogerson, J. M. (2021). Tourism business responses to South Africa's COVID-19 pandemic emergency. *Geo Journal of Tourism and Geosites*, *35*(2), 338–347. doi:10.30892/gtg.35211-657

Soehardi, S., Untari, D., & Raya, U. J. (2020). The effect of the COVID-19 pandemic on hotel employees, hotel occupancy rates and hotel tax income in Jakarta, Indonesia. *Systematic Reviews in Pharmacy*, *11*, 964–972.

South African Tourism. (2021). *South African tourism annual report 2020/21*. https://live.southafrica.net/media/299000/south-african-tourism-annual-report-2020-21.pdf

Statistics South Africa. (2021). *SA tourism industry struggles amidst COVID-19 pandemic*. https://www.statssa.gov.za/?p=14281

Sucheran, R. (2021b). Preliminary Economic Impacts of the COVID-19 Pandemic on the Hotel Sector in South Africa. *African Journal of Hospitality, Tourism and Leisure*, *10*(1), 115–130. doi:10.46222/ajhtl.19770720-90

Šuligoj, M. (2022). Work in hotels under conditions related to the COVID-19 disease: Views of Slovenian employees. *Tourism (Zagreb)*, *70*(2), 285–297. doi:10.37741/t.70.2.9

UNWTO. (2020). *Worst Year in Tourism History with 1 Billion Fewer International Arrivals. 2021*. Available online: https://www. unwto.org/news/2020-worst-year-in-tourism-history-with-1-billion-fewer-international-arrivals

van der Merwe, P., Saayman, A., & Jacobs, C. (2021). Assessing the economic impact of COVID-19 on the private wildlife industry of South Africa. *Global Ecology and Conservation*, *28*, e01633. doi:10.1016/j.gecco.2021.e01633 PMID:34541259

Williamson, D., Rasmussen, E., & Ravenswood, K. (2017). Power in the darkness: Taking a Historical and Critical Employment Relations Approach in Hospitality. *Journal of Hospitality and Tourism Management*, *33*, 134–141. doi:10.1016/j.jhtm.2017.10.014

WTTC. (2020). *More than 197m travel and tourism jobs will be lost due to prolonged travel restrictions, according to new research from WTTC*. https://wttc.org/News-Article/More-than-197m-Travel-Tourism-jobs-will-be-lost-due-to-prolonged-travel-restrictions

Yacoub, L., & El Hajjar, S. (2020). How do hotels in developing countries manage the impact of COVID-19? The case of Lebanese hotels. *International Journal of Contemporary Hospitality Management*, *33*(3), 929–948. doi:10.1108/IJCHM-08-2020-0814

Zschiegner, A. K. (2021). Tourism in protected and conserved areas amid the COVID-19 pandemic. *Parks*, (27), 103–118.

Chapter 8 Disaster Management Activities in an Indian Engineering Institute: A Case Study of Gharda Institute of Technology

Sunil Jayant Kulkarni
https://orcid.org/0000-0002-5988-3448
Gharda Institute of Technology, India

Vivek Digambar Talnikar
Annasaheb Dange College of Engineering and Technology, Ashta, India

ABSTRACT

Appropriate and fast response to the emergency is the most important aspect of disaster management. Gharda Institute of Technology (GIT) is situated near Chiplun, Maharashtra, India. The predominant natural disasters which are frequent in the nearby vicinity like floods and water logging are not seen in the institute area due to its elevation above sea level. The possible disasters are chemical spillage, fire, and snake related issues (snake bite). Periodic mock drills and demo sessions are conducted for preparedness related to these disasters. The snake bite injections are made available at the nearby hospital. The disaster management cell of GIT has handled the COVID-19 situation with appropriate precaution. The cell conducts frequent checks on the preparedness of the concern taskforce to combat any untoward incident. The display of contact numbers of hospital, police, district level authorities brings about the safe feeling among the students and staff. The institute has chemical department and hence handling of chemicals is an important aspect.

DOI: 10.4018/979-8-3693-2721-0.ch008

INTRODUCTION

Disaster is a sudden unfortunate event or emergency that needs immediate response to minimize the damage and get back to normalcy. The disasters can be classified as natural and man-made. Third one is hybrid and is combination of both, natural and man-made disasters. According to Comfort (2005), a network of the companies, firms, managements that are committed to effective disaster management through continual inquiry, informed response and adaptive to learning new things is far better approach than conventional disaster management through administrative structure (Comfort, 2005). Ethics in disaster management is a topic of discussion and needs to be given due consideration while preparing disaster management policy. Disaster is most of the time a situation that leads to monetary and human casualties so appropriate and fast response to the emergency is the most important aspect of disaster management. In many organizations, the allocation of fund for disaster management is part of the financial planning. The most important aspects of disaster management are cautiousness, bravery, fair-mindedness, watchfulness, toughness, communication and effective implementation of plan (Geale, 2012). While allocating resources to a certain task, the risks and uncertainties involved must be given due consideration (Roland et al., 2010). Findings by Sawalha indicated that the disaster management in academics and society at large still depends on traditional method (Sawalha, 2020). Coping with disruptive challenges has led to human evolution. Changes in the disaster management strategies and methodologies to meet the new challenges have always been the way forward for effective disaster management (O Brian et al., 2010). Disaster preparedness and disaster management are the terms seldom used in the academics and training of the physicians (Guerdan, 2009). Disaster preparedness includes understanding the level of knowledge and legal aspects involved that can overlap with medical and social emergency. Adequate training to medical, communication and incident controllers is an important part of preparedness. Disaster can be looked upon as an imbalance between the availability of resources and the allocation, especially of medical care facilities. It is estimated that in the past quarter century, 3400000 lives have been lost in disasters(Guerdan, 2009). Recent data on Covid -19 indicates that more than 6.5 million peoples have lost their lives due to coronavirus between 2020 to 2022(World Disasters Report, 2022). On the economic front, trillions of dollars have been lost/spend on these disasters.

Motivation

The author, while working as a member secretory of disaster management cell, realized that there is lack of awareness and negligence about disaster management.

The activities carried out by disaster management cell during Corona/Covid-19 period underlined important role of institutional disaster management cell. The book chapter provides insight into the activities carried out by the Cell to tackle the disaster management issues in the institute and other attempts by the university to include disaster management at various learning stages for budding engineers.

Problem Statement

Disaster management is an ignored area in Indian education system. Even in engineering colleges, small incidences like spillage and burning injuries occur while dealing with chemicals. Improper handling due to lack of awareness may multiply the problems. These incidences can be used for students' advantage by making them aware of disaster management practices.

Outline of the Book Chapter

The book chapter is a case study about the disaster management related activities in an engineering institute. Initially the concept and classification of disaster is explained. Various laws related to disaster management are explained preceded by literature survey on disaster management related education. Then the introduction about the Gharda Institute of Technology is followed by the disaster management related activities. Two activities carried out at the institute are explained with the pictures and procedures.

Aim and Objectives

Aim of this chapter is to discuss disaster management preparedness for an engineering institute and related activities with case studies India. In addition to academic activities, institutes are involved in educating the students about civil and industrial disasters. It outlines the measures taken in the education system by virtue of various mandatory committees and inclusion of subjects at undergraduate level to create awareness about emergency preparedness.

CLASSIFICATIN OF DISASTERS

Classification of disasters is shown in figure 1. Three main classes of disasters are natural, manmade and hybrid, based on the cause of the incidents. Geological or meteorological events that are capable of causing loss of property and life are termed as natural disasters. Disasters due to human intent, interference or errors are manmade disasters. Hybrid disasters occurs because of combination of the first two or cascading effect of first two types.

Figure 2 depicts general flowchart of disaster management activities. Five stages of disaster management are prevention, mitigation, preparedness, response and recovery(University of central Florida, no date). In colleges and corporate offices, the prevention measures like proper evacuation plan and availability of required tools, can help in preventing the disaster. Mitigation measures like adopting effective evacuation plan and training the staff about the first aid measures can help in minimizing the losses. Preparedness includes the mock drills and training sessions for disaster management. Proper communication plan leads to effective response to the disaster. The disaster-prone area should be provided with back up for the emergency services like water, electricity and sanitation. The normal services should be restored rapidly leading to recovery from the disaster.

Figure 1. Classification of disasters (Below et al., 2009; Teh & Khan, 2021)
Compiled by the Authors, 2024

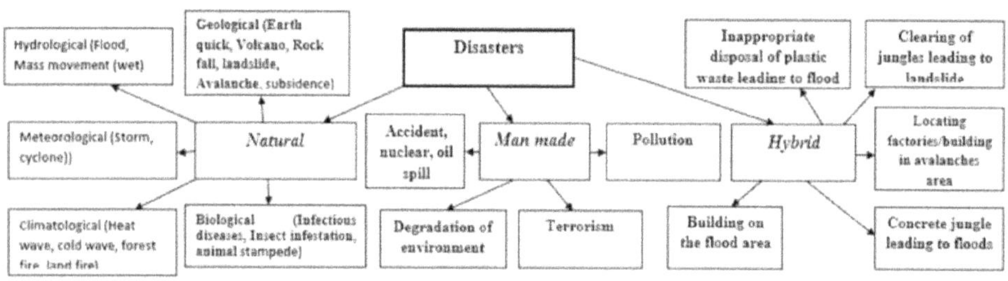

Figure 2. General flowchart of disaster management activities (Nath, 2008; Beg et al., 2013)
Compiled by the Authors, 2024

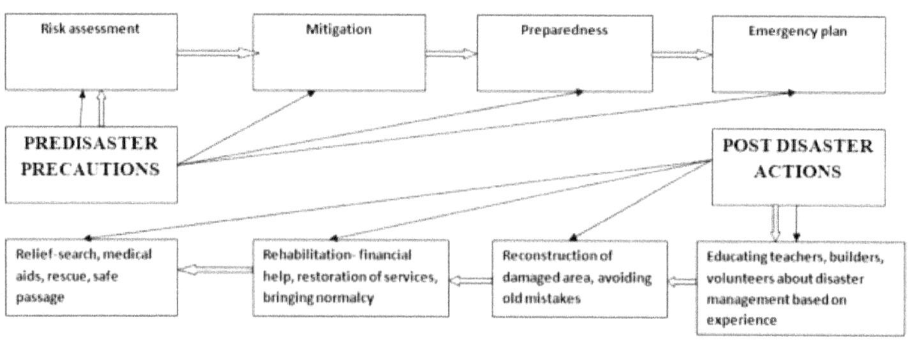

ABOUT GHARDA INSTITUTE OF TECHNOLOGY: ITS LAYOUT AND DISASTER-PRONE AREAS

Gharda Institute of Technology is situated near Lote Parshuram MIDC(Maharashtra Industrial Development Corporation) in Chiplun, Maharashtra, India. It is focused on creating the engineering professional that are competent to face the challenges in chemical industry through innovation and problem-solving ability. From disaster point of view, the predominant natural disasters which are frequent in the nearby vicinity like floods and water logging are not seen in the area due to its elevation above the sea level. The institute is based in Lavel, the village that is well above the sea level and is at sufficient elevation than nearby cities Chiplun and Khed. The snake bite is an occasional problem which is addressed by the college authorities pretty well. The security personnel are acquainted with snakes, their types and snake catching methods. The Institute is founded and run by Gharda Foundation(GF). Gharda Foundation has its own Hospital. The vaccines required for snake bites, dog bites are readily available in this hospital that is adjacent to the institute. Also, periodic health checks ups are done free of cost for Gharda Foundation(GF) stakeholders. In handling the Covid-19 situation, the hospital played a major role. Also, for the snake catching and facing any critical situations, the emergency contact numbers along with contact number of snake catchers, are displayed at prominent places in the campus. The assembly points are located for each of the building. The institute has a Disaster management cell that suggests the precautionary measures to avoid disasters and handles the emergency situations. The laboratories in chemical engineering department handle considerable amount of chemicals.

The institute has the Chemical engineering branch that provides the skilled manpower for nearby chemical industries, moreover, the Gharda Foundation is the initiative by the owners of Gharda Chemicals Limited, the parent company of the trust that run the Institute. The stock checking is done periodically and safety precautions about handling of chemicals are displayed in each laboratory. The possible hazards in the institute are listed in table 1. Possibility of fire and explosion due to chemical engineering laboratories is considered while preparing the disaster management plan. The Konkan, area in Maharashtra(India) is prone to floods due to heavy rains. Training the staff to handle the flood situation is an important part of disaster preparedness. The institute provides medical insurance facilities for staff and students to avoid the financial insecurity among the stakeholders. Also, special discount is given to the GF stakeholders for treatment in the hospital founded and managed by Gharda foundation.

Table 1. List of possible hazards in the institute (Price, 2008; Islam et al., 2016)

SR. NO	ACCIDENT SCENARIO	CAUSES	ACTION REQUIREMENT
1	Fire & Explosion due to leak of hydrogen	Improper operation, Handling, Failure of operational / process control	Better operation control, Periodical calibration and inspection of hydrogen detection system, Training to lab assistants regarding better operational control, Isolate of flammable material from laboratory, firefighting without any risk.
2	Heavy rain fall/ Flooding	Natural calamity	Laboratory assistants are trained to handle flood like situation, safe storage of reactive chemicals.

Compiled by the Authors, 2024

LITERATURE REVIEW

Disaster Management and Engineering Education

Almost all the engineering fields are prone to the disaster and hence the people working in engineering sector must be acquainted with disaster management skills (Lawrence & Suresh, 2012). Facilities, support systems, resources and modeling are focus areas of disaster management research(Rebeeh et al., 2019). There is the need to have a disaster management subject in the syllabus of all the engineering branches in India. Mechanical, chemical and civil engineering students are exposed to various risk due to the nature of their jobs. Also, computer and electronics engineers work in the crowded localities and high storied building where fire related incidents may occur. In Indian perspective, 70% of the land is prone to drought, 60% is prone to frequent earthquake, 12% to flood and 8% to cyclones (Lawrence & Suresh, 2012). Two approaches can be used for disaster awareness. Disaster management can be added as a separate subject and each of the subjects can contain the possible disasters arising from that subject related reactions and equipment.

The other approach is to provide separate practical training after learning the disaster management course/subject. In the branches like chemical engineering special attention should be give to toxic exposure (Lawrence & Suresh, 2012). Built environment related disciplines have a major role to play in disaster management. Constructional engineers must possess the disaster management knowledge (Amaratunga et al., 2011). It is envisaged that the urban infrastructure should be built considering broad spectrum of topics namely social, economic and environmental issues (Afkhamiaghda, & Elwakil, 2023). Data mining-based methods can be used for future disaster predictions. Building and geographical information systems can be integrated for better disaster management(Cao et al., 2023). In case of regional disaster management, collaborative approach can help in sharing the experience, tools and expertise(Wahyuli et al., 2023). Smart emergency crises tools encompass application of artificial intelligence in construction and disaster management(Pan & Zhang, 2023; Rane, 2023; Cao, 2023; Ahmed et al., 2023; Bannour et al., 2023). For hazard manager, it is very important to possess qualities and expertise that are beyond the response to disaster. They must be able to learn from the disaster, create the awareness that includes social science and technology for handling disasters(Blanchard and Branch, 2003). Solving social, technical and economical problem during disaster situation is important for disaster management professional (Moseley, 2004).

In disaster management, intercultural and interdisciplinary training is required. The gaps in communication command and control and lack of standardization cause major deficiency in the disaster management training (Khorram-Manesh et al., 2015). It is important to include these aspects in disaster management education. United States has experienced 1.5 trillion losses from 2003 to 2013 due to disasters in universities and educational institutes. In 2014 to 2015, 92 billion dollars was the loss (Onigbinde, 2018). This indicates that educational sector is vulnerable to disasters and its accompanying perceptions. It is responsibility of the institutions to provide disaster management facility and program. Lack of knowledge, resources, experience and teaching materials are the barriers in effective implementation of the disaster management education (Spiekermann et al. 2015, Nahayo, et al., 2018, Nipa et al., 2020). Studies indicate that fifty percent of the students are not prepared for facing possible disaster (Tan et al.,2017). Another issue is teachers approach towards disaster management and their inability to motivate the students to get trained for disaster management (Nahayo et al., 2018). Social sustainability plays a major role in disaster management at all levels of disaster management (Syed Zakaria et al., 2020). In service learning, the social and community services are included in the educational method(Furco. & Billig, 2002, Eyler, 2002).

The self-learning and service leaning approach can be used in disaster management education, predominantly disaster prevention. The demand for engineers in oil producing countries has given way to easy and lenient admission procedure. Inadequate knowledge can be dangerous while handling volatile substances. It is necessary to carefully educate the students for facing the disaster management challenges including earthquake since the area is prone to it(Baytiyeh & Naja, 2014). Virtual environment can be used for more realistic and practical education of disaster management(Tichon et al., 2003). Artificial intellence is being looked upon as a tool that can increase the speed of the response. Also, big data can be utilized for developing AI adviser platform(Lee et al., 2023). A deep learning approach can be used for improving disaster management in emergency situations(Galera-Zarco & Floros,2023).Islam et al. have proposed a combined convolutional neural network (CNN) and sorting algorithm for disaster management(Islam et al., 2023).

The industries that handle the dangerous waste more than threshold limit are known as maximum hazard units (Bhardwaj et al., 2007). India has been vulnerable to various chemical disasters. Leakage of Ammonia during the cyclone in Orissa and release of acrylonitrile during an earthquake in Gujarat are examples of this (Bhardwaj et al., 2007). Bhopal gas tragedy in which 16000 human casualties occurred was prominent disaster in India (Abraham, 1988). Post Bhopal gas tragedy, initiatives were taken by various agencies, government and non-government organizations for minimization of such incidents. In India, the concept of smart cities has become very popular. Smart devices are also being considered for safety enhancement (Aboualola et al., 2023). Various tools like internet of things, edge sensing, media analytics and big data are being explored. Social media analytics and artificial intelligence have big impact on disaster and emergency management(Pachar, 2023). Real time crises administration can be improved with the help of predictive analytics for allocation of resources during disaster situation.

Disaster Management Related Laws/Acts in India

Government of India brought the Disaster management act in 2005. This law envisaged creation of National disaster management authority. Various laws related to the handling of disaster situation are tabulated in table 2. The civil defense act defines the meaning of civil defense and underlines the powers that are immediately delegated to various agencies upon emergency situations. Under disaster management act, 2005, the disaster management fund was set up by government of India. This act describes powers and functions of state and district level executive committees in the event of disaster situation. The national institute of disaster management was set up under The National Institute of Disaster Management Regulations, 2006 by Government of India. The Notification of National Disaster Response Force (NDRF) Rules, 2008 ensured the formation of a team of personnel trained for disaster management with trainings related to response to rescue situations and disasters.

POSSIBLE DISASTERS IN AN ENGINEERING EDUCATIONAL INSTITUTE

An engineering institute handles thousands of students. The chemistry and chemical engineering laboratories need handling of some reactive and inflammable substances. Awareness and training for the staff and students is carried out by theoretical as well as experiential learning. Theoretical learning includes the lectures of safety experts from nearby industry, courses and workshops on safety and disaster management.

Table 2. Disaster management related laws in India (Saha, & Chowdhury, 2020; Gupta, 2018)

Sr. No.	Law/Act/Rule
1	The Civil Defence Act, 1968 with amendments 1975 and 2010
2	The Disaster Management Act, 2005.
3	The Civil Defence Regulations, 1968 with amendments 1975, 1973 & 1971
4	The Disaster Management (Annual Report of National Authority) Rules, 2006.
5	The Disaster Management (National Institute of Disaster Management) Rules, 2006.
6	The Disaster Management National Executive Committee (Procedure and Allowances etc) Rules, 2006.
7	The National Institute of Disaster Management Regulations, 2006.
8	The Notification of National Disaster Response Force (NDRF) Rules, 2008.

Compiled by the Authors, 2024

Mock drills and demos are carried out periodically to make students and staff competent enough to face any emergency.

The disaster management committee is formed in the institute as envisaged by All India Institute of Technical Education. The committee meets twice in the semester and takes review of the disaster management preparedness and any additional measures required. Review of electrical safety is taken twice in a semester. The period from July to December is considered as odd semester and January to June, even semester. The electrical wiring and its possible sparking scenarios are discussed. Inspection of the Electrically susceptible areas are done frequently to avoid any short circuit like situation.

EMERGENCY PREPAREDNESS PLAN AT GHARDA INSTITUTE OF TECHNOLOGY

Gharda Institute of Technology(GIT) has emergency preparedness plan for handling disaster situations. Fig.3 depicts the steps involved in preparation of the disaster management plan. The disaster management committee is headed by the Registrar or the Principal, who is assisted by the member secretary of the disaster management cell. For the preparation of the disaster management plan, the disaster management draft team interacted with various departments. The risk is identified based on the laboratory data, history, geographical data and inputs from the local community. Then disaster management tools are identified for the disaster. The areas prone to disaster are surveyed and elimination of disaster situation is facilitated by providing sufficient refugee area near the laboratories, electric installations and structures. The location of fire hydrants and smoke detectors is reviewed. The communication, incident control and evacuation teams are prepared. The disaster management plan prepared was put before various stakeholders and feedback and suggestions were incorporated.

The incident controllers are the heads of various departments in the institute. The communication team contains 3 to 4 members. Emergency team takes the immediate initiatives and evacuation teams facilitates evacuation of the people and property to safe zone. Standard operating procedure for the disaster management is displayed on the institute website. The institute has seven departments namely chemical, civil, mechanical, electronics- telecommunication, computer, Artificial intelligence and Applied Science-Humanities. Chemical handling is required in one laboratory (environmental) in civil engineering, one laboratory in Applied Science and Humanities (chemistry) and three laboratories namely chemical reaction engineering, mass transfer operations and chemical technology in chemical engineering. In these chemical department laboratories, all the instructions are displayed at prominent locations. The hazardous chemicals are stored separately and in appropriate quantities. All the laboratories are equipped with the fire extinguishers, smoke detectors and first aid boxes. Laboratory assistants are trained to handle any emergency situation. The demonstration of chemical spillage is carried out to acquaint the students with handling of such incidences. Emergency exits are clearly indicated with arrows. Assembly points are provided at strategic locations. The institute has four wings. A wing is located at the entrance and contains administrative offices. B wing contains chemical department and examination cell. C wing contains computer and electronics and telecommunication department. Central library is located near the civil engineering department in D wing. Three hostels, two for boys and one for girls, fulfills the need of accommodation. All these places are provided

Figure 3. Steps in preparation on disaster management plan (Miller et al., 2006; Mohanty et al., 2020)

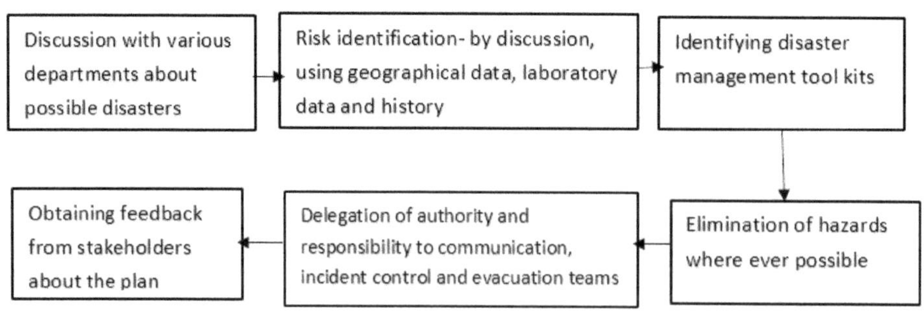

with fire fighting systems, emergency exits and assembly points. Also, the Bai Ratanbai Gharda Hospital is located adjacent to the college in case of emergency. Also, there are four buildings in the GIT colony. Principal bungalow, 2 Bed room hall kitchen(BHK) flats, self-contain rooms and 1 BHK. All these locations are provided with the fire extinguishers, emergency exits and assembly points. Also, two canteens of the hostel and one canteen of the college are provided with the emergency preparedness measures.

DISASTER MANAGEMENT ACTIVITIES

Disaster Management Cell Activities

Mock drill of chemical spillage

Objectives of the mock drill were;

a.To increase safety awareness of student and staff
b.To increase awareness about safe handling of chemicals
c.To guide the students about chemical spillage response planning.

The students were told about past examples of accidents due to chemical spillage. The properties of chemicals generally used in the laboratories were discussed with students and students were told about safe handling of these chemicals. Emphasis was on increasing awareness about storage and handling. Also, classification of chemicals and their properties were discussed. Following points were discussed

a. appropriate evacuation zones and procedures

b.availability and adequacy of fire suppression equipment
c. disposal containers for spill clean-up materials
d. the first aid procedures that might be required.

Safety aspects for chemical spillage, including assessment of the risk, selection of personal protective equipment, confinement of the spill, stopping the source, evaluation of the incident and implement clean-up, decontaminating the site and completion of required report were discussed thoroughly in disaster management cell meeting. First, students were informed about response to spillage. Then a spillage situation was created in Mass Transfer Operation laboratory by using a chemical. The windows were opened immediately to vent off fume generated. The laboratory was evacuated. The spillage was isolated and then washed off. Students became aware of safety precautions and steps to be taken in case of spillage and fume generation. Disaster management cell of Gharda institute of Technology also organizes lectures for the staff and student to make them competent in handling disaster situations in the organization. Figure 4 depicts the structure of disaster management cell and its information flow. Considering the importance of disaster management and safety in chemical and allied industries, the lecture on industrial safety was organized by disaster management cell in association with the Chemical engineering department of Gharda Institute of Technology. The guest lecture was aimed at understanding basic principles of safety. Various aspects like near miss, accident prevention methods and possibility of career in disaster and risk management were discussed. Senior safety official from H and R Johnson was invited for this online talk. Other safety awareness programs were conducted with the help of security personnel. Moreover, some subjects are there in chemical engineering like Industrial safety that make students competent enough to face emergency arising in chemical industry. Also, elective subjects like disaster management in the final year of four-year degree course makes students aware of disaster and the appropriate response. Final year projects for the students ensure that students are familiar with various unit operations and hazards arising due to inappropriate handling of equipments and chemicals. Material science and engineering is a subject that ensures basic knowledge of materials properties is available with the students. In first year, there are subjects on electrical engineering which makes students aware of electrical properties and fundamentals related to electrical engineering. Concepts of static electricity, electrical hazards are better understood by the students because of such subjects. Also subjects related to environmental management work as a tool for understanding the risk arising due to inappropriate disposal and handling of waste. Activities like chemical spillage mock drill (Figure 5) and fire extinguisher demo (Figure 6) ensures disaster preparedness in staff and students.

Fire Extinguisher Demo

Objectives of the activity were

a. To increase safety awareness about fire disaster among student and staff
b. To demonstrate use of fire extinguishers and to educate the workers/housekeeping staff about operation of fire extinguishers.

Fire was generated using Wood fuel (or fuel wood). Then staff was instructed to gather in safe place. The security personal demonstrated the use of fire extinguisher. The fire was brought under control within few minutes. Some staff members also handled fire extinguishers. It was good to see the awareness among staff members.

Figure 4. Structure of disaster management cell and emergency preparedness plan at Gharda Institute of Technology

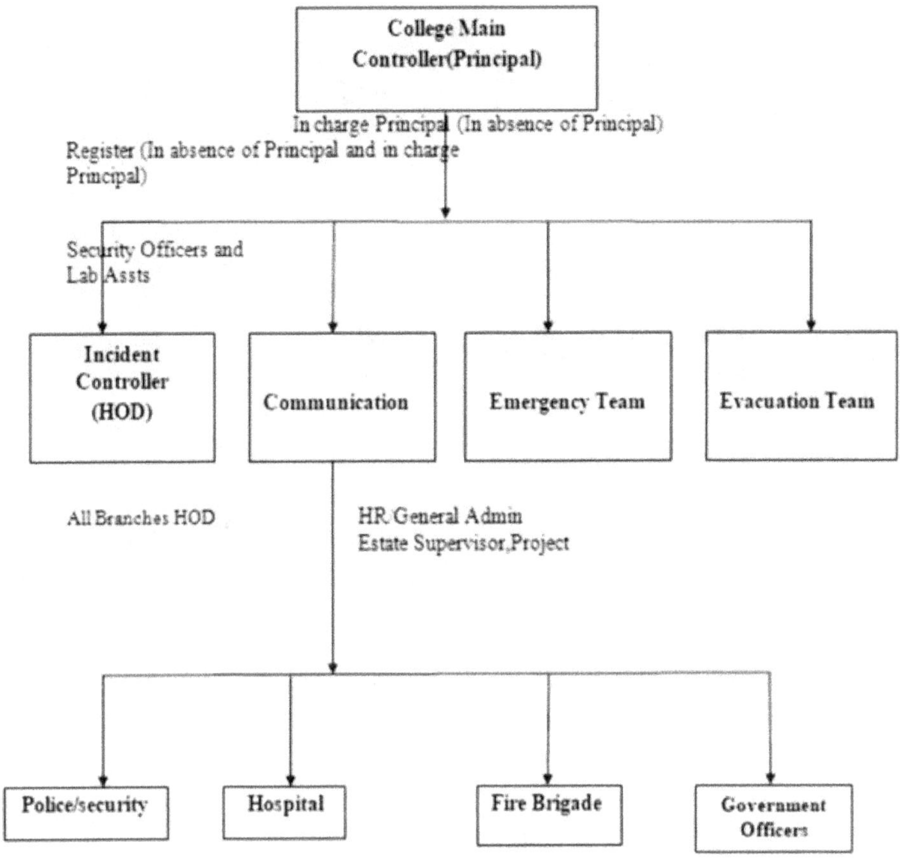

Staff(Teaching, nonteaching, housekeeping) became aware of fire safety precautions They became aware of steps to be taken in case of fire. Few staff members now can operate fire extinguishers in case of fire. Figure 6 is pictorial representation of fire extinguisher demonstration.

Disaster Management Cell Activities Related to COVID-19 Outbreak

Covid Appropriate Measures

During Corona break up in 2020, the disaster management cell of Gharda Institute of Technology played significant role. It prepared standard operating procedure

Figure 5. Mock Drill of Chemical Spillage. Original photographs

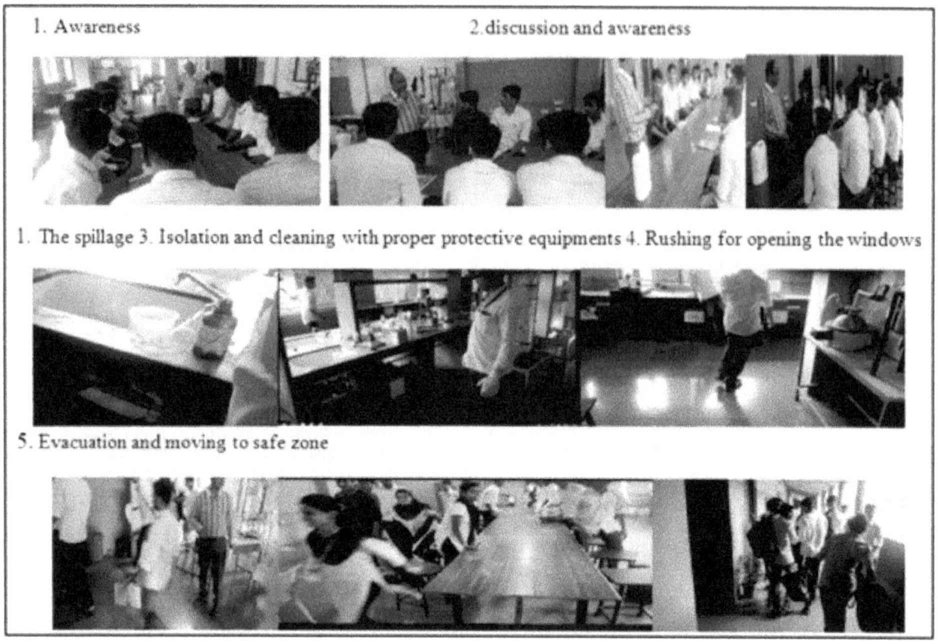

to deal with any covid- 19 infected person in the campus. Also post Covid- 19 precautions were prepared by the Disaster management cell. The cell took periodic review of the situation and suggested the measures for combating the situation. The sanitizer stands were manufactured and put into use as per following schedule. On 12 February, - Total 12 Nos. of Sanitizer Stands were manufactured and put into use. On 16 February, Total 20 nos. of Hand Wash Bottle Containers were manufactured. On 26th February, 02 Nos. of Sanitizer stand were manufactured and put into use. By 20 the February 25 Hand Wash Bottle Containers were put into use. It was decided to make the masks available for the students at the entry points when students physically start attending the college. It was decided to ensure refilling of sanitizer and handwash bottles. In addition to handling the situation in the institute the staff and students of Gharda Institute of Technology took part in the activities in nearby vicinities like vaccination, helping the patients, providing the emergency help like ambulance, accommodation and medication. The Bai Ratanbai Gharda Hospital has played a proactive role in combating the Covid 19 outbreak. The Gharda Institute of Technology played a key role by providing its hostels for Covid center during the pandemic. To avoid any situation arising due to personal issues and mental health of the students, the institute has special counselor. Counselor service can be availed by the students as and when required.

Figure 6. Pictorial representation of fire extinguisher demo. Original photographs

During Covid -19, student face the problems related to mental health due to the trauma faced by them. Many relatives of the students hospitalized, succumb to the disease which affect them emotionally and mentally. Counseling them and motivating them to study in physical or offline mode was also important. Covid-19 disaster has left the footprints on students' mentality related to learning ability which needs consultations and counselling.

FINDINGS

Disaster management activities during Covid-19 indicated that proper planning and monitoring is an important part of the disaster management. During covid-19 pandemic the disaster management cell suggested number of measures which helped in protecting the health of the staff and students. The proper planning about the location of sanitizers, temperature detectors and handwash ensured that the staff and student feel safe during the working hours. The activities like chemical spillage are important for the budding chemical engineers. The mock drills can help in inculcating the safety culture among the students that can help them in their professional life. Snake bite handling and related first aid treatment is life long learning for the staff and students. Fire extinguisher demo is common disaster management activity and important from domestic and industrial perspectives. The chemical industries are prone to the fire disasters due to the high temperature reactions, volatile compounds and flammable materials. The emergency preparedness of the engineers to possible fire disasters can help them address any incident in their professional career.

The locations of fire extinguishers in GIT were decided based on possible fire types. Carbon dioxide(CO_2) type fire extinguishers were located near the electrical rooms as the extinguishing agent in these extinguishers do not cause damage to the equipments and installations. Carbon dioxide extinguishers cannot be used for flammable gases and in confined space as it may cause breathing problem. Dry chemical powder type (DCP) fire extinguishers are used for almost all types of fires including carbonaceous material fire, electrical fire and fire due to liquid. Fig.7 depicts the positions of fire extinguishers in GIT. It can be seen that, the near chemical and chemistry department adequate number of DCP extinguishers were installed as CO_2 fire extinguishers cannot be used for confined space and flammable gases. CO_2 fire extinguishers are located near computer, electronics and related department. Other areas were equipped with both type of extinguishers in adequate numbers. Hostel has adequate open space and large area compared to various other departments. CO_2 fire extinguishers were provided more in hostel compared to other areas.

CONCLUSION

Appropriate and fast response to the emergency is the most important aspect of disaster management. In many organizations, the allocation of fund for disaster management is ignored part of financial planning. The most important virtues of disaster management are cautiousness, bravery, fair-mindedness, watchfulness, toughness, communication and effective implementation of plan. Response and recovery are two important phases of disaster management that need these virtues. While allocating resources to a certain task, the risks and uncertainties involved must be given due consideration. Gharda Institute of Technology has emergency preparedness plan for handling emergency situations. The disaster management cell of Gharda Institute of Technology has handled the Covid-19

Figure 7. Fire extinguisher types and their locations in GIT

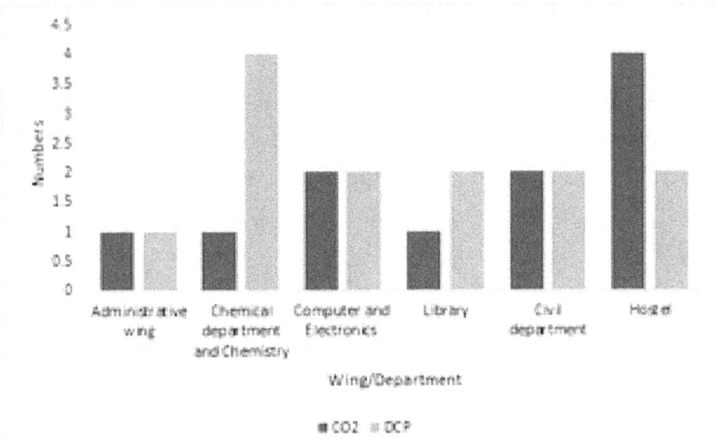

situation with appropriate precaution. The cell conducts frequent checks on the preparedness of the concern taskforce to combat any untoward incident. The display of contact numbers of hospital, police, district level authorities brings about the safe feeling among the students and staff. The institute has chemical department and hence handling of chemicals is important aspect. The disaster management cell collaborates with various departments to review possible hazards and takes appropriate training and awareness programs. The cell also conducts programs for industrial disaster management.

ACKNOWLEDGMENT

The authors are grateful to the Disaster management cell of Gharda Institute of Technology for its support and cooperation.

REFERENCES

Aboualola, M., Abualsaud, K., Khattab, T., Zorba, N., & Hassanein, H. S. (2023). *Edge Technologies for Disaster Management: A Survey of Social Media and Artificial Intelligence Integration*. IEEE. doi:10.1109/ACCESS.2023.3293035

Abraham, M. (1988). The Bhopal Tragedy—A Community Action Perspective. *Community Development Journal: An International Forum*, *23*(4), 229–234. https://www.jstor.org/stable/44258578. doi:10.1093/cdj/23.4.229

Afkhamiaghda, M., & Elwakil, E. (2023). Challenges review of decision making in post-disaster construction. *International Journal of Construction Management*, *23*(14), 2409–2418. doi:10.1080/15623599.2022.2061751

Ahmed, A. A. A., Mahalakshmi, A., ArulRajan, K., Alanya-Beltran, J., & Naved, M. (2023). Integrated artificial intelligence effect on crisis management and lean production: Structural equation modelling frame work. *International Journal of System Assurance Engineering and Management*, *14*(1), 220–227. doi:10.1007/s13198-022-01679-1

Amaratunga, R. D. G., Siriwardena, M. L., Malalgoda, C. I., Pathirage, C. P., & Thayaparan, M. (2011). *Lifelong learning needs for disaster management education in the built environment*. https://salford-repository.worktribe.com/output/1449597/lifelong-learning-needs-for-disaster-management-education-in-the-built-environment

Bannour, W., Maalel, A., & Ben Ghezala, H. H. (2023). Emergency management case-based reasoning systems: A survey of recent developments. *Journal of Experimental & Theoretical Artificial Intelligence*, *35*(1), 35–58. doi:10.1080/0952813X.2021.1952654

Baytiyeh, H., & Naja, M. K. (2014). Revolutionizing engineering education in the Middle East region to promote earthquake-disaster mitigation. *European Journal of Engineering Education*, *39*(5), 573–583. doi:10.1080/03043797.2014.895705

Beg, K., Bhadra, B. K., Sharma, J. R., Punia, M. P., & Chaurey, R. (2013). Seismic Data Analysis for Disaster Management in Jodhpur District of Rajasthan Using GIS, GPS and Remote Sensing Techniques. *Jour. Ind. Geol. Cong*, *5*, 1.

Below, R., Wirtz, A., & Guha-Sapir, D. (2009). *Disaster category classification and peril terminology for operational purposes*. No. UCL-Université Catholique de Louvain.

Bhardwaj, J. R., Chawla, R., & Sharma, R. K. (2007). *Chemical disaster management: Current status and perspectives*. https://nopr.niscpr.res.in/handle/123456789/1140

Blanchard, B. W., & Branch, R. (2003). *Outlines of competencies to develop successful 21st century hazard or disaster or emergency or hazard risk managers*. Presentation at the *Natural Hazards Workshop*, Boulder, Colorado.

Bull, M. T., Sanhueza, F., Oyarzo, C., & Cea, P. (2017). *Towards A Culture Of Prevention: Including Disaster Management In Engineering Undergraduate Programs*. https://www.wcee.nicee.org/wcee/article/16WCEE/WCEE2017-2893.pdf

Cao, L. (2023). AI and data science for smart emergency, crisis and disaster resilience. *International Journal of Data Science and Analytics*, *15*(3), 231–246. doi:10.1007/s41060-023-00393-w PMID:37035277

Cao, Y., Xu, C., Aziz, N. M., & Kamaruzzaman, S. N. (2023). BIM–GIS Integrated Utilization in Urban Disaster Management: The Contributions, Challenges, and Future Directions. *Remote Sensing (Basel)*, *15*(5), 1331. doi:10.3390/rs15051331

Comfort, L. K. (2005). Risk, security, and disaster management. *Annual Review of Political Science*, *8*(1), 335–356. doi:10.1146/annurev.polisci.8.081404.075608

Eyler, J. (2002). Reflection: Linking service and learning – Linking students and communities. *The Journal of Social Issues*, *58*(3), 517–534. doi:10.1111/1540-4560.00274

Furco, A., & Billig, S. H. (Eds.). (2002). *Service learning: The essence of the pedagogy*. IAP.

Galera-Zarco, C., & Floros, G. (2023). A deep learning approach to improve built asset operations and disaster management in critical events: An integrative simulation model for quicker decision making. *Annals of Operations Research*, 1–40. doi:10.1007/s10479-023-05247-z

Guerdan, B. R. (2009). Disaster preparedness and disaster management. *The American Journal of Chinese Medicine*, *6*, 32–40.

Gupta, A. K. (2018). Disaster governance and legal systems in India. *Disaster risk governance in India and cross cutting issues*, 39-60.

Islam, M. A., Rashid, S. I., Hossain, N. U. I., Fleming, R., & Sokolov, A. (2023). An integrated convolutional neural network and sorting algorithm for image classification for efficient flood disaster management. *Decision Analytics Journal*, *7*, 100225. doi:10.1016/j.dajour.2023.100225

Islam, R., Kamaruddin, R., Ahmad, S. A., Jan, S., & Anuar, A. R. (2016). A review on mechanism of flood disaster management in Asia. *International Review of Management and Marketing*, *6*(1), 29–52.

Kathleen Geale, S. (2012). The ethics of disaster management. *Disaster Prevention and Management: An International Journal, 21*(4), 445-462. doi:10.1108/09653561211256152

Khorram-Manesh, A., Ashkenazi, M., Djalali, A., Ingrassia, P. L., Friedl, T., Von Armin, G., Lupesco, O., Kaptan, K., Arculeo, C., Hreckovsi, B., Komadina, R., Fisher, P., Voigt, S., James, J., & Gursky, E. (2015). Education in disaster management and emergencies: Defining a new European course. *Disaster Medicine and Public Health Preparedness*, *9*(3), 245–255. doi:10.1017/dmp.2015.9 PMID:25781077

Lawrence, S., & Suresh, E. S. M. (2012). The need for integration of disaster management into engineering curriculum. *International Journal of Modern Engineering Research, 2*(6), 4524-4526.

Lee, M., Mesicek, L., Bae, K., & Ko, H. (2023). AI advisor platform for disaster response based on big data. *Concurrency and Computation*, *35*(16), e6215. doi:10.1002/cpe.6215

Miller, H. E., Engemann, K. J., & Yage, R. R. (2006). Disaster planning and management. *Communications of the IIMA*, *6*(2), 4.

Mohanty, M. P., Mudgil, S., & Karmakar, S. (2020). Flood management in India: A focussed review on the current status and future challenges. *International Journal of Disaster Risk Reduction*, *49*, 101660. doi:10.1016/j.ijdrr.2020.101660

Moseley, L. (2004). Educational needs for disaster management. *Australian Journal of Emergency Management*, *19*(4), 28–31.

Nahayo, L., Li, L., Habiyaremye, G., Richard, M., Mukanyandwi, V., Hakorimana, E., & Mupenzi, C. (2018). Extent of disaster courses delivery for the risk reduction in Rwanda. *International Journal of Disaster Risk Reduction*, *27*, 127–132. doi:10.1016/j.ijdrr.2017.09.046

Nath, S. K., Roy, D., & Thingbaijam, K. K. S. (2008). Disaster mitigation and management for West Bengal, India–An appraisal. *Current Science*, 858–864.

Nipa, T. J., Kermanshachi, S., Patel, R., & Tafazzoli, M. (2020, September). Disaster preparedness education: Construction curriculum requirements to increase students' preparedness in pre-and post-disaster activities. In *Associated Schools of Construction (ASC) International Conference* (pp. 142-131). Academic Press.

O'Brien, G., O'keefe, P., Gadema, Z., & Swords, J. (2010). Approaching disaster management through social learning. *Disaster Prevention and Management*, *19*(4), 498–508. doi:10.1108/09653561011070402

Onigbinde, L. (2018). *The impacts of natural disasters on educational attainment: Cross-country evidence from macro data*. https://repository.usfca.edu/thes/1078/

Pachar, S., Dudeja, D., Batra, N., Tomar, V., Bhimavarapu, J. P., & Singh, A. K. (2023). Recent Developments in Machine Learning Predictive Analytics for Disaster Resource Allocation. *Engineering Proceedings*, *59*(1), 19. https://www.mdpi.com/2673-4591/59/1/19

Pan, Y., & Zhang, L. (2023). Integrating BIM and AI for smart construction management: Current status and future directions. *Archives of Computational Methods in Engineering*, *30*(2), 1081–1110. doi:10.1007/s11831-022-09830-8

Price, R. K., & Vojinovic, Z. (2008). Urban flood disaster management. *Urban Water Journal*, *5*(3), 259–276. doi:10.1080/15730620802099721

Rane, N. (2023). Integrating Building Information Modelling (BIM) and Artificial Intelligence (AI) for Smart Construction Schedule. *Cost, Quality, and Safety Management: Challenges and Opportunities.*

Rebeeh, Y. A., Pokharel, S., Abdella, G. M., & Hammuda, A. S. (2019). Disaster management in industrial areas: Perspectives, challenges and future research. *Journal of Industrial Engineering and Management*, *12*(1), 133–153. doi:10.3926/jiem.2663

Rolland, E., Patterson, R. A., Ward, K., & Dodin, B. (2010). Decision support for disaster management. *Operations Management Research, 3*, 68-79. https://link.springer.com/article/10.1007/s12063-010-0028-0

Saha, A., & Chowdhury, S. (2020). Legal Framework of Disaster Management in India along with the Functioning of NDRF. *Int'l JL Mgmt. &. Human.*, *3*, 529.

Sawalha, I. H. (2020). A contemporary perspective on the disaster management cycle. *Foresight, 22*(4), 469-482.

Spiekermann, R., Kienberger, S., Norton, J., Briones, F., & Weichselgartner, J. (2015). The DisasterKnowledge Matrix – Reframing and evaluating the knowledge challenges in disaster risk reduction. *International Journal of Disaster Risk Reduction*, *13*, 96–108. doi:10.1016/j.ijdrr.2015.05.002

Syed Zakaria, S. A., Majid, A. T., Lau, T. L., & Azimi, M. A. (2020). Exploration of Training and Education for Attaining Disaster Resilience from Sustainability Perspective Among Civil Engineering Students. In *Proceedings of AICCE'19: Transforming the Nation for a Sustainable Tomorrow 4* (pp. 361-376). Springer International Publishing.

Tan, Y., Liao, X., Su, H., Li, C., Xiang, J., & Dong, Z. (2017). Disaster Preparedness Among University Students in Guangzhou, China: Assessment of Status and Demand for Disaster Education. *Disaster Medicine and Public Health Preparedness*, *11*(3), 310–317. doi:10.1017/dmp.2016.124 PMID:27666525

Teh, D., & Khan, T. (2021). Types, definition and classification of natural disasters and threat level. In *Handbook of Disaster Risk Reduction for Resilience: New Frameworks for Building Resilience to Disasters* (pp. 27–56). Springer International Publishing. doi:10.1007/978-3-030-61278-8_2

Tichon, J., Hall, R., Hilgers, M., Leu, M., & Agarwal, S. (2003). Education and training in virtual environments for disaster management. In *EdMedia+ Innovate Learning* (pp. 1191-1194). Association for the Advancement of Computing in Education (AACE). https://www.learntechlib.org/p/13974/

University of Central Florida. (n.d.). *The Disaster management cycle: 5 stages and how leadership can help prepare. Leadership and management*. https://www.ucf.edu/online/leadership-management/news/the-disaster-management-cycle/

Wahyuli, S., Nurfarhaty, N., & Haeril, H. (2023). Natural Disaster Management in Local Government from the Perspective of Collaborative Governance. *Jurnal Studi Ilmu Pemerintahan*, *4*(1), 93–102.

Chapter 9
Fostering a Sustainable and Disaster-Resilient Transportation Infrastructure

Syed Mohd. Arif
https://orcid.org/0000-0002-0821-317X
Department of Civil Engineering, Chandigarh Engineering College, CGC Jhanjeri, Mohali, India

Shalom Akhai
https://orcid.org/0000-0002-7533-457X
Department of Mechanical Engineering, Maharishi Markandeshwar University (deemed), Ambala, India

ABSTRACT

The chapter addresses resilience engineering's role in sustainable and disaster-resistant transportation infrastructure. Integrating autonomous cars, intelligent transportation systems (ITS), and network reconfiguration methodologies improves transportation network flexibility and recovery. The technology may decrease traffic, increase safety, and improve inhabitants' commuting experiences. Integration requires stakeholder involvement, infrastructure development, and strategic planning, according to the research. Contraflow, crossing removal at junctions, optimisation models, traffic assignment problem (TAP), and gradient projection increase traffic flow, congestion, and evacuation efficiency in resilient engineering. It strengthens transportation systems' ability to adapt, respond, and recover from disturbances to protect communities and ensure infrastructure sustainability. Finally, resilience engineering reduces transportation system disruptions and makes infrastructure more adaptive, durable, and sustainable.

DOI: 10.4018/979-8-3693-2721-0.ch009

INTRODUCTION

Automobiles that operate autonomously and smart cities that decrease congestion are intended to improve urban transportation and mobility by enhancing energy efficiency, resident satisfaction, and safety (Butler et al., 2020). Participation of stakeholders, infrastructure development, and strategic planning are all critical components in ensuring a successful integration process (Shareeda et al., 2021). Transportation resilience engineering focuses on methods to assist transport networks in recovering from natural disasters while constructing and maintaining infrastructure that is resilient to sustainability and disasters (Serdar et al., 2022). Disaster mitigation and evacuation require robust infrastructure, network reconfiguration, and evacuation readiness (Heaslip et al., 2009; Freckleton et al., 2012). It is imperative to comprehend the impact of infrastructure disruptions on vulnerable communities in the event of catastrophes in order to establish infrastructure resilience that prioritises human needs (Tachaudomdach et al., 2021). Critical infrastructure, social capital, nutritional security, information and communication technology, and institutional integrity are all assets that can aid in mitigating catastrophe losses (Radvanovsky et al., 2018).

This chapter aims to bridge this gap by examining the integration of autonomous vehicles and ITS technologies within the context of resilience engineering for sustainable and safe urban transportation. By exploring the strategies for network reconfiguration and their implementation challenges, this research seeks to provide insights into how transportation systems can enhance their resilience to disruptions and ensure the safety and well-being of communities during emergencies (Raadsen et al., 2020; Nipa & Kermanshachi, 2021).

The objectives of this chapter are twofold: firstly, to assess the current state of knowledge regarding the integration of autonomous vehicles and ITS technologies in urban transportation; and secondly, to investigate the strategies for network reconfiguration in resilience engineering and their potential benefits and implementation challenges.

Through this chapter, we aim to contribute to the existing body of literature by providing actionable insights for policymakers, transportation professionals, and urban planners to enhance the resilience and sustainability of transportation infrastructure in the face of evolving challenges and uncertainties.

In the subsequent sections, we reviewed the existing literature on intelligent transportation systems, autonomous vehicles, and resilience engineering. We also discussed the conceptual framework guiding this study and outline the research methodology employed. Finally, we will present the findings of our analysis and discuss their implications for theory, practice, and future research directions.

Integration of Autonomous Vehicles and Intelligent Transportation Systems

Intelligent transport systems (ITS) and autonomous vehicles have the capacity to fundamentally transform urban mobility (Aldakkhelallah and Simic 2021). Cities have the ability to furnish their residents with a streamlined and effective mode of conveyance by incorporating these technologies (Meneguette 2018 and Seuwou 2020). In the context of a smart city, ITS technologies, including communication networks, traffic management systems, and real-time data collection, have the potential to improve the coordination and operation of autonomous vehicles (Mahrez et al., 2021). Residents may experience reduced traffic congestion, enhanced safety, increased energy efficiency, and an overall improvement in their transportation experiences as a result of this integration (Kuru and Khan 2020; Iliopoulou et.al, 2020).

Strategies for Network Reconfiguration in Resilience Engineering

The modern transportation management method of resilience engineering influences network reconfiguration to minimise congestion, increase evacuation efficiency, and shorten travel time for all users (Xu et al., 2021). This approach uses traffic flow changes, roundabouts, traffic signal optimisations, mathematical optimisation, TAP, and gradient projection (Gholamhosseinian & Seitz 2022). Using these methods, resilience engineering strengthens transport infrastructure against interruptions, improving everyday operations and emergency response (Zou & Chen 2019). Resilience engineering includes reversing traffic flow on roads or highways to speed emergency evacuations. The method needs real-time traffic monitoring and management, public awareness initiatives, and strategic planning. Technology, public relations, and strategic planning are needed to execute contraflow (Renne et al., 2022). Eliminating intersection crossings is another novel way to improve traffic flow and safety. Restructuring road networks reduces congestion, improves safety, and strengthens the transportation system (Nugmanova et al., 2019). Optimisation models need careful planning, stakeholder collaboration, and powerful traffic modelling and simulation tools to reconfigure systems efficiently.

Traffic Assignment Problem (TAP) models distribute traffic flow to diverse routes to reduce travel time while retaining transportation system efficiency. Successful implementation needs advanced traffic modelling and simulation tools, real-time traffic data gathering and analysis, and adaptive traffic signal control systems (Patriksson 2015). The advanced Gradient Projection approach in resilience engineering shifts traffic flow from non-shortest pathways to the shortest path between origin-destination pairs, encouraging user equilibrium traffic assignment.

Due to its computing resources, implementation may be difficult (Tap et al., 2023). These methods must be integrated into a comprehensive and adaptable framework encompassing government agencies, local authorities, emergency services, and the public to improve transportation infrastructure resilience. In emergencies, stakeholders must be aware and able to cooperate via effective communication protocols and information transmission. Artificial intelligence and machine learning can improve resilience engineering (Saravi et al., 2019). Real-time data analytics and predictive modelling help make proactive decisions and allocate resources more efficiently during routine operations and emergencies related to traffic patterns (Pérez et al., 2018).

Table 1 provides an overview of specific strategies for network reconfiguration in resilience engineering, along with their implementation challenges, potential benefits, and best practices.

Resilience Engineering for Sustainability in Transportation Infrastructure

Consideration of elements that boost a system's capacity to tolerate and recover from disturbances is an important part of resilience engineering, which contributes to the sustainability of the transportation infrastructure (Raadsen et at., 2020). This comprises the planning and management of the infrastructure with robustness, redundancy, flexibility, and adaptation as primary considerations (Sun et at., 2020). Transportation systems may better absorb shocks, minimise interruptions, and preserve performance if they include these concepts, which will eventually contribute to the long-term sustainability of the infrastructure (Lu, X., Liao, W., Fang, D., Lin, K., Tian, Y., Zhang, C., Zheng, Z., & Zhao, P. 2020).

Resilience Engineering for Safety and Well-Being of Communities

Building and maintaining infrastructures that are able to resist and recover from disruptions brought on by natural disasters or other severe occurrences is the primary emphasis of resilience engineering (Nipa & Kermanshachi 2021). This kind of engineering makes a significant contribution to the health and safety of communities (Akhai et al., 2016; Akhai et al., 2020). It is alike providing human comfort through engineering technology in buildings like providing air conditioning (Akhai et at., 2021; Tanwar & Akhai 2017; Kumar & Akhai 2022). It encompasses ideas and practises that strengthen the capacity of transportation systems, infrastructure, and organisations to successfully adapt, react, and recover from disruptions. Resilience engineering aims to protect the safety and well-being of communities, minimise

Table 1. Strategies for network reconfiguration in resilience engineering

Strategies	Descriptions	Implementation Challenges	Potential Benefits	Best Practices
Contraflow or lane reversal	This approach includes modifying traffic flow on selected lanes or highways to facilitate emergency evacuation traffic.	Requires real-time traffic monitoring and control systems	Improved evacuation times during emergencies	Simulations can assess contraflow's efficacy under various traffic scenarios. Use visible signs and public awareness efforts to warn drivers of contraflow practises.
Crossing elimination at intersections	By reconfiguring intersections, such as implementing roundabouts or traffic signal optimization, the efficiency and safety of traffic flow can be improved, reducing congestion and enhancing resilience	May require significant infrastructure changes	Reduced congestion and improved traffic flow	- Conduct a thorough analysis of existing traffic patterns and identify intersections that would benefit from reconfiguration. - Engage with the community and stakeholders to address potential concerns about intersection reconfiguration projects.
Optimization models	Mathematical optimization models can be used to determine the optimal system reconfiguration that minimizes total travel time for all users, including evacuees and regular commuters	Requires careful planning and coordination with stakeholders	Enhanced resilience to disruptions	- Utilize advanced traffic modeling and simulation tools to calibrate and validate optimization models. - Establish clear communication protocols and procedures for sharing data and coordinating reconfiguration decisions among stakeholders.
Traffic assignment problem (TAP)	TAP models can be used to allocate traffic flow to different routes, ensuring that users can minimize their own travel time while maintaining overall system efficiency	May require advanced traffic modeling and simulation tools	More efficient use of transportation infrastructure	- Develop real-time traffic data collection and analysis capabilities to inform TAP models. -Implement adaptive traffic signal control systems that can adjust signal timings based on real-time traffic conditions.
Gradient projection method	This method can be employed to shift traffic flow from non-shortest paths to the shortest path between origin-destination pairs, promoting user equilibrium traffic assignment	May require significant computational resources	Improved user experience	Implement efficient algorithms and optimisation strategies to minimise gradient projection method computational load. Combine gradient projection with modern traffic modelling and simulation techniques to boost its efficacy.

(Iliopoulou et al., 2020; Arif, S. M., & Gupta, S. 2022; Zhang et al., 2019; Ahmed & Dey 2020; Jawad & Liu 2020) disruptions to daily activities, and ensure the continuous functionality of critical systems (Arif et at., 2021). This is accomplished by designing infrastructure to withstand hazards, implementing effective emergency response plans, and promoting adaptive capacity (Achillopoulout al., 2020).

How Can Transportation Systems Ensure That They Are Inclusive and Accessible to all Populations During Evacuations?

To minimise disaster damage, evacuation transportation must be inclusive and accessible to everyone. Comprehensive preparation, accessible choices, public awareness, agency collaboration, and real-time evacuation order updates are strategies. Transportation for elderly homes, isolated areas, and low-income populations must be prioritised. Emergency responders and transportation professionals need cultural sensitivity training. Community engagement, particularly for underprivileged communities, is encouraged. Regular accessibility audits and transportation infrastructure assessments are needed. Safety and effectiveness depend on developing and adapting evacuation plans and tactics based on prior catastrophes and community needs. Complete evacuation plans must include the requirements of all demographic groups, including those with impairments, elderly persons, restricted mobility, and language or cultural difficulties. Paratransit, accessible public transportation, and disabled automobiles are available. Public outreach and education advise vulnerable people on evacuation methods, routes, and transit alternatives (Ebrahimnejad et al., 2023; Wang et al., 2023). Emergency response agencies, transportation companies, and community organisations work together to evacuate. Closed captioning, sign language interpretation, and multilingual communication give accessible information. Transportation workers and emergency responders get cultural sensitivity and awareness training for vulnerable communities (Cadwell 2021). Communities are urged to participate, and accessibility audits and evaluations identify and remove impediments. Table 2 outlines resilience engineering network reconfiguration strategies.

How Resilience Engineering Helps Transportation Systems Minimize Disruptions

To minimise transportation delays, resilience engineering is essential. It comprises optimising network reconfiguration, enhancing infrastructure design and maintenance, and using ICT for real-time monitoring and communication during disturbances. Adaptability requires planning and ongoing learning. Authorities, emergency responders, community organisations, and IT companies must collaborate. Risk

Table 2. Strategies for network reconfiguration in resilience engineering

Strategy	Description
Comprehensive Planning and Preparedness	Develop detailed evacuation plans considering the needs of all population groups, including those with disabilities, older adults, individuals with limited mobility, and those with linguistic or cultural barriers. Ensure widespread dissemination and accessibility of these plans.
Accessible Transportation Options	Provide a variety of transportation options to accommodate diverse needs, including accessible public transportation, specialized vehicles for people with disabilities, and paratransit services. Establish accessible pickup points and evacuation routes.
Public Outreach and Education	Conduct targeted outreach campaigns to inform vulnerable populations about evacuation procedures, routes, and available transportation options. Provide information in multiple languages and formats for accessibility.
Coordination and Collaboration	Establish clear lines of communication and coordination among emergency response agencies, transportation providers, and community organizations to ensure a coordinated and effective evacuation response.
Accessible Communication and Information	Provide real-time updates and information through accessible channels such as closed captioning, sign language interpretation, and multilingual communication regarding evacuation orders, road closures, and transportation options.
Vulnerable Populations Identification	Identify and prioritize the needs of vulnerable populations, including those in isolated areas, nursing homes, and low-income communities, ensuring their access to transportation during evacuations.
Cultural Sensitivity and Awareness	Train transportation personnel and emergency responders to be culturally sensitive and aware of the unique needs of different population groups during evacuations.
Community Engagement and Participation	Involve community members, especially those from vulnerable populations, in planning and decision-making processes related to evacuation plans and transportation strategies.
Accessibility Audits and Assessments	Regularly conduct accessibility audits and assessments of transportation infrastructure and evacuation routes to identify and address barriers hindering accessibility for all users.
Continuous Improvement and Adaptation	Continuously evaluate and improve evacuation plans and transportation strategies based on lessons learned from past disasters and evolving community needs.

(Zhang et al., 2019; Raadsen et al., 2019; Achillopoulout al., 2020; Zhang et al., 2019; Achillopoulout al., 2020; Muzafar et al., 2020; Zhou et al., 2019; Hollnagel 2013; Hollnagel 2013; Raharjo et al., 2022; Raharjo et al., 2022)

management and vulnerability assessments identify disturbances and implement remedies to heal and learn. Reconfiguring transport networks to eliminate counterflow and crossings at junctions improves flexibility during interruptions. Infrastructure design and maintenance use durable materials and hazard-resistant design for robustness, redundancy, flexibility, and adaptability. ICT allows real-time monitoring, data collecting, and communication amid disturbances. A culture of readiness, emergency response training, and continual learning and development create adaptive capability in transportation organisations. Coordination and collaboration among stakeholders are necessary for disruption reduction and recovery. Risk management and vulnerability assessments detect possible disruptions and reduce their effect. After a disturbance, learning from earlier disruptions improves readiness and reaction (Ganin et al., 2019; Zhou et al., 2019). Table 3 outlines transportation resilience strategies.

Table 3. Transportation resilience strategies

Strategy	Description	Strategies	Considerations
Network Reconfiguration	Optimization of transportation networks for increased adaptability during disruptions.	Contraflow (lane reversal), Crossing elimination at intersections, Optimization models for effective system reconfiguration	Requires careful planning and coordination with emergency response agencies
Infrastructure Design and Maintenance	Emphasis on designing and maintaining infrastructure with robustness, redundancy, flexibility, and adaptability.	Use of durable materials, Incorporation of redundancy in critical components, Design for hazard resistance	May require physical infrastructure changes, such as roundabout construction
Information and Communication Technologies (ICT)	Leveraging ICT for real-time monitoring, data collection, and communication during disruptions.	Real-time monitoring, data collection, and communication during disruptions	Requires computational resources and expertise to develop and implement optimization models
Adaptive Capacity Building	Promotion of adaptive capacity within transportation organizations through a culture of preparedness.	Personnel training in emergency response procedures, Encouragement of continuous learning and improvement	Requires accurate traffic data and models to effectively allocate traffic flow
Collaboration and Partnerships	Emphasis on collaboration and partnerships among stakeholders.	Collaboration and partnerships among stakeholders	Fosters a coordinated approach to disruption mitigation and recovery
Vulnerability Assessment and Risk Management	Conducting vulnerability assessments to identify potential disruptions and implementing risk management strategies to mitigate their impact.	Assessing the resilience of critical infrastructure components, Evaluating potential hazards	Involves assessing the resilience of critical infrastructure components and evaluating potential hazards
Post-Disruption Recovery and Learning	Emphasis on learning from past disruptions to improve future preparedness and response capabilities.	After-action reviews, Identification of areas for improvement, Incorporation of lessons learned into future plans and procedures	Includes after-action reviews, identification of areas for improvement, and incorporation of lessons learned into future plans and procedures

(Ahmed & Dey 2020; Jawad & Liu 2020; Raadsen et al., 2020; Zhang et al., 2019; Zhang et al., 2019; Achillopoulou et al., 2020; Achillopoulou et al., 2020; Achillopoulou et al., 2020; Shareeda et al., 2021)

Challenges and Future Scope

Stakeholder collaboration, infrastructure development, public awareness and communication, advanced technology deployment, and computational resource management are issues for transportation infrastructure resilience engineering techniques. Traffic modelling tools, real-time data gathering systems, and adaptive traffic signal control systems are needed to implement these techniques. Public awareness and comprehension of resilience engineering methods like contraflow need improved PR.

Advanced technologies like artificial intelligence and machine learning need significant investment, training, and careful deployment to succeed. Gradient Projection is promising but requires a lot of processing resources. Emerging technology, inclusion in evacuation planning, ongoing improvement, integration with smart city projects, and worldwide cooperation and information sharing are the future of resilience engineering in transportation infrastructure. Infrastructure development is essential for success, and stakeholders must collaborate to achieve smooth coordination. To help people adapt to traffic changes and collaborate in crises, public relations should be improved. So, resilience engineering for transport infrastructure has great promise despite its difficulties. Addressing these difficulties and investigating future horizons may strengthen, maintain, and disaster-resistant transportation networks.

CONCLUSION

The article highlights the important role that resilience engineering plays in the process of developing ecologically friendly and disaster-resistant transportation infrastructure. The capacity of transportation networks to adapt and recover from disruptions may be improved using technologies such as autonomous cars, intelligent transportation systems, and efficient strategies for network reconfiguration. During evacuations, the problems of inclusiveness and accessibility of transportation continue to be of the utmost importance, highlighting the need for extensive preparation, coordination, and cultural sensitivity. The ongoing improvement of evacuation plans via the application of lessons learned in the past is another factor that helps to the protection of communities. When seen in the larger perspective of the goals, resilience engineering appears as a basic strategy to minimising disruptions, increasing sustainability, and safeguarding the future of transportation infrastructure.

Recommendations

The review outlines resilience engineering suggestions for sustainable and disaster-resistant transportation infrastructure research, policy, and implementation. These proposals involve multidisciplinary research cooperation between academics, engineers, urban planners, and AI/data science professionals. Government, local authorities, emergency services, technology providers, and the public should be more involved. Governments and commercial organisations should combine to finance and promote technology integration research for resilience.

Community awareness campaigns on resilience engineering in transportation, including contraflow, crossing removal, and emergency procedures, are recommended. Transportation workers, emergency responders, and stakeholders should also get resilience engineering training. Pilot programmes and case studies should evaluate resilience engineering techniques in real-world situations.

International knowledge exchange platforms could disseminate resilience engineering in transportation best practices, case studies, and research. Smart city policy integration should include network reconfiguration and intelligent transport systems to improve urban resilience. Resilience methods should be monitored and evaluated regularly using robust frameworks.

Using lessons from prior catastrophes, community input, and technology advances, governments and authorities update and improve evacuation plans over time. Exit planning includes inclusive accessible measures for all people, particularly disadvantaged ones. Resilience engineering behavioural research examines how people respond to resilience techniques. Emergency risk communication techniques emphasise the need to invest in communication infrastructure. Technology suppliers should collaborate to speed the deployment of cutting-edge technologies and integrate resilience engineering principles. Sustainability is also considered in resilience engineering, including eco-friendly infrastructure, renewable energy, and sustainable materials.

REFERENCES

Achillopoulou, D. V., Mitoulis, S. A., Argyroudis, S. A., & Wang, Y. (2020). Monitoring of transport infrastructure exposed to multiple hazards: A roadmap for building resilience. *The Science of the Total Environment*, *746*, 141001. doi:10.1016/j.scitotenv.2020.141001 PMID:32795756

Ahmed, S., & Dey, K. (2020). Resilience modeling concepts in transportation systems: A comprehensive review based on mode, and modeling techniques. *Journal of Infrastructure Preservation and Resilience*, *1*(1), 1–20. doi:10.1186/s43065-020-00008-9

Akhai, S., Mala, S., & Jerin, A. A. (2020). Apprehending air conditioning systems in context to COVID-19 and human health: A brief communication. *International Journal of Healthcare Education & Medical Informatics, 7*(1&2), 28-30.

Akhai, S., Singh, V. P., & John, S. (2016). Investigating Indoor Air Quality for the Split-Type Air Conditioners in an Office Environment and Its Effect on Human Performance. *Journal of Mechanical Civil Engineering*, *13*(6), 113–118.

Akhai, S., Thareja, P., & Singh, V. P. (2017). Assessment of Indoor Environment Health Sustenance in Air Conditioned Class Rooms. *Advanced Research in Civil and Environmental Engineering*, *4*(1&2), 1–9. doi:10.24321/2393.8307.201701

Aldakkhelallah, A., & Simic, M. (2021). Autonomous vehicles in intelligent transportation systems. In *Human Centred Intelligent Systems: Proceedings of KES-HCIS 2021 Conference* (pp. 185-198). Springer Singapore.

Arif, S. M., & Gupta, S. (2022). Assessment of A Higher Water Tank Taking into Account the Capacity and Foundation Wall in Earthquake Regions. In *2022 11th International Conference on System Modeling & Advancement in Research Trends (SMART)* (pp. 708-712). IEEE. 10.1109/SMART55829.2022.10047220

Borowska-Stefańska, M., Goniewicz, K., Grama, V., Hornak, M., Masierek, E., Morar, C., ... Wiśniewski, S. (2023). Evaluating approaches to wartime mass evacuation management in eastern NATO territories: a literature review. *Safety &Defense, 1.*

Butler, L., Yigitcanlar, T., & Paz, A. (2020). Smart urban mobility innovations: A comprehensive review and evaluation. *IEEE Access : Practical Innovations, Open Solutions*, *8*, 196034–196049. doi:10.1109/ACCESS.2020.3034596

Cadwell, P. (2021). Translation and interpreting in disaster situations. The Routledge handbook of translation and health, 253-268.

Ebrahimnejad, S., Villeneuve, M., & Tavakkoli-Moghaddam, R. (2023). An optimization model for evacuating people with disability in extreme disaster conditions: A case study. *Scientia Iranica*, *30*(4), 1498–1517.

Freckleton, D., Heaslip, K., Louisell, W., & Collura, J. (2012). Evaluation of resiliency of transportation networks after disasters. *Transportation Research Record: Journal of the Transportation Research Board*, *2284*(1), 109–116. doi:10.3141/2284-13

Ganin, A. A., Mersky, A. C., Jin, A. S., Kitsak, M., Keisler, J. M., & Linkov, I. (2019). Resilience in intelligent transportation systems (ITS). *Transportation Research Part C, Emerging Technologies, 100*, 318–329. doi:10.1016/j.trc.2019.01.014

Gholamhosseinian, A., & Seitz, J. (2022). A comprehensive survey on cooperative intersection management for heterogeneous connected vehicles. *IEEE Access : Practical Innovations, Open Solutions, 10*, 7937–7972. doi:10.1109/ACCESS.2022.3142450

Heaslip, K., Louisell, W., &Collura, J. (2009). *A methodology to evaluate transportation resiliency for regional networks* (No. 09-3194).

Hollnagel, E. (Ed.). (2013). *Resilience engineering in practice: A guidebook*. Ashgate Publishing, Ltd.

Iliopoulou, C., Konstantinidou, M. A., Kepaptsoglou, K. L., & Stathopoulos, A. (2020). ITS technologies for decision making during evacuation operations: A review. *Journal of Transportation Engineering. Part A, Systems, 146*(4), 04020010. doi:10.1061/JTEPBS.0000329

Islam, R., Jones, S., & Hudnall, M. (2023). Transportation system functions during hurricane response: A systematic review of modes and methods. *Transportation Research Interdisciplinary Perspectives, 18*, 100786. doi:10.1016/j.trip.2023.100786

Jawad, S., & Liu, J. (2020). Electrical vehicle charging services planning and operation with interdependent power networks and transportation networks: A review of the current scenario and future trends. *Energies, 13*(13), 3371. doi:10.3390/en13133371

Kumar, P., & Akhai, S. (2022). Effective energy management in smart buildings using VRV/VRF systems. *Additive Manufacturing in Industry 4.0: Methods, Techniques, Modeling, and Nano Aspects*, 27.

Kuru, K., & Khan, W. (2020). A framework for the synergistic integration of fully autonomous ground vehicles with smart city. *IEEE Access : Practical Innovations, Open Solutions, 9*, 923–948. doi:10.1109/ACCESS.2020.3046999

Lu, X., Liao, W., Fang, D., Lin, K., Tian, Y., Zhang, C., ... Zhao, P. (2020). Quantification of disaster resilience in civil engineering: A review. *Journal of Safety Science and Resilience, 1*(1), 19-30.

Mahrez, Z., Sabir, E., Badidi, E., Saad, W., & Sadik, M. (2021). Smart urban mobility: When mobility systems meet smart data. *IEEE Transactions on Intelligent Transportation Systems, 23*(7), 6222–6239. doi:10.1109/TITS.2021.3084907

Meneguette, R. I., De Grande, R., & Loureiro, A. A. (2018). *Intelligent transport system in smart cities*. Springer International Publishing. doi:10.1007/978-3-319-93332-0

Nipa, T. J., & Kermanshachi, S. (2021). Dimensions of Resilience Measurement in Critical Transportation Infrastructure. In *International Conference on Transportation and Development 2021* (pp. 302-312). 10.1061/9780784483541.028

Nugmanova, A., Arndt, W. H., Hossain, M. A., & Kim, J. R. (2019). Effectiveness of ring roads in reducing traffic congestion in cities for long run: Big Almaty ring road case study. *Sustainability (Basel)*, *11*(18), 4973. doi:10.3390/su11184973

Patriksson, M. (2015). *The traffic assignment problem: models and methods*. Courier Dover Publications.

Pérez, J. L., Gutierrez-Torre, A., Berral, J. L., & Carrera, D. (2018). A resilient and distributed near real-time traffic forecasting application for Fog computing environments. *Future Generation Computer Systems*, *87*, 198–212. doi:10.1016/j.future.2018.05.013

Raadsen, M. P., Bliemer, M. C., & Bell, M. G. (2020). Aggregation, disaggregation and decomposition methods in traffic assignment: Historical perspectives and new trends. *Transportation Research Part B: Methodological*, *139*, 199–223. doi:10.1016/j.trb.2020.06.008

Radvanovsky, R. S., & McDougall, A. (2018). *Critical infrastructure: homeland security and emergency preparedness*. CRC Press.

Raharjo, E. P., Sarjana, S., & Safitri, M. (2022). Transportation infrastructure planning in supporting disaster mitigation: Case study in Mount Gamalama. *Jamba (Potchefstroom, South Africa)*, *14*(1), 1123. doi:10.4102/jamba.v14i1.1123 PMID:35401943

Renne, J., Wolshon, B., Pande, A., Murray-Tuite, P., & Kim, K. (2022). *Creating Resilient Transportation Systems: Policy, Planning, and Implementation*. Elsevier.

Rouhanizadeh, B., Kermanshachi, S., & Dhamangaonkar, V. S. (2020, March). Reconstruction of critical and interdependent infrastructure due to catastrophic natural disasters: lessons learned. In *Construction Research Congress 2020* (pp. 895-904). Reston, VA: American Society of Civil Engineers. 10.1061/9780784482858.096

Saravi, S., Kalawsky, R., Joannou, D., Rivas Casado, M., Fu, G., & Meng, F. (2019). Use of artificial intelligence to improve resilience and preparedness against adverse flood events. *Water (Basel)*, *11*(5), 973. doi:10.3390/w11050973

Serdar, M. Z., Koç, M., & Al-Ghamdi, S. G. (2022). Urban transportation networks resilience: Indicators, disturbances, and assessment methods. *Sustainable Cities and Society*, *76*, 103452. doi:10.1016/j.scs.2021.103452

Seuwou, P., Banissi, E., &Ubakanma, G. (2020). The future of mobility with connected and autonomous vehicles in smart cities. *Digital twin technologies and smart cities*, 37-52.

Shareeda, A., Al-Hashimi, M., & Hamdan, A. (2021). Smart cities and electric vehicles adoption in Bahrain. *Journal of Decision Systems*, *30*(2-3), 321–343. doi: 10.1080/12460125.2021.1911024

Sun, W., Bocchini, P., & Davison, B. D. (2020). Resilience metrics and measurement methods for transportation infrastructure: The state of the art. *Sustainable and Resilient Infrastructure*, *5*(3), 168–199. doi:10.1080/23789689.2018.1448663

Syed Arif, S. M., Pandey, V., Dhiman, S., & Sharma, P. (2021). Conceptualization of zero energy building coupled with seismically resistant structure. *International Journal of Mechanical Engineering, 6*, 238.

Tachaudomdach, S., Upayokin, A., Kronprasert, N., & Arunotayanun, K. (2021). Quantifying road-network robustness toward flood-resilient transportation systems. *Sustainability (Basel)*, *13*(6), 3172. doi:10.3390/su13063172

Tanwar, N., & Akhai, S. (2017). Survey Analysis for Quality Control Comfort Management in Air Conditioned Classroom. *Journal of Advanced Research in Civil and Environmental Engineering*, *4*(1&2), 20–23. doi:10.24321/2393.8307.201702

Tao, X., Cheng, L., Zhang, R., Chan, W. K., Chao, H., & Qin, J. (2023). Towards Green Innovation in Smart Cities: Leveraging Traffic Flow Prediction with Machine Learning Algorithms for Sustainable Transportation Systems. *Sustainability (Basel)*, *16*(1), 251. doi:10.3390/su16010251

Wang, C., Yin, F., Zhao, Y., & Yin, L. (2023). Making Transportation Systems in US Cities Smarter and More Inclusive: A Synthesis of Challenges and Evaluation of Strategies. *ISPRS International Journal of Geo-Information*, *12*(2), 72. doi:10.3390/ijgi12020072

Xu, X., Chen, A., Xu, G., Yang, C., & Lam, W. H. (2021). Enhancing network resilience by adding redundancy to road networks. *Transportation Research Part E, Logistics and Transportation Review*, *154*, 102448. doi:10.1016/j.tre.2021.102448

Zhang, X., Mahadevan, S., & Goebel, K. (2019). Network reconfiguration for increasing transportation system resilience under extreme events. *Risk Analysis*, *39*(9), 2054–2075. doi:10.1111/risa.13320 PMID:31039286

Zhou, Y., Wang, J., & Yang, H. (2019). Resilience of transportation systems: Concepts and comprehensive review. *IEEE Transactions on Intelligent Transportation Systems*, *20*(12), 4262–4276. doi:10.1109/TITS.2018.2883766

Zou, Q., & Chen, S. (2019). Enhancing resilience of interdependent traffic-electric power system. *Reliability Engineering & System Safety*, *191*, 106557. doi:10.1016/j.ress.2019.106557

Chapter 10
Pandemic Resilience, Crisis Management and COVID-19 Health and Safety Responses in the Hospitality Industry

Reshma Sucheran
Durban University of Technology, South Africa

ABSTRACT

The COVID-19 pandemic has impacted the hospitality industry, prompting the implementation of health and safety measures. This study examines COVID-19 health and safety measures used in South Africa's accommodation sector. Data was obtained from 477 establishments through online surveys. The findings reveal a robust commitment to health and safety practices driven by adherence to regulatory standards. Disparate levels of social distancing measures highlight several challenges in implementation. A strong association with star grading indicates the use of measures tailored to each type of establishment. High-end establishments are more inclined to implement comprehensive measures. Larger facilities use floor marks more frequently, whereas smaller ones rely on personal communication. The number of employees influences the adoption of measures, highlighting the importance of ongoing training and communication on health issues. The study advocates for proactive health practices, customised strategies, financial support mechanisms, and a commitment to long-term preparedness.

DOI: 10.4018/979-8-3693-2721-0.ch010

INTRODUCTION

The COVID-19 pandemic, recognized globally as a significant health crisis (Fong et al., 2020; Zenker & Kock, 2020), originated in Wuhan, China, in December 2019, spreading rapidly to over 200 countries and regions. By July 18, 2020, China had reported more than 85,000 confirmed cases, with the global count surpassing 14,000,000 (World Health Organization, 2020). This pandemic profoundly impacted the hospitality and tourism industry, particularly hotels and public establishments reliant on interactions between staff and guests for service delivery (Sharma et al., 2021). Due to the transmission risks associated with COVID-19, many hotels faced complete or partial shutdowns, with some experiencing clusters of outbreaks. This dual impact has significantly affected the hotel industry at both macro and micro levels, influencing sustainability and regular operations (Zenker & Kock, 2020; Baum & Hai, 2020; Jiang & Wen, 2020). The essence of hospitality services lies in creating unique experiences through close customer interactions and personalized engagement with staff and facilities. However, the pandemic has forced accommodation establishments into uncharted territory, necessitating a complete restructuring of their value-creation process to adhere to new health standards. In response to the uncertainty, hotels were compelled to become contact-free or minimize contact, swiftly adapting service procedures to align with the evolving situation (Sigala, 2020).

While the global hospitality industry has faced disruptive events and crises in the past (Gossling et al., 2021), the unprecedented impact of the COVID-19 pandemic required extraordinary measures for recovery. Accommodation facilities have had to implement high-level safety leadership strategies and operational health and safety standards to navigate these challenging times. Recent literature draws attention to the fundamental role of COVID-19 preventive measures in revitalizing the hospitality industry amid the pandemic (Prentice et al., 2021). Victor et al. (2021) and Wei et al. (2021) suggest that these measures rebuild customer trust and, ultimately, their intention to revisit hospitality businesses during the reopening period. Gupta and Sahu (2021) maintain that these measures contribute to the revival of the hospitality sector after the pandemic's aftermath. Moreover, Kim, Han, and Ariza-Montes (2021), as well as Kim, Kim, and Wang (2021), confirm that proper implementation of preventive measures could enhance tourist intention to visit and revisit destinations during the pandemic.

While existing research in the hospitality and tourism literature predominantly concentrates on crisis management and coping strategies at the macro-level, such as destination or industry-wide analyses (Prayag, 2018; Faisal et al., 2020; Radic et al., 2020), there is a noticeable gap in studies on crisis management strategies at the micro-level, particularly within organizations. The imperative for research examining the impacts of hotel characteristics on implementing health and safety

COVID-19 response measures in the South African accommodation sector is undeniably crucial, especially considering the notable absence of existing studies within this specific context. The COVID-19 pandemic has triggered unprecedented challenges, compelling the hospitality industry to adapt swiftly to ensure the safety of guests and staff. However, the diverse nature of accommodation establishments, varying in size, rating, ownership, and type, necessitates a thorough investigation into how these specific characteristics influence the adoption and efficacy of health and safety measures. Understanding how factors such as hotel star grading, ownership structures, establishment size, workforce size and type of facility contribute to implementing health and safety response measures is fundamental for developing tailored strategies and guidelines. This research is not merely an academic pursuit; it holds practical implications for industry stakeholders, policymakers, and public health authorities. It provides insights to inform effective decision-making, resource allocation, and formulating sector-specific guidelines. As the hospitality sector grapples with ongoing uncertainties, a comprehensive study in this unexplored domain becomes essential to enhance preparedness, resilience, and safety within the South African accommodation sector.

LITERATURE REVIEW

The literature on the hospitality sector extensively addresses the sector's vulnerability to disasters and crises, considering both natural and human-induced factors (Ritchie, 2004; Faulkner, 2001; Lo et al., 2006). Faulkner (2001) defines disasters as unforeseeable catastrophic changes beyond an organisation's control, requiring reactive responses through contingency plans. In contrast, crises are disruptions hindering operations, threatening viability, or impacting stakeholders, potentially being anticipated through proper management (Ritchie, 2004). Kim et al. (2005) highlight that disasters stem from external factors, while crises involve internal issues. The COVID-19 outbreak, due to its rapid and uncontrollable global spread, is classified as a disaster due to the limited control over its impacts. Crisis management, designed to mitigate damage from crises or disasters, follows a systematic procedure (Ritchie, 2004). Crises vary in magnitude, originating from internal or external factors within organizations (Pearson & Clair, 1998; Ritchie, 2004; de Sausmarez, 2004). The pre-crisis, crisis, and post-crisis framework guides crisis management approaches, emphasizing risk assessment and environmental scanning (Meyers & Holusha, 1986; Faulkner, 2001; Ritchie, 2004). The relationship between crises and tourism, particularly the impact on destinations' reputation and attractiveness, has become a significant focus (de Sausmarez, 2007; Santana, 2004).

Preventive measures against the COVID-19 pandemic played a crucial role in safeguarding public health and impacted international tourist arrivals, leading to a substantial decline worldwide (Kim et al., 2021; UNWTO, 2021). The hospitality sector faced unprecedented challenges, with projected losses exceeding 2.7 trillion US dollars, prompting operational closures and unemployment (WTTC, 2020). Recommendations for hotel owners included reducing capacity, prioritizing delivery services, rigorous cleaning, Personal Protective Equipment (PPE) usage, promoting personal hygiene, and specific protocols for guests exposed to the virus (Davahli et al., 2020). Studies, such as Diayudha's (2020) on Jakarta hotels affiliated with the Omega chain, emphasize the importance of SOP guidelines, health assessments, disinfection, isolation rooms, entry protocols, and complete PPE adherence. Implementing COVID-19 preventive measures extends beyond public health considerations to become integral to business operations, especially within the hospitality industry (Garrido-Moreno et al., 2021; Amani & Ismail, 2022; Wei et al., 2021). The crisis management literature emphasizes including these measures in corporate rebranding strategies during the industry's reopening (Victor et al., 2021; Tevi & Otubanjo, 2012).

METHODOLOGY

The methodological approach in this study commenced with an extensive phase of desktop research, focusing on examining research reports and academic sources. The dynamic and unprecedented nature of the COVID-19 pandemic and the rapid dissemination of information across various online platforms during the research period necessitated a thorough examination. The target population for this study included managerial staff within accommodation sector establishments across all nine provinces in South Africa. The accommodation establishments included hotels, guesthouses, bed and breakfasts, backpackers and hostels, caravan and campsites, country houses, game lodges, nature lodges, and self-catering establishments. Both graded and ungraded establishments were included in the research, with the list of these accommodation establishments being in the public domain and sourced from various relevant websites. To ensure a representative sample, a minimum of 360 establishments was targeted.

Online and electronic surveys were employed to elicit views and perceptions from accommodation establishments throughout South Africa, forming a substantial component of the primary data collection for this research. The distribution of online survey questionnaires unfolded over four weeks, from July to August 2020. The development of the online, self-administered questionnaire drew upon the foundational work conducted by various researchers, including AlBattat and Som (2014 & 2015) and Hung et al. (2018). Notwithstanding these influences, the survey deployed for this research was unique and was implemented in its current format for the first time. The voluntary nature of the questions, except for the one requesting consent, allowed respondents to omit specific inquiries. Notably, the questions exclusively pertained to business matters, devoid of exploring potentially triggering personal or psychological factors. The questionnaire comprised three distinct sections. Section one concentrated on the characteristics of accommodation establishments, exploring features such as star grading, provincial location, type of ownership, size (number of guestrooms), number of employees, and the facilities offered. Section two examined the financial impact of COVID-19 on accommodation establishments, emphasizing aspects such as the operational status during the lockdown, the vulnerability of employees and guests, cancellations of bookings, staffing challenges, and the assistance required during the pandemic. The final section of the questionnaire delved into COVID-19 preparedness and response measures, covering policy and planning, staffing considerations, education and training initiatives, communication and knowledge management strategies, economic measures, and COVID-19. The analysis of collected data was conducted utilising the Statistical Package for the Social Sciences (SPSS), version 26. SPSS, which offered a systematic exploration and interpretation of the gathered information. The data analysis process involved various statistical techniques within SPSS.

RESULTS

Characteristics of Accommodation Establishment

The study's objective was to examine the attributes of accommodation establishments, including their star grading, establishment size, staff count, facility type, duration of operation and ownership type. The study will examine the impact of these attributes on the extent and type of health and safety response measures adopted by businesses.

A minority of establishments classified as 1-star (0.2%) and 2-star (2.1%) are scarcely represented in the data. The mid-range category 3-star (36.7%) and 4-star (40.0%) establishments dominate the sample. Premium 5-star establishments are less represented at 8.2%. Furthermore, a significant proportion of establishments (12.8%) are not star-graded. The dataset also sheds light on the diversity of accommodation establishments based on their type. Guesthouses emerge as the predominant category, capturing a significant market share (37.9%), followed by self-catering facilities (17.2%). Hotels, resorts and lodges (bush/country) contribute comparably to the market, representing 10.3% and 10.7%, respectively. Bed and breakfast establishments hold a substantial fraction (16.1%), while game lodges (2.1%), country

Table 1. Characteristics of accommodation establishments

Characteristic		Frequency	Percent
Star Grading	1 star	1	0.2
	2 star	10	2.1
	3 star	175	36.7
	4 star	191	40.0
	5 star	39	8.2
	Not graded	61	12.8
	Total	477	100.0
Type of Facility	Hotel/Resort	49	10.3
	Lodge (bush/country)	51	10.7
	Guesthouse	181	37.9
	Self-Catering	82	17.2
	Game Lodge	10	2.1
	Country House	9	1.9
	Bed and Breakfast	77	16.1
	Caravan and Camp Site	7	1.5
	Backpackers and Hostel	9	1.9
	Motel/Inn	2	0.4
	Total	477	100.0
Ownership Type	Independently owned, self-managed	464	97.3
	Chain-owned, managed by chain	11	2.3
	Unsure	2	0.4
	Total	477	100.0
Facility Size	Less than ten rooms	261	54.7
	11 to 30 rooms	162	34.0
	31 to 60 rooms	30	6.3
	61 to 100 rooms	15	3.1
	More than 100 rooms	9	1.9
	Total	477	100.0
Duration of Operation	Less than five years	61	12.8
	5 to 10 years	88	18.4
	11 to 15 years	130	27.3
	16 to 20 years	84	17.6
	21 to 25 years	63	13.2
	More than 25 years	51	107
	Total	477	100.0

continued on following page

Table 1. Continued

Characteristic		Frequency	Percent
Number of Employees	Less than ten employees	333	69.8
	11 to 30 employees	100	21.0
	31 to 60 employees	27	5.7
	61 to 100 employees	7	1.5
	101 to 130 employees	4	0.8
	131 to 150 employees	1	0.2
	More than 150 employees	5	1.0

Source: Author's own compilation, 2022

houses (1.9%), caravan and campsites (1.5%), backpackers and hostels (1.9%), and motels/inns (0.4%) constitute smaller segments within the dataset. The ownership and management structure of accommodation establishments reveals autonomy, with 97.3% being independently owned and self-managed. Regarding the size of facilities, more than half (54.7%) of establishments have fewer than ten rooms, while 34% have room counts ranging from 11 to 30. Establishments with a capacity of 31 to 60 rooms (6.3%), 61 to 100 rooms (3.1%) and more than 100 rooms (1.9%) represent a comparatively small fraction of the overall sample. Approximately 12.8% of establishments have operated for less than five years, while 18.4% have operated for 5 to 10 years. Notably, establishments operating for 11 to 15 years represent a significant proportion at 27.3%. Established businesses with operational histories spanning 16 to over 25 years collectively represent a significant percentage (30.8%), with 10.7% operating for over 25 years. Regarding workforce size, most (69.8%) of the accommodation businesses have less than ten staff. Establishments with 11 to 30 employees account for 21.0% and 5.7% have between 31 to 60 employees. Establishments with more than 60 employees accounted for 3.5% of the sample.

FACTOR ANALYSIS

The research instrument employed for this study comprised 14 items designed to assess the health and safety response measures implemented in the accommodation sector during the COVID-19 pandemic. To ensure the precision and reliability of the instrument, the study highlighted the assessment of reliability and validity. A reliability coefficient of 0.60 or higher was considered "acceptable" for a newly developed construct. Table 2 presents the Cronbach's alpha scores for all questionnaire items, with each section demonstrating reliability scores surpassing the recommended Cronbach's alpha value. This outcome affirms the instrument's reliability, indicating consistent and dependable measurement across the various factors of the research, thereby enhancing the methodological strength of the study.

For determining common variables, exploratory factor analysis was conducted to confirm interdependent correlations among several variables. Extraction of the factors was done with the help of Principle Component Analysis. The Kaiser Meyer Olkin measure (Table 3) examined the adequacy of the sample. At p value <0.001, Bartlett's test of sphericity, chi-square value shows that exploratory factor analysis is appropriate and all the conditions are satisfied for factor analysis.

Exploratory factor analysis (EFA) with PCA through SPSS 28.0 was conducted to examine the reliability and validity of questionnaire data (Tables 3 and 4). The EFA results showed that Cronbach's α for Health Monitoring and Sanitation Practices (0.906) and Social Distancing and Space Management (0.804) were all over 0.8, indicating that the questionnaire items had reasonably good internal consistency. The KMO index of these constructs was 0.898, the factor loading coefficients of each item were above 0.5 and Bartlett's Test of Sphericity sig. value was less than 0.05, suggesting a sound validity structure of the questionnaire. The principal component analysis was used as the extraction method, and the rotation method was Varimax with Kaiser Normalization. It is noted that the variables that constituted this section loaded along two components (sub-themes). Component 1 focused on Health Monitoring and Sanitation Practices and has higher loadings for items directly monitoring health symptoms and providing sanitization materials. It emphasizes the active efforts taken by an establishment to detect and prevent the spread of COVID-19 through health monitoring and maintaining cleanliness. Component 2 focuses on

Table 2. Cronbach's alpha

		N of Items	Cronbach's Alpha
Component 1	Health Monitoring and Sanitation Practices	10	0.906
Component 2	Social Distancing and Space Management	4	0.804
All items included		14	0.886

Source: Author's own compilation, 2023

Table 3. KMO and Bartlett's test

Kaiser-Meyer-Olkin Measure of Sampling Adequacy.		0.898
Bartlett's Test of Sphericity	Approx. Chi-Square	3916.525
	df	91
	Sig.	< 0.001

Table 4. Rotated component matrix

Measures	Component	
	1	2
Provide alcohol-based hand sanitiser that contains at least 60% alcohol for all guests and staff.	.650	.069
Provide approved PPE to employees at no cost.	.625	.248
Post appropriate bulletins and posters on COVID-19 operational instructions for staff and guests.	.501	.451
Monitor employees at least once before starting their shift for fever or respiratory symptoms.	.744	.337
Monitor guests as they arrive for fever or respiratory symptoms.	.617	.407
Stock enough cleaning and sanitising products for the size of your operation.	.748	.037
Maintain records that will help you trace who has been in contact with any infected individuals who have been to your establishment.	.808	.199
Keep employees informed on outbreaks and relevant healthcare precautions.	.809	.199
Routinely clean all frequently touched surfaces in the establishment to limit the spread of the virus.	.842	.190
Identify sources for timely and accurate pandemic information	.739	.188
Ensure markings on the floor are maintained to maintain social distancing where needed.	.086	.679
Due to Social Distancing norms, allocate alternate rooms or leave two rooms vacant in between based on occupancy levels	.182	.668
Reduce the number of tables in restaurants to maintain Social Distancing norms.	.187	.858
Seating for restaurant tables is to be reduced to half capacity.	.205	.845

Extraction Method: Principal Component Analysis. Rotation Method: Varimax with Kaiser Normalization. a. Rotation converged in 3 iterations.

Source: Author's own compilation, 2022

Social Distancing and Space Management and is characterized by higher loadings for items on space management and social distancing guidelines.

Health and Safety Response Measures

As observed in Table 5, the 'Health Monitoring and Sanitation Practices' subtheme reveals a comprehensive and proactive response to the COVID-19 pandemic within accommodation establishments. Across all measures, a significantly higher proportion of establishments have implemented health measures, which is evident in the 'have done' category. In particular, the scoring patterns indicate a strong commitment to providing hand sanitizer (91.4%), the availability of approved PPE (78.8%), posting operational guidelines (65.4%), health monitoring of employees (77.8%) and guests (73.2%), availability of cleaning supplies (85.5%), record-keeping for contact tracing (83.2%), employee communication (84.5%), routine cleaning of high-touch surfaces (86.8%), and identification of accurate pandemic information sources (81.1%).

Table 5. Implementation of health monitoring and sanitation practices

Measures	Have done	Have not done, but plan to do	Do not plan to do
Provide alcohol-based hand sanitiser that contains at least 60% alcohol for all guests and staff.	91.4%	7.5%	1.0%
Provide approved PPE to employees at no cost.	78.8%	17.4%	3.8%
Post appropriate bulletins and posters on COVID-19 operational instructions for staff and guests.	65.4%	25.4%	9.2%
Monitor employees at least once before starting their shift for fever or respiratory symptoms.	77.8%	18.4%	3.8%
Monitor guests as they arrive for fever or respiratory symptoms.	73.2%	21.4%	5.5%
Stock enough cleaning and sanitizing products for the size of your operation.	85.5%	12.4%	2.1%
Maintain records that will help you trace who has been in contact with any infected individuals who have been to your establishment.	83.2%	14.9%	1.9%
Keep employees informed on outbreaks and relevant healthcare precautions.	84.5%	13.0%	2.5%
Routinely clean all frequently touched surfaces in the establishment to limit the spread of the virus.	86.8%	10.7%	2.5%
Identify sources for timely and accurate pandemic information	81.1%	15.5%	3.4%

Source: Author's own compilation, 2022

Several studies provide robust support for prioritising the physical well-being of staff and guests within the hospitality industry, particularly in response to the ongoing COVID-19 pandemic. These studies confirm that hotels have taken a proactive stance by implementing stringent measures such as hygiene control, provision of protective equipment, online medical consultations, staff training in epidemic detection (Hao et al., 2020; Lai & Wong,2020) regular temperature checks, more frequent cleaning and disinfection in common areas and guest rooms, installation of hand sanitizing stations, promotion of social distancing measures, contactless check-in processes, and dissemination of information about COVID-19 prevention measures to guests (Smart et al., 2021; 2021; Rady & Atia, 2020; Ramírez et al., 2021). In line with these findings, Rady and Atia (2024) emphasized the significance of hotels actively tracking the contacts of an injured person during the COVID-19 crisis and subsequently tracing guests to link the contact during their stay. This thorough approach highlights the industry's commitment to health and sanitation practices.

One significant driver behind this strong commitment to health monitoring and sanitation practices is the requirement for regulatory compliance. Governments and health authorities worldwide have imposed stringent regulations to curb the spread of the virus, compelling accommodation providers to adhere to prescribed health and safety measures (Ramírez et al., 2021; Lai & Wong, 2020; Fu, 2020). Failure to comply could result in legal consequences, financial penalties, and reputational damage, motivating establishments to take a proactive stance. Moreover, this shift in consumer priorities towards an increased emphasis on health and safety considerations highlights the strategic insight recognized by accommodation providers, who acknowledge the advantage of positioning themselves as secure havens during public health crises. Direct quotes from respondents further inform this sentiment, with one expressing, *"We are very concerned about the health of the guests and staff. The pandemic is spreading like fire, putting human beings at risk,"* and another emphasizing, *"This pandemic is real; we must maintain social distancing, and wash our hands, wear masks, and follow the safety measures"*. Adopting rigorous health measures is perceived as necessary to increase competitiveness and appeal to safety-conscious guests. The proactive stance of certain managers is also evident, as highlighted by a respondent who articulated, *"I did intensive research on the Health and Safety Protocols of other countries, and as a team, we underwent online COVID-19 training endorsed by the WHO. I wrote protocols and have a proper risk management strategy."*

While there are establishments in the 'have not done but plan to do' category, their percentages are considerably lower than those in the 'have done' category. One respondent noted several reasons for this: *"The lack of advance notice about restrictions being lifted puts immense pressure on businesses to implement efficient and effective training to staff regarding new legislation, cleaning regimes and provide time to implement comprehensive and effective occupational health and safety regimes"*.

The findings from the 'Social Distancing and Space Management' response measures reveal that floor markings for social distancing are not uniformly adopted across establishments. The establishments that have implemented floor markings (38.6%) are possibly driven by regulatory compliance and a commitment to guest safety. On the other hand, 26.6% of establishments in the planning phase might

Table 6. Implementation of social distancing and space management measures

Measures	Have done	Have not done but plan to do	Do not plan to do
Ensure markings on the floor are maintained to ensure social distancing where needed.	38.6%	26.6%	34.8%
Allocate alternate rooms or leave two rooms vacant in between.	51.6%	19.9%	28.5%
Reduce the number of tables in restaurants to maintain Social Distancing norms.	57.9%	16.1%	26.0%
Seating for restaurant tables is to be reduced to half capacity.	57.4%	15.3%	27.3%

Source: Author's own compilation, 2022

reflect a strategic approach, taking time to assess the most effective implementation methods. A substantial proportion (34.8%) of establishments not planning to make changes may be due to challenges associated with physical space reconfiguration or financial constraints. Just over half (51.6%) of establishments implemented alternate or two rooms in between for room allocations. The 28.5% not planning such changes may indicate a calculated decision based on their specific operations. Regarding restaurant operations, most establishments have reduced tables (57.9%) and seating capacity (57.4%), aligning with social distancing norms. A significant proportion (26.0% for tables and 27.3% for seating capacity) not planning to reduce capacity may be influenced by factors such as maintaining revenue levels or perceiving that their current setups adequately meet safety requirements. These varied responses highlight the complexities inherent in adapting hospitality spaces to ensure health and safety during the pandemic. However, one respondent stated, "Yes, *my establishment is taking huge financial strain due to lockdown, and it risks the livelihood of our continued existence, but lockdowns, reducing exposure and social distancing is still the correct way to save lives*".

These findings corroborate with similar studies. Lai & Wong (2020) and Fotiadis et al. (2021) indicate that the key strategy adopted by hotels is the reconfiguration of public spaces and common areas to allow for greater distance between individuals. This may involve rearranging furniture in lobbies, restaurants, and conference rooms to promote social distancing. Moreover, hotels have re-evaluated room allocation and occupancy by leaving rooms vacant for a certain period between guest stays to allow for thorough cleaning and disinfection. Some hotels have even designated specific floors or wings for guests who may have been exposed to COVID-19 or are in quarantine, ensuring the safety of other guests and staff members (Lai & Wong, 2020; Ramírez et al., 2021). Smart et al. (2021) also discovered that methods of conveying social distancing measures and updated service protocols to guests varied, including dissemination through diverse channels such as the hotel website, brochures, and signage. On the other hand, while social distancing and space management strategies are essential in hotels during the COVID-19 pandemic, some researchers argue that these measures may negatively impact the traditional hospitality experience and hinder the personal touch that is often valued by guests and can create a sense of impersonality and isolation, as guests may feel distanced from staff and other guests (Lai & Wong, 2020), and leaving rooms vacant between stays can reduce hoteliers' revenue opportunities (Fotiadis et al., 2021; Chan et al., 2021).

Star grading. The chi-square analysis on the relationship between star grading and COVID-19 health and safety response measures has yielded essential findings with several statistically significant associations. These included the provision of 60% alcohol-based sanitiser for all guests and staff ($p = 0.004$), providing approved Personal Protective Equipment (PPE) at no cost to employees ($p = 0.029$), posting

Table 7. Chi-square analysis: Health and safety response measure and accommodation establishment characteristics

Measures	1	2	3	4	5	6
Provide alcohol-based sanitiser for guests and staff.	.004*	.145	.987	.571	.095	.173
Provide approved PPE at no cost to employees	.029*	.007*	.803	.208	.425	.162
Post bulletins & posters on COVID-19 measures	.041*	.083	.107	.000*	.055	.001*
Monitor employees for fever and respiratory symptoms	.008*	.043*	.635	.028*	.840	.073
Monitor guests on arrival for fever and symptoms	.006*	.000*	.771	.033*	.100	.056
Stock enough cleaning and sanitizing products	.022*	.001*	.522	.929	.849	.258
Keep records to trace contact with infected persons	.006*	.517	.916	.155	.074	.228
Keep employees informed on outbreaks and precautions	.006*	.144	.932	.459	.082	.377
Routinely clean all frequently touched	.000*	.261	.958	.234	.060	.185
Identify sources for accurate pandemic information.	.009*	.781	.619	.239	.256	.081
Ensure markings on the floor to maintain social distancing	.022*	.004*	.130	.000*	.045*	.000*
Allocate alternate rooms or leave vacant rooms in between	.102	.001*	.459	.003*	.319	.035
Reduce tables in restaurants for social distancing norms	.103	.000*	.481	.042*	.337	.000*
Seating for restaurants reduced to half-capacity	.002*	.000*	.775	.016*	.161	.000*

*Significant values (p<0.05) Chi-square analysis. 1: Star grading; 2: Type of establishment; 3: Ownership type; 4: Size of the establishment (number of rooms); 5: Duration of operation; 6: Number of employees

Source: Author's own compilation, 2022

bulletins and posters on COVID-19 operational instructions for staff and guests (p = 0.041), health monitoring for both employees (p = 0.008) and guests on arrival (p = 0.006), the routine cleaning of frequently touched surfaces to limit the spread of the virus (p = <.001) and reducing restaurant seating to half capacity (p = 0.002). The research findings are aligned with previous studies, disclosing a noticeable trend in the health and safety responses to the COVID-19 pandemic based on hotel star grading. Higher-graded establishments, often associated with implementing comprehensive health and safety protocols, prioritise measures to reassure guests and foster a sense of security (Rady & Atia, 2020; Sharma et al., 2021). Studies highlight that 5-star hotels focus on essential measures like contact tracing and installing reception barriers to ensure physical distancing, showcasing a commitment to stringent safety practices (Bonfanti et al., 2021).. This trend aligns with premium establishments' elevated service standards and hygiene expectations, emphasising their commitment to guest safety and well-being. Rady and Atia (2020) found that the impact of hotel star grading goes beyond visible protocols, extending to investments in staff training for advanced cleaning technologies and strict adherence to local health regulations. This commitment is also evident in the guest experience, as top-rated establishments offer contactless amenities like digital room keys and in-room dining options, demonstrating adaptability in meeting evolving guest needs amid the pandemic (Lai & Wong, 2020).

Conversely, lower-graded and ungraded establishments may face financial constraints preventing them from adopting specific health and safety measures. One respondent noted that this might not be possible for facilities with limited budgets, *"I think the government should provide relief on the PPE and sanitizing costs and more financial support for the tourism/hospitality sector"*. Another respondent mentioned, "It would have been nice to have some assistance from the *government regarding municipality bills and cost of PPE"*. However, despite resource constraints, lower-star establishments have displayed significant efforts in prioritizing health and safety during the pandemic. Budget hotels and establishments with fewer resources than their higher-graded counterparts, proactively implement strict cleaning protocols and invest in staff training for proper sanitization practices (Lai & Wong, 2020). These initiatives confirm that hotel star grading alone does not determine the commitment to ensuring a safe environment. Instead, it emphasizes how each establishment optimally leverages available resources to effectively respond to public health needs during challenging times (Jiang & Wen, 2020; Chan et al., 2021).

Type of establishment. The chi-square analysis examining the association between the type of establishment and the implementation of COVID-19 health and safety measures reveals variations in practices across different accommodation types. Significant correlations were found with monitoring employees ($p=.043$) and guests ($p = <.001$) for fever and respiratory symptoms, stocking enough cleaning and sanitizing products ($p = .001$) and ensuring markings on the floor to maintain social distancing and the type of establishment ($p = .004$). Most hotel/resorts, lodges, guesthouses and bed and breakfasts have successfully implemented floor markings. On the other hand, self-catering and caravan/camp sites showed a lower need for strict social distancing measures in these settings, which may be because guests have more control over their environments and interactions, or these establishments may have outdoor facilities that naturally lend themselves to distancing. The allocation of alternate rooms or leaving vacant rooms between guests exhibits a highly significant association with the type of establishment ($p = .001$). Hotels/resorts and lodges (bush/country) showcase relatively high compliance with this measure, potentially influenced by their larger capacities and more spacious layouts conducive to such practices. In contrast, guesthouses and bed and breakfasts demonstrate moderate adherence, while caravan/camp sites display lower levels of compliance. These variations may be influenced by space constraints, layout configurations, and the outdoor-focused nature of activities associated with caravan/campsites. A quote from a respondent further supports this trend: *"My units are far away from each other, and so guests and staff can easily maintain social distancing."*

Reducing the number of tables (p=<.001) and seating capacity (p=<.001) in restaurants to maintain social distancing norms showed a highly significant association with the type of establishment, with higher compliance in hotels/resorts, lodges, guesthouses, bed and breakfasts and game lodges. At the same time, unique operational structures and physical layouts may influence lower compliance in self-catering, game lodges, and backpackers/hostels. Past studies concur that facilities such as hotels and resorts that have a large number of rooms and shared facilities like dining areas or swimming pools may face more significant challenges in ensuring social distancing and maintaining high levels of cleanliness and hygiene (Jiang & Wen, 2020; Ramírez et al., 2021). On the other hand, smaller establishments such as guesthouses or bed and breakfasts may have fewer guests and more personalised services, making it easier to control and monitor health and safety measures (Yang & Mao, 2020).

The study found no significant associations between the type of accommodation establishment and keeping employees informed on outbreaks and health care precautions (p = 0.144), routinely cleaning all frequently touched surfaces to limit the spread of the virus (p = 0.261), identifying sources for timely and accurate pandemic (p = 0.781), keeping records to facilitate contact tracing in the event of exposure to the virus (p=0.517), the provision of 60% alcohol-based sanitiser for all guests and staff (p = 0.145), and posting bulletins and posters on COVID-19 operational instructions for staff and guests (p = 0.083). These findings imply that the accommodation type does not play a significant role in these specific health and safety response aspects, and establishments adopt similar protocols regardless of their classification.

Type of Ownership. The data reveals no significant associations between ownership type and all health and safety response measures, indicating uniformity in adopting these measures across different ownership structures within accommodation establishments. However, existing research opposes these findings. For example, Hao et al. (2020) found that chain hotels, often backed by larger corporate entities, may have more centralized health and safety protocols and resources. This can lead to more standardized and rigorous response measures across their properties. On the other hand, with their smaller and more independent nature, non-chain hotels may have the flexibility to adapt quickly to changing circumstances. This could allow them to implement customized health and safety measures tailored to their property and clientele. While chain hotels may be bound by standardized protocols, non-chain hotels can leverage their local knowledge and relationships to effectively respond to health and safety challenges posed by the COVID-19 pandemic (Hao et al., 2020; Lai & Wong, 2020).

Size of Establishment. Significant associations were found between the size of the accommodation establishment and the posting of bulletins and posters on

COVID-19 operational instructions for staff and guests (p=<0.001); monitoring employees (p=.028) and guests (p=.033) for fever and respiratory symptoms (p=.028); ensuring floor markings for social distancing (p=<.001); allocating alternate rooms or leave vacant rooms in between (p=.003); reducing the number of tables in restaurants (p=.042); and seating for restaurants reduced to half capacity (p=.016). The decision not to implement employee monitoring measures in some smaller establishments may be influenced by resource constraints, perceived lower risk due to smaller staff numbers, or reliance on other preventive measures. Larger establishments, particularly those with more than 100 guestrooms, exhibit a higher percentage of having implemented floor markings for social distancing. This finding suggests that larger establishments, having more expansive public spaces and higher foot traffic, implement visual cues to guide guests and maintain safe distances. Conversely, smaller establishments with less than ten guestrooms show a lower percentage of having done floor markings. This could be attributed to their more private settings, where direct communication and personal attention may be more feasible in ensuring social distancing without explicit floor indicators. Fittingly, one respondent stated, *"We as small businesses are far more capable of keeping to strict hygiene and COVID-19 protective measures than large corporations and hotels. The public is safer in our establishments than they are going to supermarkets etc. as hundreds pass through those doors yet we are the last permitted to operate"*.

Larger establishments, specifically those with more than 30 guestrooms, demonstrate a higher commitment in allocating alternate rooms or leaving vacant rooms in between and reducing the number of tables in restaurants. The reasons for these findings are likely tied to operational considerations, space constraints, and perceptions of risk associated with different-sized establishments. For example, larger establishments, possibly with expansive dining areas, prioritize reducing seating capacity to half to adhere to social distancing guidelines effectively. However, Hao et al. (2020) argue that large hotels' diverse amenities and facilities, such as restaurants, spas, and event spaces, present additional complexities in maintaining a hygienic environment across all areas.

There were no significant correlations between the size of the accommodation establishment and the provision of alcohol-based sanitiser (p=.571); the provision of PPE to employees (p=.208); the stocking of cleaning and sanitizing products(p=.929); keeping records for contact tracing (p=.155); keeping employees informed on outbreaks and health care precautions (p=.459); the routine cleaning of frequently touched surfaces (p=.234); and identifying sources for timely pandemic information (p=.239). These findings suggest that the size of the establishment does not strongly influence the adoption of these measures, implying that whether an accommodation establishment is a smaller boutique hotel or a larger chain, the findings indicate a similar commitment to implementing these health and safety response measures.

Duration of operation. The findings suggest that the number of years an accommodation establishment has been in operation does not strongly correlate with adopting specific health and safety COVID response measures. The decision to implement these measures appears to be influenced by factors beyond the establishment's operational history. Contrary to these findings, previous research presents a different perspective. According to Shin and Kang (2020), older hotels, equipped with established protocols and experience in managing diverse health and safety challenges, may be advantageous in responding to the pandemic. In contrast, newer hotels may leverage modern technologies and innovative solutions specifically tailored to address the unique challenges posed by the pandemic. Ramírez et al. (2021) further posit that a hotel's operational years can shape the mindset and preparedness of its staff. While older hotels may possess established protocols, they may also be hindered by outdated technologies and traditional approaches ill-suited to handle the distinctive challenges brought about by the pandemic (Lai & Wong, 2020). Long-standing hotels cultivate a deeply ingrained culture of vigilance and attention to health and safety details. At the same time, newer establishments may offer a fresh perspective and adapt more readily to rapidly evolving safety protocols. However, the flip side is that newer hotels may lack experience in dealing with health and safety challenges, potentially rendering them less prepared to respond effectively to the pandemic.

Number of employees. The only significant association was between the number of employees and the posting of bulletins and posters on COVID-19 operational instructions for staff and guests ($p=.001$). Establishments with a higher number of employees showed the highest implementation rate of this measure. Past research has presented opposing findings on the impact of workforce size. Kim et al. (2021) highlighted that a smaller workforce foster streamlined communication and coordination, resulting in the more efficient implementation of safety protocols. Conversely, Kim et al. (2021) argue that a larger workforce may provide enhanced flexibility and specialization, enabling effective management of various health and safety aspects. Ramírez et al. (2021) found that more employees can facilitate the implementation of stricter social distancing measures by providing more staff for the management and enforcement of these protocols. Additionally, as Wong et al. (2021) noted, a larger workforce contributes to increased vigilance in monitoring and enforcing hygiene practices, including regular cleaning and disinfection of common areas. This array of contrasting findings underscores the nuanced relationship between workforce size and the execution of health and safety measures.

CONCLUSION

The research findings regarding COVID-19 health and safety response measures within the South African accommodation sector have substantial implications for industry stakeholders and policymakers. This study's comprehensive and proactive approach to 'Health Monitoring and Sanitation Practices' denotes a robust commitment within establishments, demonstrating adherence to regulatory standards. The driving force behind this commitment lies in the imperative of regulatory compliance, as stringent measures imposed by governments and health authorities globally shape the industry's response. In terms of 'Social Distancing and Space Management,' the findings reveal disparate levels in the adoption of health and safety response measures, specifically in aspects like floor markings for social distancing, which exhibit variability. This diversity may be influenced by regulatory compliance and strategic considerations, with certain establishments taking time to assess effective implementation methods. Moreover, challenges related to physical space reconfiguration, financial constraints, and considerations regarding room allocations and restaurant operations underscore the complexities inherent in adapting hospitality spaces during the pandemic.

The statistically significant associations between star grading and specific COVID-19 health and safety response measures emphasize the differentiated approaches accommodation establishments adopt based on their grading. The provision of 60% alcohol-based sanitizer, free Personal Protective Equipment (PPE) for employees, posting operational instructions, health monitoring for employees and guests, routine cleaning of high-touch surfaces, and reduced restaurant seating capacity demonstrate a discernible trend in the industry. The findings suggest that premium establishments are more likely to invest in comprehensive health and safety protocols, aligning with their commitment to reassuring guests and creating a secure environment, reflecting the industry's adaptability and responsiveness to public health challenges. More extensive facilities, especially those with over 100 guestrooms, were more likely to implement measures like floor markings. On the contrary, smaller establishments with fewer than ten guestrooms show a lower propensity for floor markings, possibly due to their more intimate and private settings, where direct and personal communication may suffice. The absence of significant correlations with some health and safety measures implies a consistent commitment across accommodation establishments of varying sizes. Hotels/resorts, lodges, guesthouses, and bed and breakfasts strongly adhere to social distancing measures, while self-catering and caravan/camp sites lower the need for strict protocols. Allocation of alternate rooms or leaving vacant rooms is significantly associated with the type of establishment. However, certain aspects, such as employee information dissemination and routine cleaning, show no significant associations, suggesting uniformity across accommodation types in these areas. The absence of significant associations between ownership types and health and safety response measures suggests a consistent commitment to adopting these measures across diverse ownership structures within accommodation establishments. The duration of operation does not strongly correlate with the adoption of specific health and safety measures, suggesting that establishments, regardless of their operational history, are equally responsive to external factors such as health guidelines, industry standards, and current pandemic conditions. Notably, the number of employees within an establishment emerges as a factor influencing the adoption of health and safety measures, with establishments boasting more employees showing a higher implementation rate of measures such as posting operational instructions, possibly due to larger management teams and more sophisticated communication channels. Conversely, smaller establishments, facing resource constraints, may find it challenging to allocate resources for comprehensive communication materials.

The comprehensive analysis of COVID-19 health and safety response measures in the South African accommodate sector leads to several key recommendations. Firstly, sustaining the industry's proactive commitment to health monitoring and sanitation practices is necessary, emphasizing continuous compliance with evolving regulatory standards. The correlation between the number of employees and the posting of operational instructions highlights the significance of ongoing employee training and clear communication. Tailored approaches for different accommodation types are crucial, recognizing variations in practices. For instance, caravan and campsite facilities may require specific outdoor safety measures, while larger hotels should maintain comprehensive safety protocols. Governments and industry associations are urged to provide financial support for smaller establishments facing constraints to effectively implement necessary health and safety measures. Lastly, maintaining a long-term perspective in preparedness is essential for all accommodation providers, irrespective of operational duration. Investing in infrastructure and staff training for sustained readiness in potential future health crises is vital. Implementing these recommendations will enhance the South African accommodation sector's response to COVID-19 and related crises, ensuring the safety of guests and staff and contributing to the overall resilience of the industry.

Limitations

The research method employed in this study exhibits several limitations that warrant consideration. Firstly, the focus on managerial staff within the accommodation sector may introduce sampling bias, potentially overlooking the perspectives of other stakeholders. The four-week data collection timeframe may not fully capture the dynamic and evolving nature of the COVID-19 pandemic's impact, and the exclusion of personal or psychological factors in the questionnaire limits a holistic understanding of the situation. Lastly, the study's generalizability beyond South African accommodation establishments is questionable, emphasising the need for caution when extrapolating the findings to other regions or industries.

ACKNOWLEDGMENT

This work is based on research supported by the National Institute for the Humanities and Social Sciences (REF NO: BRI22/1215).

REFERENCES

Albattat, A., & Som, A. P. M. (2015). Emergency Preparedness of the Hotel Industry: The Case of Jordan. *Advances in Environmental Biology*, *9*(3), 19–22.

Albattat, A. R. S., & Som, A. P. M. (2013). Employee Dissatisfaction and Turnover Crises in the Malaysian Hospitality Industry. *International Journal of Business and Management*, *8*(5), 62–71.

Amani, D., & Ismail, I. J. (2022). Investigating the predicting role of COVID-19 preventive measures on building brand legitimacy in the hospitality industry in Tanzania: Mediation effect of perceived brand ethicality. *Future Business Journal*, *13*(1), 1–14. doi:10.1186/s43093-022-00128-6

Baum, T., & Hai, N. T. T. (2020). Hospitality, tourism, human rights and the impact of COVID-19. *International Journal of Contemporary Hospitality Management*, *32*(7), 2397–2407. doi:10.1108/IJCHM-03-2020-0242

Bonfanti, A., Vigolo, V., & Yfantidou, G. (2021, April 1). The impact of the Covid-19 pandemic on customer experience design: The hotel managers' perspective. *International Journal of Hospitality Management*, *94*, 10287. doi:10.1016/j.ijhm.2021.102871 PMID:34866744

Chan, I. C. C., Ma, J., Ye, H., & Law, R. (2021, January 1). A Comparison of Hotel Guest Experience Before and During Pandemic: Evidence from Online Reviews. *Springer eBooks*, 549-556.

Garrido-Moreno, A., García-Morales, J., & Martín-Rojas, R. (2021). Going beyond the curve: Strategic measures to recover hotel activity in times of COVID-19. *International Journal of Hospitality Management*, *96*, 102928. doi:10.1016/j.ijhm.2021.102928 PMID:36540887

Gossling, S., Scott, D., & Hall, C. M. (2021). Pandemics, tourism and global change: A rapid assessment of COVID-19. *Journal of Sustainable Tourism*, *29*(1), 1–20. doi:10.1080/09669582.2020.1758708

Kim, J., Kim, J., & Wang, Y. (2021). Uncertainty risks and strategic reaction of restaurant firms amid COVID-19: An assessment of antecedents, consequences and intermediate. *International Journal of Hospitality Management*, *95*, 102894.

Kim, J. J., Han, H., & Ariza-Montes, A. (2021). The impact of hotel attributes, well-being perception, and attitudes on brand loyalty: Examining the moderating role of the COVID-19 pandemic. *Journal of Retailing and Consumer Services*, *62*, 102634. doi:10.1016/j.jretconser.2021.102634

Lo, A., Cheung, C., & Law, R. (2006). The survival of hotels during disaster: A case study of Hong Kong in 2003. *Asia Pacific Journal of Tourism Research*, *11*(1), 65–79. doi:10.1080/10941660500500733

Ltifi, M., & Hichri, A. (2022). The effects of corporate governance on the customer's recommendations: A study of the banking sector at the time of COVID-19. *Journal of Knowledge Management*, *26*(1), 165–191. doi:10.1108/JKM-06-2020-0471

Napierała, T., Leśniewska-Napierała, K., & Burski, R. (2020). Impact of Geographic Distribution of COVID-19 Cases on Hotels' Performances: Case of Polish Cities. *Sustainability (Basel)*, *12*(11), 4697–4697. doi:10.3390/su12114697

Pearson, C. M., & Clair, J. A. (1998). Reframing crisis management. *Academy of Management Review*, *23*(1), 59–76. doi:10.2307/259099

Prentice, C., Altinay, L., & Woodside, A. G. (2021). Transformative service research and COVID-19. *Service Industries Journal*, *41*(1-2), 1–8. doi:10.1080/02642069.2021.1883262

Santana, G. (2004). Crisis management and tourism. *Journal of Travel & Tourism Marketing*, *15*(4), 1–8. doi:10.1300/J073v15n04_05

Smart, K., Ma, E., Qu, H., & Ding, L. (2021, April 1). COVID-19 impacts, coping strategies, and management reflection: A lodging industry case. *International Journal of Hospitality Management*, *94*, 102859–102859. doi:10.1016/j.ijhm.2021.102859 PMID:34785846

Teng, X., Teng, Y., Wu, K., & Chang, B. (2021). *Corporate Social Responsibility in Public Health During the COVID-19 Pandemic: Quarantine Hotel in China.* doi:10.3389/fpubh.2021.620930

Yang, Y., & Mao, Z. (2020). Location advantages of lodging properties: A comparison between hotels and Airbnb units in an urban environment. *Annals of Tourism Research*, *81*. doi:10.1016/j.annals.2020.102861

Chapter 11
Disaster Risk Management Challenges Towards Integrated Disaster Risk Management at the South African Municipalities:
What Is Left Untouched?

Mosekama Osia Mokhele
https://orcid.org/0000-0003-3299-2799
Nelson Mandela University, South Africa

Alice Ncube
https://orcid.org/0000-0001-7440-5308
University of the Free State, South Africa

Olivia Kunguma
University of the Free State, South Africa

ABSTRACT

Although there has been significant global recognition of the South African disaster management legislation and the policy framework, its implementation approaches are flawed across spheres of government, with municipalities being the worst. Municipalities face various obstacles to effectively integrating disaster risk management into their development planning, leaving marginalized populations such as undocumented immigrants vulnerable during disasters. This study examines the challenges faced by municipalities in integrating disaster risk management into their development planning. A qualitative case-study approach following

DOI: 10.4018/979-8-3693-2721-0.ch011

desktop research was used in this study. The secondary data was collected from peer-reviewed articles and government documents such as the city of eThekwini integrated development plan 2023-2024 to 2027-2028 and the disaster management sector plan. Using two case studies, notable challenges were revealed, such as financial and capacity constraints (technical and non-technical), incomplete or not approved plans inertia, public awareness deficits, absence of a disaster management advisory forum, communication breakdowns, and the often-overlooked impacts of sand mining and undocumented immigration on disaster resilience. The study recommends joint strategic interventions to strengthen disaster risk management capabilities and create a more resilient social fabric, prioritizing marginalized populations. This study further highlights the urgency of a paradigm shift in disaster risk management that transcends conventional silos and reactive but embraces joint planning and community-centered approaches to disaster risk reduction. By championing inclusive approaches that prioritize these marginalized populations' unique needs and challenges, municipalities can promote greater resilience and equity in an increasingly uncertain future characterized by escalating climate risks and socio-economic inequalities.

INTRODUCTION

Disasters, whether natural or human-induced, threaten communities, economies, and ecosystems worldwide. The process of thoroughly and credibly predicting and controlling risks from many synergistic sources, known as integrated disaster risk management, is a challenge to the scientific (in the literature) and public sector institutions (practices) (Bean, 2021). As human activities lead to increased disaster losses, integrated disaster risk management gets more complicated, especially in municipalities (Munsamy, 2018). For predicting possible human, economic, and environmental losses, integrated disaster risk management necessitates thorough disaster risk studies that incorporate numerous hazards, risks, and their causes (Peduzzi, 2019). Moreover, an integrated evaluation of disaster risk considers not just prospective losses but also how they are spread among regions and at-risk populations and how they may affect the marginalised populations differently, such as people with low incomes, children, women, people with disabilities, the elderly, immigrants, indigenous people, and LGBTIQA + community (World Health Organization, 2017; United Nations, 2019; Human Rights Campaign; 2021). These marginalised populations face discrimination, stigmatisation, and exclusion, which can be exacerbated during disasters, resulting in increased vulnerabilities and challenges in accessing appropriate support and resources (Meyer, 2015; Logie and Turan, 2020).

Due to combining science, technology, and policy into a strong confluence, integrated disaster risk management is quickly gaining prominence in the disaster risk management fraternity (Moşteanu, Faccia, and Cavaliere, 2020). At the same time, several studies have been conducted on integrated disaster risk management from the perspective of development, structure, performance, and institutional processes (Aldunce, Beilin, Howden, and Handmer, 2015; Verbano and Venturini, 2011). This chapter argued that integrated disaster risk management necessitates a thorough and multi-factor risk assessment and procedures and institutions that proactively decrease the risks through understanding, translating, and interpreting disaster risk management policies and plans to reflect local conditions (Van Niekerk, 2014; Munsamy, 2018). Thus, the availability of legislation and policy frameworks, including policy measures and risk information, are useless unless they are well-understood and effectively implemented (Sithole, 2014; Van Niekerk, 2014).

Both the Hyogo Framework for Action 2005–2015 (2005) and its successor, the Sendai Framework for Disaster Risk Reduction 2015-2030, emphasised the need to make disaster risk reduction a national and local priority with a solid institutional foundation for the execution of integrated disaster risk management (Van Niekerk, 2015; Van Niekerk, Coetzee, and Nemakonde, 2020). However, even with a comprehensive evaluation, there will be divergent perspectives about the gravity of the threats, their acceptability, and the economic and ecological costs of mitigation (Belle, 2016). For municipalities to retain legitimacy in the eyes of an increasingly critical public realm, policymakers must progressively involve the affected individuals to participate in the design process and implementation of disaster risk reduction policies, plans, and standards. Thus, integrated disaster risk management requires a joint-planning approach that includes academics from various disciplines such as engineering, social sciences, urban and land-use planning, economics and finance, practitioners from multiple fields, national and local policymakers, and community and NGO representatives. Unfortunately, while many nations and circumstances have progressed toward integrated disaster risk management, the usual response to disasters remains reactive, a similar case in the South African disaster risk management fraternity, especially in municipalities (Sithole, 2014; Van Niekerk and Wisner, 2014; Oosthuizen, 2018).

Municipalities play a crucial role in disaster risk management, striving to enhance resilience, minimise vulnerabilities, and safeguard the well-being of their constituents. However, several challenges persist amidst the complex landscape of disaster risk management, hindering the effective integration of risk reduction measures into municipal development planning processes (Sithole, 2024; Sibanda, 2016). Among these challenges, the impacts of sand mining activities and the presence of undocumented immigrants have emerged as significant yet often overlooked contributors to vulnerability and resilience in disaster-prone areas.

Therefore, the chapter explores South African municipalities' diverse challenges in pursuing integrated disaster risk management. By examining the factors left untouched or inadequately addressed, we uncover critical insights into the gaps, vulnerabilities, and opportunities for enhancing disaster resilience at the local level. Through a comprehensive analysis of these challenges, we aim to inform policy discussions, guide decision-making processes, and catalyse action towards a more robust and sustainable disaster risk management framework in South Africa.

LITERATURE REVIEW

Legislative Framework for Disaster Risk Management in South Africa

The Disaster Management Act 57 of 2002 (DMA), as amended by Act 16 of 2015, serves as the foundational legislation guiding disaster risk management practices across spheres of government in South Africa, such as national, provincial, and local governments. This comprehensive Act establishes the necessary institutional capacity across all spheres of government to proactively address and mitigate the impacts of various disasters (Republic of South Africa, 2002). In Chapter 5, Section 42 of the Act, specific responsibilities are assigned to metropolitan and district municipalities to develop integrated disaster management frameworks (Republic of South Africa, 2002). These frameworks serve as strategic blueprints for coordinating disaster preparedness, response, and recovery efforts within their jurisdictions. Importantly, these frameworks must align with the national and provincial disaster management frameworks to ensure coherence and synergy in disaster management strategies (Duze and Reddy, 2020).

Furthermore, Section 42 of the DMA also mandates metropolitan and district municipalities to establish disaster management centers within their administrative structures (Republic of South Africa, 2002). The centers serve as a nerve for disaster management activities and are pivotal in coordinating emergency responses. While the primary responsibility for establishing these centers lies with metropolitan and district municipalities, the DMA allows for partnerships with local municipalities to optimise resource allocation and enhance operational effectiveness (Republic of South Africa, 2002; 2015). The responsibilities assigned to municipal disaster management centers, outlined in Sections 43 to 55 of the Act, entail a comprehensive range of functions to strengthen disaster resilience and response capabilities within local communities (Republic of South Africa, 2002). These responsibilities encompass a broad spectrum of activities, including but not limited to:

- Championing an integrated approach to disaster management that fosters collaboration and synergy among stakeholders representing diverse sectors.
- Providing strategic guidance on funding mechanisms and advocating for legislative measures to bolster disaster risk reduction efforts.
- Cultivating a culture of volunteerism and community engagement through capacity-building programs.
- Serving as conduits for disseminating timely guidance, information, and best practices to pertinent stakeholders.
- Vigilantly monitoring and evaluating the efficacy of disaster management plans and initiatives.
- Orchestrating swift and coordinated response actions in the event of disaster occurrences.

Furthermore, the Act stipulates oversight and accountability mechanisms, including the municipal council appointing heads for municipal disaster management centres. These appointed individuals are entrusted with the authority to exercise powers, make decisions in alignment with council directives, and ensure the effective implementation of disaster management strategies within their respective municipalities (Republic of South Africa, 2002).

INTEGRATED DISASTER RISK MANAGEMENT IN MUNICIPALITIES: A FRAMEWORK FOR SOUTH AFRICA

In South Africa, disaster management is a collaborative effort involving various spheres of government (Van der Berg, 2015). While much of the focus tends to be on the roles played by national and provincial entities in disaster preparedness and response, it is vital to recognise the significant role that municipalities also play in this process due to their proximity to localities. Despite the overarching framework provided by national and provincial bodies, municipalities are granted the flexibility to develop their disaster risk management capacities and structures (Van Niekerk and Wentink, 2017). This autonomy allows municipalities to design strategies and initiatives tailored to address their marginalised populations' unique needs and challenges. By empowering municipalities this way, South Africa ensures a more targeted and effective disaster response, ultimately enhancing the nation's resilience (Sithole, 2014; Sibanda, 2016).

Ensuring municipalities have the authority and resources to take proactive measures is crucial for bolstering disaster resilience (Van Niekerk, 2015). This empowerment encompasses various actions, including establishing disaster management centres or units, implementing customised risk reduction initiatives, and integrating disaster risk management into broader development planning processes (Oosthuizen, 2018). Establishing dedicated disaster management centres or units within municipalities allows for a focused and coordinated approach to disaster preparedness and response. These units serve as central hubs for disaster-related activities, facilitating efficient communication, resource allocation, and decision-making during crises.

Integrating disaster risk management into broader development planning processes is essential for promoting sustainable and resilient development (Duze and Reddy, 2020). By embedding disaster risk management principles into integrated development plans, land-use planning strategies, and infrastructure projects, municipalities can minimise exposure to hazards and enhance adaptive capacity (Ngcamu, 2022). This requires a shift towards holistic, multi-sectoral approaches considering disaster risks in all development planning and decision-making aspects. Comprehensive risk assessments form the cornerstone of this integration. Municipalities must conduct thorough evaluations to identify and understand potential hazards and vulnerabilities (Duze, 2016). This information serves as the basis for developing strategies to reduce risk and enhance resilience. Once risks are identified, municipalities can integrate disaster risk reduction measures into their land-use planning strategies (Ngcamu, 2022). This includes implementing zoning regulations to limit development in high-risk areas and ensuring new infrastructure projects are designed to withstand potential hazards.

Furthermore, it is crucial to integrate disaster risk management into infrastructure projects. Municipalities should design infrastructure resilient to potential hazards, such as implementing building codes that require structures to withstand high winds or earthquakes (Ngcamu, 2022). Additionally, nature-based solutions can be incorporated to help mitigate risks, such as green roofs or permeable pavements.

Community engagement is another vital aspect of integrating disaster risk management into development planning. Involving marginalised populations in decision-making processes helps identify local risks and priorities and builds local capacity for disaster preparedness and response (Oosthuizen, 2018). Empowering marginalised populations to engage actively in decision-making processes and resilience-building initiatives is essential for effective disaster risk management at the local level (Nkombi and Wentink, 2022). Furthermore, active community involvement is indispensable in cultivating a sense of ownership and commitment to disaster preparedness and response efforts. Community-based disaster management programs represent a cornerstone of this approach (Reddy, 2010). These programs facilitate the engagement of residents in various critical activities, including risk assessments, planning, and capacity-building exercises. Involving community members directly in these initiatives allows municipalities to harness invaluable local knowledge and resources, thereby enhancing the overall effectiveness of disaster risk management efforts (Sithole, 2014; Nkombi and Wentink, 2022).

Moreover, fostering collaboration between municipalities and marginalised populations fosters trust and strengthens social cohesion, both pivotal for resilience in the face of disasters. Working with marginalised populations provides municipalities with insights into the unique vulnerabilities and needs of different neighborhoods and populations, enabling them to tailor their disaster risk management strategies accordingly. Through these community-driven approaches, South Africa can develop more robust and responsive disaster management frameworks firmly rooted in local realities and help build stronger, safer, and more resilient communities capable of effectively withstanding and recovering from disasters.

INTEGRATED DISASTER RISK MANAGEMENT: POTENTIAL CHALLENGES FOR MUNICIPALITIES

Disaster risk management is essential for safeguarding communities against the impacts of natural and man-made disasters. However, municipalities in South Africa face several challenges in effectively integrating disaster risk management into their development planning processes (Botha et al., 2011; Munsamy, 2018). These challenges encompass financial constraints, limited human resources, lack of political will, lack of awareness, and poor communication (Mamabolo and Sebola, 2021). Each of these challenges presents unique obstacles that hinder the ability of municipalities to prepare for, mitigate, and respond to disasters effectively. These challenges will be discussed below:

FINANCIAL CONSTRAINT

Financial constraints pose significant challenges for municipalities in managing disaster risks effectively. These constraints manifest in various ways and can severely impact the ability of municipalities to prepare for, mitigate, and respond to disasters. One consequence of financial constraints is the limited funding for critical infrastructure upgrades to reduce disaster risks (Atkinson, 2021). Projects such as building flood defenses, retrofitting buildings for earthquake resistance, and constructing emergency shelters require substantial investment. However, without adequate funding, municipalities may struggle to prioritise these infrastructure needs, leaving marginalised populations vulnerable to the impacts of natural hazards (Oosthuizen, 2018).

Moreover, financial limitations impede the implementation of proactive risk reduction measures, including hazard mapping, early warning systems, and community resilience programs (Mamabolo and Sebola, 2021). These initiatives require investment in data collection, technology, and community outreach efforts. Yet, without sufficient resources, municipalities may be unable to undertake these activities, leaving marginalised populations ill-prepared to anticipate and mitigate the impacts of disasters.

Poor funding affects investment in disaster preparedness and response efforts. Training emergency responders, stockpiling emergency supplies, and developing contingency plans are essential components of effective disaster response (Ngcamu, 2022). Without proper funding, emergency services may lack the resources and capacity to respond effectively to disasters, leading to delays in rescue operations and increased casualties. Furthermore, smaller municipalities may struggle to access external funding sources from national or international agencies, NGOs, and donor organisations. Complex application processes, stringent eligibility criteria, and reporting requirements can hinder accessing external funding. As a result, municipalities may miss out on opportunities to secure additional resources and technical assistance for disaster risk management initiatives. The impacts of financial constraints are often felt most acutely by vulnerable, marginalised populations, exacerbating existing social and economic inequalities (Sibanda, 2016).

Marginalised populations, such as low-income households and informal settlements, are disproportionately affected by disasters due to their lack of resources and limited access to social safety nets. Without targeted interventions, these marginalised populations remain at heightened risk of displacement, loss of livelihoods, and long-term recovery challenges. Inadequate funding can undermine the long-term sustainability of disaster risk management efforts. Short-term funding cycles and uncertain budget allocations make it difficult for municipalities to undertake comprehensive, multi-year initiatives aimed at building resilience. As a result, investments in disaster risk reduction may be sporadic and reactive rather than strategic and proactive, compromising the overall effectiveness of disaster risk management initiatives.

LIMITED HUMAN RESOURCES

Limited human resources present a significant challenge for municipalities in effectively managing disaster risks within their jurisdictions. This challenge encompasses various aspects that hinder municipalities' ability to prepare for, mitigate, and respond to disasters (Mamabola and Sebola, 2021; Mkhwanazi and Mpanza, 2020). Shortage of trained personnel is one of the primary consequences of limited human resources, including emergency responders and disaster management specialists. Insufficient staffing levels can compromise municipalities' capacity to respond to disasters effectively. Emergency services may be overwhelmed, leading to delays in rescue and relief efforts and potentially increasing the severity of the disaster's impact (Atkinson, 2021). This can also result in a lack of specialised expertise in disaster risk management. Without personnel trained and experienced in disaster preparedness, mitigation, and response, municipalities may struggle to develop and implement effective strategies for reducing disaster risks, leaving marginalised populations vulnerable to the impacts of natural hazards (Mkwanazi and Mpanza, 2020).

Additionally, it can impede the conduct of comprehensive risk assessments, which are essential for identifying and understanding potential hazards and vulnerabilities. Municipalities may not clearly understand their marginalised populations' risks without adequate personnel to collect and analyse data, making prioritising and implementing risk reduction measures challenging (Van Niekerk and Wentink, 2017). Furthermore, efforts to engage marginalised populations in disaster preparedness and response activities can be hindered. Community engagement is critical for building resilience and fostering a culture of readiness, as these disaster events occur locally (Atkinson, 2021; Nkombi and Wentink, 2022). However, without sufficient personnel to coordinate community outreach efforts, municipalities may struggle to communicate effectively with residents and involve them in decision-making processes. Human resource constraints can also limit municipalities' capacity to build the skills and knowledge of their staff in disaster risk management (Atkinson, 2021). However, without adequate resources for training and development, municipalities may find it challenging to enhance the capabilities of their workforce (Atkinson, 2021). Moreover, it can exacerbate the vulnerability of marginalised populations. These marginalised populations may already face barriers to accessing information and resources, and a lack of trained personnel to support them in disaster preparedness and response efforts can further compound their challenges.

Addressing the challenge of limited human resources for disaster risk management requires a multifaceted approach. Municipalities can explore strategies such as capacity-building programs, partnerships with academic institutions and NGOs, and the use of technology to enhance their workforce's effectiveness. By investing in training and development, fostering collaboration, and leveraging available resources, municipalities can improve their capacity to manage disaster risks and build resilient communities.

LACK OF POLITICAL WILL

The lack of political will at the local level is a significant barrier to effective disaster risk management (Van Niekerk, 2018). This happens when political leaders don't give enough attention to disaster risk management because they focus on other immediate issues or short-term goals. For example, they might prioritise economic development or social welfare over investing in long-term risk reduction measures (Riet, 2009). Additionally, there's often a lack of awareness among leaders about the importance of disaster risk management until a major disaster occurs. This means that disaster risk management may not get the funding or support it needs to be effective. As a result, disaster risk management efforts may be inconsistent or poorly integrated into broader development planning processes, leaving marginalised populations vulnerable to disasters (Botha et al., 2011; Sibanda, 2016).

Also, short-term political considerations exert a notable influence on the decision-making of political leaders, particularly in the context of disaster risk management. Elected officials are frequently driven by the need to secure short-term gains, such as winning reelection or achieving rapid policy victories, which can shape their priorities and actions (Mkhwanazi and Mpanza, 2020). In pursuing immediate success, leaders may prioritise projects or policies that promise immediate benefits over long-term risk reduction measures. Disaster risk management efforts, by their nature, demand sustained commitment and investment over time to yield tangible results in risk reduction and community resilience. However, these long-term objectives may not align with the shorter electoral cycles or the limited tenure of political leaders. As a result, disaster risk management initiatives may struggle to garner consistent support and resources from policymakers, leading to a lack of continuity and coherence in disaster risk management policies and programmes.

Focusing on short-term political gains can also create a bias towards projects or policies that offer visible and easily quantifiable benefits within a short timeframe. This may lead to the neglect of disaster risk management initiatives, often involving intangible or longer-term outcomes that are not immediately apparent. Consequently, disaster risk management efforts may be sidelined or relegated to secondary importance in political agendas despite their critical role in reducing disaster risks and protecting marginalised populations. Furthermore, the transient nature of political leadership can contribute to a lack of continuity in disaster risk management policies and programs (Wentink and Niekerk, 2017). Changes in political leadership, whether due to elections, administrative reshuffles, or other factors, can disrupt ongoing disaster risk management initiatives and hinder the implementation of long-term strategies. This lack of continuity can undermine the effectiveness of disaster risk management efforts, as progress made under previous administrations may be undone or neglected by subsequent leaders.

LACK OF PUBLIC AWARENESS

The lack of awareness regarding disaster risk management disaster risk management is a significant challenge that can impede effective preparedness, response, and recovery efforts. Lack of awareness refers to a limited understanding among individuals, communities, and decision-makers about the risks they face, the measures available to mitigate them, and the importance of taking proactive steps to prepare for disasters (Atkinson, 2021). This lack of awareness can manifest in various ways and across different levels of society, posing barriers to building resilience and increasing vulnerability to disasters (Mamabola and Sebola, 2021; Kunguma, Ncube, and Mokhele, 2021).

The lack of public awareness in disaster risk management presents challenges across various levels of society and governance structures. At its core, this challenge arises from a deficiency in understanding and acknowledging the risks posed by natural and man-made disasters and the necessary measures to mitigate and respond effectively (Atkinson, 2021). Limited awareness can manifest in various forms at the individual and community levels. Firstly, there may be a lack of knowledge regarding the specific hazards and vulnerabilities within a particular area. Without this understanding, individuals may underestimate the potential impact of disasters on their lives, livelihoods, and marginalised populations (Botha et al., 2011).

Consequently, they may fail to take appropriate preventive actions or adequately prepare for emergencies, leaving themselves and their marginalised populations more susceptible to harm. Moreover, cultural beliefs, social norms, and historical experiences can influence risk perceptions and shape disaster preparedness behavior (Fatti and Patel, 2013). Misconceptions or outdated beliefs about disasters may persist, leading individuals to downplay the importance of preparedness measures or resist adopting new approaches to risk management. Language barriers, literacy levels, and access to information can further exacerbate the lack of awareness, particularly among marginalised or vulnerable populations with limited resources or opportunities to access educational materials.

At the institutional and governance level, the lack of awareness poses challenges for policymakers, government agencies, and other stakeholders involved in disaster risk management. Decision-makers may be unaware of the long-term implications of inadequate risk management or prioritise competing priorities over investments in disaster preparedness and resilience-building (Fatti and Patel, 2013). This can result in insufficient funding, limited policy support, and fragmented approaches to disaster risk management, hindering the effectiveness of response efforts and exacerbating vulnerabilities. Furthermore, institutional inertia, bureaucratic barriers, and a lack of coordination among government agencies can impede efforts to address the lack of awareness and implement proactive measures (Kunguma et al., 2021). Without clear mandates, resources, and mechanisms for collaboration, stakeholders may struggle to develop and implement comprehensive disaster risk management strategies that address marginalised populations' diverse needs and priorities (Atkinson, 2021).

POOR COMMUNICATION

Communication problems in disaster risk management refer to challenges related to exchanging information, disseminating warnings, and coordinating response efforts among various stakeholders involved in disaster preparedness and response (Kunguma, 2020). This issue can manifest at multiple levels, including between government agencies, different levels of government, authorities, and the affected population (Tempelhoff et al., 2009). One aspect of the communication problem is inadequate information sharing and coordination among government agencies involved in DRM. Different departments and agencies may have their datasets, protocols, and communication channels, leading to fragmented information silos. This lack of coordination can result in duplication of efforts, gaps in coverage, and delays in response efforts (Kunguma, 2022).

Furthermore, communication breakdowns can occur between different spheres of government. Poor communication and coordination between these levels can lead to confusion, jurisdictional disputes, and delays in decision-making. For example, delays in obtaining approval or resources from higher levels of government can hamper the timely implementation of response measures at the local level. Another critical aspect of the communication problem is the challenge of effectively communicating risk information to the affected population (Solomon, 2011). This includes disseminating timely warnings, providing clear instructions on evacuation procedures, and educating marginalised populations about preparedness measures. Language barriers, literacy levels, cultural factors, and access to communication channels can all impact the effectiveness of risk communication efforts. Also, spreading misinformation and rumors during disasters can further exacerbate communication problems. False or inaccurate information circulating through social media and other channels can create confusion and panic and undermine public trust in official communication channels.

Addressing these challenges requires a concerted effort to prioritise disaster risk management within the local sphere of government in South Africa. By acknowledging and actively addressing financial constraints, limited human resources, lack of political will, lack of awareness, and poor communication, municipalities can enhance their capacity to effectively prepare for, mitigate, and respond to disasters (Van Niekerk et al., 2018). This requires strategic investments in infrastructure, training, public awareness campaigns, and interagency collaboration. Additionally, fostering a culture of resilience and accountability at all levels of government is essential for ensuring that disaster risk management remains a priority in policy and decision-making processes. Ultimately, South African municipalities can better protect their marginalised populations and build a more resilient future by overcoming these challenges and strengthening disaster risk management capabilities.

METHODOLOGY

In the context of disaster preparedness, South African municipalities face a multitude of complex challenges, including sand mining and undocumented immigration, which are often overlooked. This study explores the environmental and social implications of sand mining activities on marginalized populations' resilience and the effects of fire disasters on undocumented immigration within vulnerable communities. These topics are crucial for devising effective disaster response strategies and integrating marginalized groups into disaster risk management.

This study employed a qualitative research approach (Creswell, 2014) to investigate these issues, allowing for in-depth exploration of intricate phenomena within specific contexts (Rashid et al., 2019). Data collection involves sourcing information from academic journals, government documents, news articles, and publications to ensure a comprehensive understanding of the challenges faced by South African municipalities in implementing integrated disaster risk management.

The case study method is chosen for its reliability, robustness, and versatility in providing detailed insights into specific cases and contexts (Crowe et al., 2017). Two case studies, namely the catastrophic floods in KwaZulu-Natal in April 2022 and the devastating fire disaster in the Kennedy Road informal settlement in July 2023, are selected based on their relevance and significance in highlighting municipalities' challenges in integrating disaster risk management into their development planning.

Thematic analysis is utilised to analyse the collected data, enabling the identification of common challenges and issues across the case studies. This method facilitates the systematic organization and interpretation of data, allowing for extracting meaningful insights and generating actionable recommendations. Moreover, Sitorus's (2021) research findings regarding the complexity of case movements in important situations are considered in framing the analysis. These insights contribute to a deeper understanding of the dynamics in disaster risk management within South African municipalities.

Throughout the data collection process, duplicate records not specific to South Africa, irrelevant, not within the scope of the study, lacking sufficient information, or duplicative of another report were excluded to ensure the integrity and relevance of the analysis. Based on the analysis findings, recommendations are formulated to address the identified challenges and gaps in disaster risk management. These recommendations aim to enhance South African municipalities' disaster risk management capabilities and promote resilience and equity, particularly among marginalized populations.

Case Study 1: Devastating KwaZulu-Natal (KZN) Floods of April 2022

The flooding event that struck the KZN coastal zone in April 2022 was a catastrophic disaster with far-reaching implications for the affected communities and municipalities—triggered by a robust low-pressure weather system off the east coast of southern Africa. The event brought relentless rain to the region, exacerbated by an influx of low-level moist air from the Southern Indian Ocean (Reliefweb, 2022). Cut-off low-pressure systems delivered extremely high rainfall levels of 450

Figure 1. Flow chart of the record selection process: PRISMA approach
Source: Page et al. (2020)

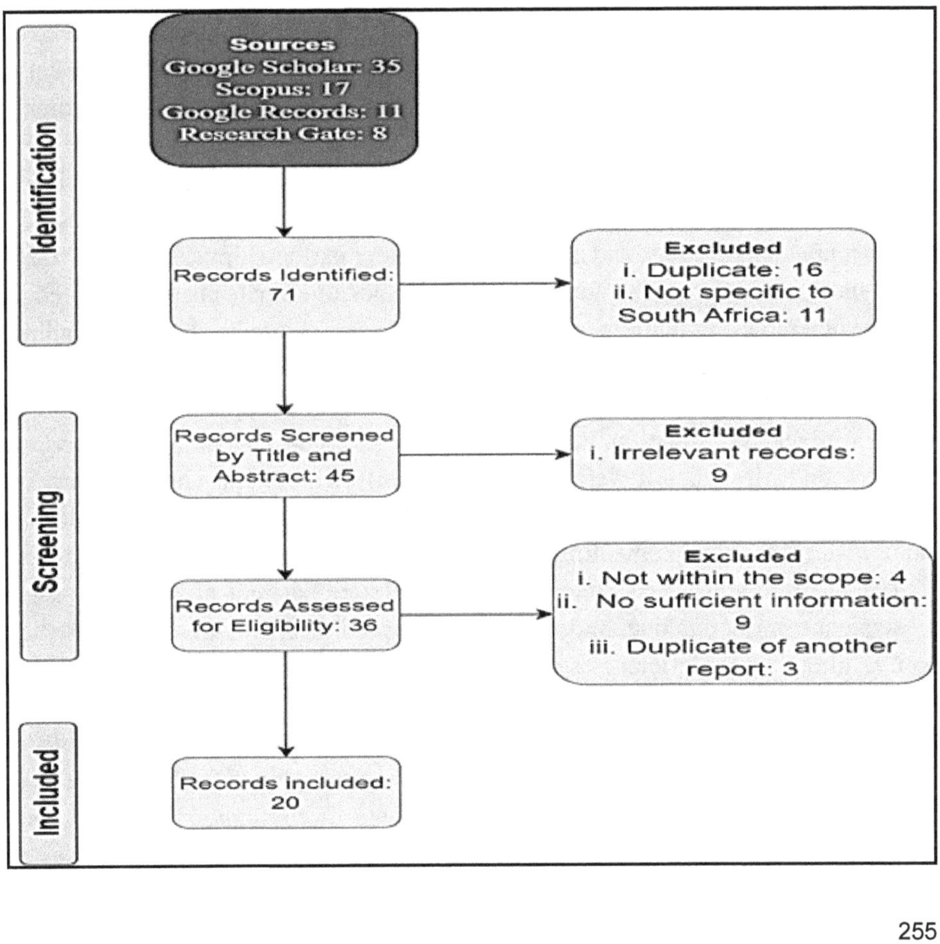

mm in some areas in just 48 hours (Engelbrecht, Le Roux, and Mbiyozo, 2022). The resulting deluge exceeded meteorological expectations, leading to widespread flooding, sinkholes, mudslides, and significant traffic disruptions. The South African Weather Service issued escalating warnings as the situation unfolded, ultimately reaching Level 9, reflecting the severity of the impending disaster. Durban, a city familiar with flooding events, recorded unprecedented rainfall levels, with some areas receiving more than four times the average monthly rainfall in just 24 hours. Rivers overflowed, inundating informal settlements and riverside communities, while critical infrastructure, including roads, bridges, and utility networks, suffered extensive damage.

The disaster had a profound human toll, resulting in the loss of 459 lives and leaving an additional 88 individuals reported missing as of May 2022. The destruction of over 4000 homes rendered tens of thousands of people homeless, while damage to businesses and infrastructure led to temporary unemployment for 45,000 individuals. Vulnerable communities bore the brunt of the devastation, underscoring existing socio-economic inequalities and vulnerabilities. Environmental impacts were equally significant, with dams overflowing beyond capacity and rivers contaminated with high levels of E. coli. The inability to operate essential infrastructure, like hydroelectric generators, highlighted the vulnerability of critical systems to extreme weather events. President Cyril Ramaphosa attributed the disaster to climate change, a sentiment echoed by scholars and experts (Kunguma, 2022; Ngcamu, 2022).

Ngcamu's review of disaster preparedness plans in two affected provinces revealed deficiencies in prioritising climate change interventions, incompetence among municipal officials, and a lack of stakeholder partnerships (Ngcamu, 2022). Inadequate training programs for officials and a lack of climate change knowledge among vulnerable communities exacerbated the disaster's effects. A notable finding was the connection between apartheid urban planning legacies and the catastrophic impacts of flooding in KZN. Poverty, land scarcity, and rapid, unplanned urbanisation also exacerbated the situation. Economic repercussions on agricultural production, livestock, and infrastructure were severe, particularly in the EC province. The lack of disaster preparedness plans was attributed to staff and resource shortages, hindering municipalities from implementing effective programs (Kunguma, 2023).

The disaster's unusual characteristics, such as a stationary cut-off low system and subsequent tropical cyclone, underscored the unpredictability of extreme weather events. Infrastructure deficiencies, including inadequate drainage systems and poorly planned developments, exacerbated flooding and mudslides. Municipalities faced criticism for neglecting infrastructure maintenance and construction, particularly in flood-prone areas. Informal settlements in vulnerable locations further heightened community vulnerability. The aftermath prompted discussions on recovery and resilience-building efforts, emphasising the urgent need for comprehensive disaster risk management strategies to safeguard lives, livelihoods, and ecosystems amid escalating climate risks (Reuters, 2022; eNCA, 2022).

The flooding revealed significant challenges that require urgent attention. One major issue was the ineffective warning system, which caught health facilities unprepared for the disaster's severity. Furthermore, a significant portion of the affected population, particularly Black South Africans, live in areas with inadequate infrastructure, such as poor drainage systems, worsening the flooding's impact. The government's response to the crisis was slow, leaving many individuals and communities without essential assistance in the disaster's aftermath. Additionally, socio-economic factors, including poverty and the prevalence of informal settlements, exacerbated the difficulties faced by vulnerable populations. To effectively address these issues, municipalities must prioritise comprehensive planning, improve coordination mechanisms, and provide targeted support for at-risk communities.

Uncovering Overlooked Risks: The Role of Sand Mining in Exacerbating Disaster Impacts

During the KZN 2022 disaster, numerous challenges emerged, reflecting the vulnerabilities faced by communities and municipalities. These challenges included the loss of lives, destruction of homes and infrastructure, disruptions to essential services like water and electricity, and the displacement of thousands of people. Amidst the immediate aftermath, attention was understandably focused on addressing these pressing issues and providing relief to affected populations. However, amidst these visible challenges, there remained an overlooked factor that significantly contributed to the severity of the disaster: the impact of sand mining activities on the KZN region's environmental resilience and disaster risk management capabilities.

Sand mining is a prominent activity in KwaZulu-Natal, with many mines operating in the region (Trade and Investment KwaZulu-Natal, 2020). Alarmingly, many of these operations occur within ecosystem service hotspots, posing a substantial threat to ecosystems and their ability to support human well-being (Davids et al., 2018). Sand mining results in extensive landscape degradation and contributes to pollution, land disputes, and climate change (Honke and Kranz, 2013). Research indicates that sand mining disrupts sediment reservoirs, releasing fecal organisms like E. coli into rivers and water bodies. Moreover, it disturbs biofilm growth in sand aquifer systems, potentially compromising water quality.

Additionally, mining activities impact surface and groundwater, leading to higher levels of antibiotic resistance genes and heavy metal contamination (Zou et al., 2021). The repercussions of sand mining extend beyond microbial contamination to alterations in river channels, erosion, and habitat loss (Pitchaiah, 2017). These changes increase the vulnerability of coastal areas to flooding and erosion, threatening infrastructure, livelihoods, and the tourism industry.

Furthermore, sand mining disproportionately affects ecosystem services, undermining water filtration, soil stabilisation, and biodiversity conservation (Davids et al., 2018). The massive impact of sand mining on roads and other infrastructure stems from its disruptive effects on natural landscapes and water systems. Sand mining alters river channels and increases erosion rates, leading to instability in riverbanks and surrounding areas. As a result, roads built near riverbanks or flood-prone areas become more susceptible to damage and collapse during extreme weather events like floods. The excessive removal of sand can also destabilise the soil structure, compromising the integrity of road foundations and contributing to pavement degradation and subsidence.

Moreover, the sedimentation of rivers and water bodies due to sand mining can obstruct drainage systems, exacerbating flood risks and causing water to inundate roads and infrastructure. Additionally, the loss of habitats and agricultural land due to sand mining can further exacerbate erosion and increase the likelihood of landslides, posing additional threats to roads and infrastructure. Overall, the extensive environmental alterations caused by sand mining significantly increase the vulnerability of roads and infrastructure to damage during disasters, emphasising the importance of addressing this issue within disaster risk management strategies.

Sand mining complicates disaster risk management efforts by:

- *Increasing vulnerability*: The environmental degradation caused by sand mining makes marginalised populations more vulnerable to disasters like flooding and erosion (Mngeni, Betek, Musampa, and Nakin, 2017; Mathada and Kori, 2012). With natural buffers depleted, the impact of such events is amplified, posing more significant risks to lives, property, and infrastructure.
- *Health hazards*: Microbial contamination from mining activities compromises water quality, leading to health hazards for marginalised populations relying on affected water sources (Manjuatha, Chengappa, and Chandrakanth, 2006). This complicates disaster response efforts as authorities must address the immediate impacts of disasters and the longer-term health consequences of waterborne diseases.
- *Infrastructure challenges*: The destruction of roads, bridges, and other infrastructure due to changes in river channels and erosion disrupts transportation networks and access to essential services (Ohaeri, Ogbonna, Gobo, and Ngah, 2021). This hampers the deployment of emergency response teams, evacuation efforts, and the delivery of relief supplies to affected marginalised populations.

- *Environmental degradation*: The loss of ecosystem services caused by sand mining further undermines the natural resilience of ecosystems and communities to disasters (Sankara and Sultana, 2016). Without these services, marginalised populations struggle to cope with the aftermath of disasters and face prolonged recovery periods. Recognising the impact of sand mining as a disaster risk management challenge is crucial for municipalities. The unchecked expansion of sand mining operations exacerbates environmental degradation, increases vulnerability to disasters such as flooding, and undermines efforts to build resilient communities.

Therefore, integrating measures to address unsustainable mining practices into disaster risk management strategies is essential to mitigate risks, protect marginalised populations, and promote sustainable development.

Case Study 2: 2023 Fire Incidents in Kennedy Road Informal Settlement, KwaZulu-Natal

The 2023 fire incidents in the Kennedy Road informal settlement in Durban, KwaZulu-Natal, and Johannesburg have devastated the affected communities unprecedentedly. In Kennedy Road, the fire erupted on July 16, 2023, engulfing more than 1101 houses and leaving over 1213 families homeless. Over 2357 people, including 647 children and 1710 adults, were affected and displaced (Reliefweb, 2023). Tragically, the fire resulted in two fatalities and several reported injuries. South African Red Cross Society (SARCS) and other response teams swiftly mobilised to provide aid and support to those impacted. Immediate relief efforts included the distribution of blankets, mattresses, and hot meals to the affected families. Additionally, SARCS facilitated Psychosocial Support (PSS) services and Restoring Family Links (RFL) assistance to help individuals cope with the trauma and reconnect with their loved ones (International Federation of Red Cross [IFRC], 2023).

However, despite these efforts, there were significant gaps in the response. Limited resources and minimal support from various stakeholders, including the government, businesses, and donors, posed challenges in adequately addressing the needs of the affected population. The majority of affected individuals were undocumented migrants, further complicating the situation as they faced barriers in accessing assistance and support. Additionally, there were concerns about the lack of sustainable shelter interventions and long-term solutions for the displaced families. SARCS and other humanitarian organisations continued to advocate for increased support and coordinated efforts to ensure the well-being and recovery of the affected marginalised populations (IFRC, 2023).

Unseen Challenges: Undocumented Immigration's Impact on Disaster Response

The fire incidents of 2023 in KwaZulu Natal, occurring within informal settlements, draw attention to the complexities surrounding disaster risk management, particularly in the context of illegal immigration. Given that 98% of the victims were undocumented immigrants, it becomes imperative to recognise and address the unique challenges they encounter during disaster response. South Africa is home to approximately 2.9 million immigrants, a substantial portion of whom are believed to be undocumented due to various factors, including economic hardship, political instability, and limited access to legal migration pathways (Migration Policy Institute, 2021). This dynamic adds layers of complexity to disaster management initiatives, especially at the municipal level. The informal settlements, commonly marked by overcrowding, poor infrastructure, and low-quality housing, are home to a large number of undocumented immigrants.

Unfortunately, the high density of vulnerable, marginalised populations in these areas greatly increases the potential dangers during disasters, such as fires (Singh et al., 2014). Illegal immigration worsens pre-existing vulnerabilities and introduces distinctive hurdles in disaster risk management. Undocumented immigrants, lacking formal documentation and integration into mainstream society, are especially vulnerable to the negative consequences of disasters. Their inability to access essential information, emergency services, and humanitarian aid during crises can result in delays in response operations and heightened casualties. Furthermore, the pressure on already stretched resources and infrastructure in informal settlements exacerbates the difficulties of disaster risk management. The influx of undocumented immigrants into these areas strains emergency services, shelters, and humanitarian aid groups, impeding their capacity to deliver prompt and efficient support to all inhabitants (DuBard and Massin, 2007).

The case study description observed that undocumented immigrants were excluded from receiving certain forms of support. The exclusion of undocumented immigrants from vital resources and support networks can have profound implications for both the individuals and the broader community (Mendez et al., 2020). When undocumented immigrants are denied access to essential services such as healthcare, education, and emergency assistance, they are left in vulnerable positions. This vulnerability can lead to desperation as they struggle to meet their basic needs, which may result in resorting to illegal activities to survive. As marginalised individuals are pushed to the fringes of society and denied opportunities for legitimate means of support, and they may turn to illicit means to sustain themselves and their families (Norasakkunkit et al., 2012). This could include engaging in activities such as theft, drug trafficking, or other forms of criminal behavior. As a result, crime rates in communities with high populations of undocumented immigrants may increase, further exacerbating social tensions and straining law enforcement resources.

Moreover, it can have far-reaching consequences for disaster risk management. The response efforts become less effective when these individuals cannot access timely information, emergency services, or humanitarian aid during disasters (Djalante et al., 2020). This can lead to delays in aiding affected communities, increasing the risk of casualties and exacerbating the impact of the disaster. Furthermore, the exclusion of undocumented immigrants from disaster preparedness and response efforts can perpetuate a cycle of vulnerability and marginalisation. Without access to resources and support networks, these individuals are left more susceptible to the negative impacts of disasters, making it harder for them to recover and rebuild in the aftermath. This undermines the resilience of affected marginalised populations and strains disaster risk management efforts at every level of government.

In dealing with illegal immigration in disaster risk management, we need to take a big-picture approach. This means ensuring everyone, including undocumented immigrants, is part of planning and responding to disasters. We also need to push for policies that tackle the root causes of problems like informal settlements and inequality and ensure everyone gets fair access to help when disaster strikes. By tackling both illegal immigration and disaster risk together, we can create more robust and unbiased marginalised populations. These marginalised populations will be better at handling disasters and helping everyone recover afterward. It is about making sure everyone, no matter their immigration status, gets a fair shot at safety and support.

LESSONS LEARNT ON INTEGRATED DISASTER RISK MANAGEMENT IN THE LOCAL SPHERE OF GOVERNMENT IN SOUTH AFRICA

Several key lessons can be learned regarding integrated disaster risk management from the two case studies. Both case studies underscore the imperative of embracing a holistic approach to disaster planning, integrating diverse factors such as environmental risks, socio-economic vulnerabilities, and community needs. By recognising the intricate interplay between these elements, comprehensive disaster risk management strategies can be developed, fostering resilience and effectively addressing the multifaceted challenges of disasters. This necessitates collaborative efforts among government agencies, humanitarian organisations, and non-governmental organisations, emphasising the importance of a multi-sectoral approach in mitigating vulnerabilities and enhancing preparedness.

Environmental considerations, including sand mining activities and climate change impacts, are critical factors influencing disaster risk. Strengthening environmental resilience through sustainable land use planning, ecosystem restoration, and protection of critical natural infrastructure is essential for mitigating disaster impacts and safeguarding communities against future risks. Proactive measures to address environmental concerns are vital for reducing vulnerabilities and building long-term resilience against environmental hazards. Investing in ecosystem-based approaches to disaster risk reduction offers multiple benefits, including enhanced ecosystem services, improved livelihoods, and increased social cohesion. Empowering marginalised populations to engage in environmental conservation efforts and participate in decision-making processes enhances ownership and sustainability of resilience-building initiatives. Prioritising environmental considerations in disaster risk management mitigate immediate impacts, contributes to long-term sustainability, and creates a more resilient future for future generations (Tajima, Gore, and Fischer, 2014).

Ensuring the inclusion of marginalised populations, such as undocumented immigrants, in disaster response efforts is crucial for equitable disaster risk management. By integrating equity considerations across all phases of disaster management, from risk assessment to recovery, we can guarantee that vulnerable populations receive fair access to resources and support. This approach acknowledges marginalised populations' diverse needs and challenges, fostering inclusivity and fairness in disaster preparedness, response, and recovery efforts. By actively involving marginalised populations in decision-making processes and ensuring their voices are heard, DRM initiatives can be more effective and responsive to the needs of all individuals, contributing to more resilient and cohesive communities (Meena and Hridya, 2022).

Additionally, fostering collaboration among government agencies, humanitarian organisations, and stakeholders is crucial for ensuring coordinated disaster response and recovery efforts. By establishing partnerships, sharing resources, and coordinating activities, stakeholders can enhance the effectiveness and efficiency of disaster risk management initiatives, particularly in complex emergency scenarios. This collaborative approach allows for pooling expertise, resources, and capacities, enabling a more comprehensive and timely response to disasters. Moreover, it promotes synergy among different actors involved in disaster risk management, leading to more cohesive and integrated strategies for disaster preparedness, response, and recovery.

Prioritising long-term planning and investment in resilience-building measures is fundamental to disaster risk management strategies. This approach involves allocating resources towards infrastructure upgrades, early warning systems, capacity-building programs, and community resilience initiatives (Jha et al., 2013). By investing in these measures, marginalised populations can mitigate disaster risks and enhance preparedness for future events (Mehta, 2007; Maru et al., 2014; Ton et al., 2019). Infrastructure upgrades, such as strengthening buildings and improving drainage systems, help reduce vulnerabilities to hazards like floods and earthquakes (Economic and Social Commission for Asia and the Pacific (ESCAP), 2006; 2010). Early warning systems provide timely alerts, allowing marginalised populations to take preventive actions and evacuate if necessary (Wafi et al., 2015). Capacity-building programs empower individuals and organisations to respond effectively to disasters, while community resilience initiatives foster a culture of preparedness and mutual support (Gil-Rivas and Kilmer, 2016). Overall, prioritising long-term resilience-building measures is essential for building safer and more resilient communities in the face of evolving disaster risks (Deria et al., 2020).

Embracing an adaptive management approach is equally crucial, allowing for continuous learning, innovation, and adjustment in response to evolving risk profiles. Monitoring and evaluating the effectiveness of interventions and incorporating lessons learned from past experiences are indispensable for building resilience over time, ensuring that disaster risk management efforts remain responsive and effective in mitigating disaster risks. Through collaborative action and inclusive engagement, marginalised populations can fortify their ability to navigate the complexities of disaster risk and foster collective resilience in the face of adversity.

By incorporating these lessons into disaster risk management practices, municipalities can strengthen their capacity to manage disaster risks effectively, protect vulnerable populations, and build a more resilient future for all.

OVERALL ASSESSMENT OF INTEGRATED DISASTER RISK MANAGEMENT IN SOUTH AFRICAN MUNICIPALITIES

South African municipalities face significant challenges and constraints in implementing integrated disaster risk management. These challenges include financial constraints, limited human resources, and a lack of political will. Insufficient funding and resources hinder the implementation of proactive risk reduction measures, while a shortage of trained personnel limits the capacity to respond effectively to disasters. Additionally, the lack of political will often results in disaster risk management not receiving adequate attention or resources, further complicating disaster preparedness and response efforts (Lassa et al., 2019).

There is a need for improved coordination and integration of disaster risk management efforts across different levels of government and sectors. Fragmentation in planning and implementation hinders the effectiveness of disaster risk management initiatives, leading to gaps in coverage and duplication of efforts. Enhanced coordination mechanisms and integration of disaster risk management into broader development planning processes are essential to address these challenges and ensure a more coherent and effective approach to disaster management. Strengthening partnerships among government agencies, civil society organisations, and the private sector is crucial to leveraging resources, sharing knowledge, and enhancing collaboration in disaster preparedness and response efforts (Adams et al., 2018).

Marginalised populations, including undocumented immigrants and residents of informal settlements, are disproportionately affected by disasters. These marginalised populations often face barriers to accessing essential services and resources, exacerbating their vulnerability during disasters. Addressing the needs of these marginalised populations requires targeted interventions and inclusive approaches that consider their unique challenges and vulnerabilities. Implementing policies and programs that prioritise including marginalised populations in disaster risk management planning and response efforts is essential to ensure equitable outcomes and build resilience across all segments of society.

Disaster risk management efforts in South African municipalities must adopt a multi-sectoral approach that engages various stakeholders, including government agencies, civil society, and the private sector. Collaboration among these stakeholders is essential to leverage resources, share knowledge, and build resilience at the community level. A multi-sectoral approach ensures that disaster risk management efforts are comprehensive, inclusive, and sustainable, addressing the root causes of vulnerabilities and building long-term resilience. By fostering partnerships and coordination among diverse actors, municipalities can enhance the effectiveness of disaster risk management initiatives and ensure that they are responsive to the needs and priorities of all stakeholders (Perney and D'angelo, 2023).

Building capacity and resilience at the local level is crucial for effective disaster risk management. This encompasses investing in training and development programs, fostering community engagement, and integrating disaster risk management into development planning processes. By empowering municipalities and marginalised populations to take proactive measures, South Africa can enhance its ability to prepare for, mitigate, and respond to disasters, ultimately building more resilient communities and reducing the impact of future disasters. Strengthening local capacity improves response capabilities and fosters a culture of preparedness and resilience, ensuring marginalised populations are better equipped to withstand and recover from various hazards and challenges.

FUTURE DEVELOPMENT OF INTEGRATED DISASTER RISK MANAGEMENT

To address these challenges effectively, it is recommended that national and municipalities in South Africa collaborate closely to:

- *Develop Comprehensive Policies*: National governments should lead the development of comprehensive policies and regulatory frameworks addressing sand mining, disaster risk management, and immigration. These policies should integrate environmental protection, social inclusion, and sustainable development principles to mitigate risks and build resilience effectively.
- *Facilitate Implementation*: National governments should support municipalities implementing these policies by providing guidance, technical assistance, and financial resources. This may include capacity-building initiatives, training programs, and knowledge-sharing platforms to enhance the capability of municipalities to enforce regulations and coordinate disaster response efforts.
- *Promote Collaboration*: Foster collaboration between spheres of government, as well as relevant key stakeholders such as civil society organisations, academia, and affected marginalised populations. Establishing multi-stakeholder platforms and coordination mechanisms can facilitate information sharing, decision-making, and joint action to address shared challenges.
- *Monitor and Evaluate*: Establish robust monitoring and evaluation mechanisms to assess the effectiveness of policies and interventions in achieving their intended outcomes. Regular review and adjustment of policies based on evidence and feedback can ensure their relevance and responsiveness to evolving needs and circumstances.
- *Advocate for Change*: Advocate for policy reforms and institutional changes at the national level to address underlying structural factors contributing to vulnerability, such as informal settlements, inequality, and economic marginalisation. This may involve engaging with policymakers, legislators, and other stakeholders to raise awareness, build consensus, and mobilise support for transformative action.

By working collaboratively across different levels of government and engaging with stakeholders at all levels, South Africa can effectively address the complex challenges posed by sand mining, undocumented immigration, and disaster risk management, ultimately building more resilient and inclusive communities.

CONCLUSION

This explored municipalities' complex issues in implementing integrated disaster risk management. Data collected from several secondary sources such as, among others, academic material from reliable journals, municipal integrated development plan, and disaster management sector plan of the city of eThekwini Municipality showed the progress made by the municipality in integrating disaster risk management into IDPs, still the municipality still reactive instead of proactive in its approaches to risk reduction. Thus, this chapter emphasised the significant paradigm shift in disaster risk management that transcends conventional silos and embraces joint planning and community-centered approaches in disaster risk reduction.

Drawing from the case study presented in this chapter, the study revealed some significant challenges, among others, financial and capacity constraints (technical and non-technical), incomplete or not approved plans inertia, public awareness deficits, absence of disaster management advisory forum, communication breakdowns, and often-overlooked impacts of sand mining and undocumented immigration on disaster risk management in South Africa. The findings highlight the urgent need for integrated approaches that address these complex issues within disaster preparedness, response, and recovery efforts. As revealed, sand mining exacerbates environmental degradation and increases marginalized populations' vulnerability to disasters such as flooding and soil erosion. Its far-reaching impacts on infrastructure, water quality, and ecosystem services highlight the necessity of incorporating measures to regulate and mitigate sand mining activities into disaster risk management strategies.

Similarly, the intersection of undocumented immigration with disaster response introduces unique hurdles that must be addressed. Excluding undocumented immigrants from essential resources and support networks not only deepens their vulnerability but also undermines the resilience of entire communities. Thus, this chapter contributes to the broader knowledge expansion on the need to take into consideration the integration of undocumented immigrants into the whole municipal disaster risk management system through the development of inclusive policies that ensure fair access to assistance and tackle root causes such as informal settlements and inequality are imperative for building more robust, more equitable communities capable of effectively managing disasters and facilitating recovery for all. By advocating for inclusivity in approaches that prioritises marginalised populations' unique needs and challenges, municipalities can promote greater resilience and equity in an increasingly uncertain future characterised by escalating climate risks and socio-economic inequalities.

Declaration of Competing Interest

The authors declare that they have no known competing financial interests or personal relationships that could have appeared to influence the work reported in this paper.

Authors' Contributions

Mokhele conceptualised, compiled, and proofread the chapter. Ncube and Kunguma proofread and provided an editorial as well as compilation.

Data Availability

Not available

REFERENCES

Adams, R. M., Prelip, M. L., Glik, D. C., Donatello, I., & Eisenman, D. P. (2018). Facilitating partnerships with community-and faith-based organizations for disaster preparedness and response: Results of a national survey of public health departments. *Disaster Medicine and Public Health Preparedness*, *12*(1), 57–66. doi:10.1017/dmp.2017.36 PMID:28735594

Aldunce, P., Beilin, R., Howden, M., & Handmer, J. (2015). Resilience for disaster risk management in a changing climate: Practitioners' frames and practices. *Global Environmental Change*, *30*, 1–11. doi:10.1016/j.gloenvcha.2014.10.010

Atkinson, D. (2021). Preparing for the worst: South African municipalities' readiness to manage disasters related to potential shale gas mining. *International Journal of Disaster Risk Reduction*, *65*, 102537. doi:10.1016/j.ijdrr.2021.102537

Bean, H. (2021). Researching Integrated Disaster Risk Management from an Interpretive/Critical Perspective. *Journal of Integrated Disaster Risk Management*, *11*(2), 31150. doi:10.5595/001c.31150

Belle, J. A. (2016). *The Integration of Disaster Risk Reduction and Climate Change Adaptation Strategies into Wetlands Management in the Eastern Free State, South Africa*. Unpublished Doctoral Dissertation. Bloemfontein, South Africa. University of the Free State.

Botha, D, Van Niekerk, D, Wentink, G, Coetzee, C, Forbes, K, Maartens, Y, Annandale, E, Tshona, T, & Raju, E. (2011). Disaster risk management: Status assessment at municipalities in South Africa. *The African Centre for Disaster Studies, North-West University.*

Creswell, J. (2014). *Research Design: Qualitative, Quantitative and Mixed Methods Approaches* (4th ed.). SAGE Publications. doi:10.7748/nr.12.1.82.s2

Crowe, S., Cresswell, K., Robertson, A., Huby, G., Avery, A., and Sheikh, A. (2017). The case study approach. . doi:10.1016/B978-008044910-4.00408-9

Davids, R., Rouget, M., Boon, R., & Roberts, D. (2018). Spatial analyses of threats to ecosystem service hotspots in Greater Durban, South Africa. *PeerJ*, *6*, e5723. doi:10.7717/peerj.5723 PMID:30386691

Deria, A., Ghannad, P., & Lee, Y. C. (2020). Evaluating implications of flood vulnerability factors with respect to income levels for building long-term disaster resilience of low-income communities. *International Journal of Disaster Risk Reduction*, *48*, 101608. doi:10.1016/j.ijdrr.2020.101608

Djalante, R., Nurhidayah, L., Van Minh, H., Phuong, N. T. N., Mahendradhata, Y., Trias, A., Lassa, J., & Miller, M. A. (2020). COVID-19 and ASEAN responses: Comparative policy analysis. *Progress in Disaster Science*, *8*, 100129. doi:10.1016/j.pdisas.2020.100129 PMID:34173447

DuBard, C. A., & Massing, M. W. (2007). Trends in emergency Medicaid expenditures for recent and undocumented immigrants. *Journal of the American Medical Association*, *297*(10), 1085–1092. doi:10.1001/jama.297.10.1085 PMID:17356029

Duze, M., & Reddy, P. S. (2020). Disaster Risk Management at the Local Level: The Case of Ethekwini City Council in South Africa. Advances in African Economic, Social, and Political Development. In P. S. Reddy & H. Wissink (Eds.), *Reflections on African Cities in Transition*. Springer., doi:10.1007/978-3-030-46115-7_12

Duze, M. P. (2016). *A framework for disaster risk reduction for local government: A case study of eThekwini Municipality*. Unpublished master's thesis. University of KwaZulu-Natal, Westville Campus.

Economic and Social Commission for Asia and the Pacific (ESCAP). (2006). Enhancing regional cooperation in infrastructure development, including that related to disaster management. Author.

Engelbrecht, F., Le Roux, A., & Mbiyozo, A. N. (2022). *With every second city and town projected to suffer increased flooding, South Africa must be better prepared.* Daily Maverick. https://www.dailymaverick.co.za/article/2022-05-12-poor-climate-adaptation-outdated-infrastructure-served-as-catalysts-for-kzn-floods/

Environmental Protection Agency (EPA). (2000). *Abandoned mine site characterization and cleanup handbook.* USEPA.

Fatti, C. E., & Patel, Z. (2013). Perceptions and responses to urban flood risk: Implications for climate governance in the South. *Applied Geography (Sevenoaks, England), 36*, 13–22. doi:10.1016/j.apgeog.2012.06.011

Gil-Rivas, V., & Kilmer, R. P. (2016). Building community capacity and fostering disaster resilience. *Journal of Clinical Psychology, 72*(12), 1318–1332. doi:10.1002/jclp.22281 PMID:26990644

Hönke, J., & Kranz, N. (2013). Cleaning up Their Act, or More? Mining Companies and Environmental Protection in South Africa. In T. A. Börzel & C. R. Thauer (Eds.), *Business and Governance in South Africa. Governance and Limited Statehood Series*. Palgrave Macmillan. doi:10.1057/9781137317810_10

Human Rights Campaign. (2021). *Violence against the transgender and gender non-conforming community in 2020.* Retrieved from https://www.hrc.org/resources/violence-against-the-transgender-and-gender-non-conforming-community-in-2020

Jha, A. K., Miner, T. W., & Stanton-Geddes, Z. (Eds.). (2013). *Building urban resilience: principles, tools, and practice*. World Bank Publications. doi:10.1596/978-0-8213-8865-5

Kunguma, O. (2020). *South African disaster management framework: Assessing the status and dynamics of establishing information management and communication systems in Provinces.* Unpublished Doctoral Dissertation. University of the Free State, South Africa: Bloemfontein.

Kunguma, O. (2022). *COVID-19 disaster, an opportunity for the South African disaster management legislation betterment?* Academia Letters., doi:10.20935/AL4645

Kunguma, O. (2022). *Opinion: KZN floods expose significant socio-economic and environmental vulnerabilities.* Disaster Management Training and Education Centre for Africa: University of University of the Free State, South Africa: Bloemfontein. https://www.ufs.ac.za/templates/news-archive/campus-news/2022/april/opinion-kzn-floods-expose-significant-socio-economic-and-environmental-vulnerabilities

Kunguma, O., Ncube, A., & Mokhele, M. (2021). COVID-19 disaster response: South African disaster managers' faith in mandating legislation tested? *Jamba (Potchefstroom, South Africa), 13*(1). Advance online publication. doi:10.4102/jamba.v13i1.1099 PMID:34691364

Lassa, J. A., Surjan, A., Caballero-Anthony, M., & Fisher, R. (2019). Measuring political will: An index of commitment to disaster risk reduction. *International Journal of Disaster Risk Reduction, 34*, 64–74. doi:10.1016/j.ijdrr.2018.11.006

Logie, C. H., & Turan, J. M. (2020). How do we balance tensions between COVID-19 public health responses and stigma mitigation? Learning from HIV research. *AIDS and Behavior, 24*(7), 2003–2006. doi:10.1007/s10461-020-02856-8 PMID:32266502

Mamabolo, M., & Sebola, M. (2021). *The role and adequacy of disaster management unit within the South African Municipalities*. The Business and Management Review. doi:10.24052/BMR/V12NU02/ART-09

Mandal, M., Ghosh, D., & Ghosh, B. (2016). Environmental Impact of Sand Mining: A Case Study along the Lower Reaches of Ajay River, West Bengal, India. *Journal of Environment and Sociobiology, 13*, 99–108.

Manjunatha, A., Chengappa, P., & Chandrakanth, M. (2006). Externalities Due To Sand Mining and Distillery Effluent in Water Streams of India. *Journal of Global Economics, 2*, 135–151. doi:10.1956/JGE.V2I2.209

Maru, Y. T., Smith, M. S., Sparrow, A., Pinho, P. F., & Dube, O. P. (2014). A linked vulnerability and resilience framework for adaptation pathways in remote, disadvantaged communities. *Global Environmental Change, 28*, 337–350. doi:10.1016/j.gloenvcha.2013.12.007

Mathada, H., & Kori, E. (2012). An Evaluation of Sand and Gravel Extraction in Nzhelele River Valley, Limpopo Province of South Africa. *3rd International Conference on Biology, Environment, and Chemistry IPCBEE*, 46.

Meena, H. R., & Hridya, S. (2022). Preparedness of the Marginalized People in Disaster Management During Extreme Climate Events. *Analytical Approaches for Vulnerability and Adaptation to Climate Change in Agriculture,* 170.

Mehta, M. (2007). *Gender matters: Lessons for disaster risk reduction in South Asia*. The International Centre for Integrated Mountain Development (ICIMOD), Kathmandu, Nepal. Retrieved from https://www.preventionweb.net/files/2406_GenderandDisasters.pdf

Méndez, M., Flores-Haro, G., & Zucker, L. (2020). The (in) visible victims of disaster: Understanding the vulnerability of undocumented Latino/a and indigenous immigrants. *Geoforum*, *116*, 50–62. doi:10.1016/j.geoforum.2020.07.007 PMID:32834081

Merriam, S. B. (2010). Qualitative case studies. In P. Peterson, E. Baker, & B. McGaw (Eds.), *International encyclopedia of education* (3rd ed., pp. 456–462). Elsevier Ltd. doi:10.1016/B978-0-08-044894-7.01532-3

Meyer, I. H. (2015). Resilience in the study of minority stress and health of sexual and gender minorities. *Psychology of Sexual Orientation and Gender Diversity*, *2*(3), 209–213. doi:10.1037/sgd0000132

Migration Policy Institute. (2021). *South Africa as an Immigration Destination: History, Recent Trends, and Current Challenges*. https://www.migrationpolicy.org/article/south-africa-immigration-destination-history

Mkhwanazi, N., & Mpanza, S. (2020). Examining the nature of disaster risk reduction strategies in local municipalities: a case study of Ndwedwe Local Municipality. *International Conference on Public Administration and Development Alternatives (IPADA).*

Mngeni, A., Betek, C., Musampa, C. M., & Nakin, M. D. V. (2017). The physical and environmental impacts of sand mining. *Transactions of the Royal Society of South Africa*, *72*(1), 1–5. doi:10.1080/0035919X.2016.1209701

Moşteanu, N. R., Faccia, A., & Cavaliere, L. P. L. (2020). Disaster Management, Digitalization, and Financial Resources: Key factors to keep the organization ongoing. *Proceedings of the 2020 4th International Conference on Cloud and Big Data Computing*, 118-122. 10.1145/3416921.3416930

Munsamy, L. (2018). *Development of an integrated disaster risk management model for the municipalities in the Free State Province*. Unpublished Doctoral Dissertation. Bloemfontein, South Africa: Central University of Technology, Free State.

News24. (2022). *Flood-ravaged KwaZulu-Natal still waiting on R1bn disaster funds*. News24. https://www.news24.com/news24/southafrica/news/flood-ravaged-kwazulu-natal-still-waiting-on-r1bn-disaster-funds-20220511

Ngcamu, B. (2022). *Climate change and disaster preparedness issues in Eastern Cape and Kwazulu-Natal*. Town and Regional Planning. doi:10.18820/2415-0495/trp81i1.5

Ngcamu, B. S. (2011). *Disaster risk management in local government: a case study of Foreman and Kennedy Road informal settlements, eThekwini Municipality, KwaZulu-Natal*. Unpublished Doctoral dissertation. Durban University of Technology: KwaZulu-Natal.

Nkombi, Z., & Wentink, G. J. (2022). The role of public participation in disaster risk reduction initiatives: The case of Katlehong township. *Jàmbá: Journal of Disaster Risk Studies*, *14*(1), a1203. https://doi.org/https://doi.org/10.4102/jamba.v14i1.1203

Norasakkunkit, V., Uchida, Y., & Toivonen, T. (2012). Caught between culture, society, and globalization: Youth marginalization in post-industrial Japan. *Social and Personality Psychology Compass*, *6*(5), 361–378. doi:10.1111/j.1751-9004.2012.00436.x

Ohaeri, M., Ogbonna, D., Gobo, A., & Ngah, S. (2021). Environmental Impacts of Sand Mining in Some Coastal Communities in Port Harcourt Metropolis Nigeria. *Journal of Applied Life Sciences International*. . doi:10.9734/jalsi/2021/v24i1030265

Oosthuizen, R. C. (2018). *A critical analysis of disaster risk management in local governance with reference to Sarah Baartman District Municipality*. Unpublished Doctoral Dissertation. Port Elizabeth, South Africa. Nelson Mandela University. Port Elizabeth.

Page, M. J., McKenzie, J. E., Bossuyt, P. M., Boutron, I., Hoffmann, T. C., Mulrow, C. D., Shamseer, L., Tetzlaff, J. M., Akl, E. A., Brennan, S. E., Chou, R., Glanville, J., Grimshaw, J. M., Hróbjartsson, A., Lalu, M. M., Li, T., Loder, E. W., Mayo-Wilson, E., McDonald, S., ... Moher, D. (2021). The PRISMA 2020 statement: An updated guideline for reporting systematic reviews. *British Medical Journal*, *372*(71), n71. Advance online publication. doi:10.1136/bmj.n71 PMID:33782057

Peduzzi, P. (2019). The disaster risk, global change, and sustainability nexus. *Sustainability (Basel)*, *11*(4), 957. doi:10.3390/su11040957

Perney, M. E. P., & D'angelo, G. (2023). Local Governance Support Tools for Disaster Risk Reduction and Climate Adaptation Strategies: The EU Contribution in the Case Study of the Municipality of Naples. *Sustainability (Basel)*, *15*(15), 11716. doi:10.3390/su151511716

Pitchaiah, P. (2017). Impacts of Sand Mining on Environment–A Review. *International Journal of Geoinformatics*, *4*(1), 1–6. doi:10.14445/23939206/IJGGS-V4I1P101

Pitchaiah, S. P., & Sultana, N. (2016). Impact of Sand Mining on Asian Countries - A Review. *International Journal of Geoinformatics*, *3*(5), 1–5. doi:10.14445/23939206/IJGGS-V3I5P101

Rashid, Y., Rashid, A., Warraich, M., Sabir, S., & Waseem, A. (2019). Case Study Method: A Step-by-Step Guide for Business Researchers. *International Journal of Qualitative Methods, 18*. Advance online publication. doi:10.1177/1609406919862424

ReliefWeb. (2022). *South Africa: Floods and Landslides - Apr 2022*. ReliefWeb. https://reliefweb.int/disaster/fl-2022-000201-zaf

ReliefWeb. (2023). *South Africa: Fire - DREF Operation Update (MDRZA014)*. Retrieved from https://reliefweb.int/report/south-africa/south-africa-fire-dref-operation-update-mdrza014

Republic of South Africa. (2000). *Municipal Systems Act, 2000 (Act 33 of 2000)*. Government.

Republic of South Africa. (2002). Disaster Management Act, 2002 (Act 57 of 2002). Pretoria: Government Printer.

Republic of South Africa. (2015). Disaster Management Act, 2015 (Act 16 of 2015). Pretoria, South Africa. Government Printers.

Reuters. (2022, April 14). *It is bad: survivors count costs of South Africa's devastating floods*. Reuters. https://www.reuters.com/business/environment/it-is-bad-survivors-count-costs-south-africas-devastating-floods-2022-04-14

Riet, G. (2009). Disaster risk assessment in South Africa: Some current challenges. *South African Review of Sociology*, *40*(2), 194–208. doi:10.1080/21528586.2009.10425108

Ryscavage, R. (2003). The Transition from Conflict to Peace. *Emergency Relief*, 284-294

Singh, S. R., Eghdami, M. R., & Singh, S. (2014). The concept of social vulnerability: A review from disasters perspectives. *International Journal of Interdisciplinary and Multidisciplinary Studies*, *1*(6), 71–82.

Sitorus, S. (2021). Qualitative Method (Case Study Research). *Journal of Communication Education*. . doi:10.58217/joce-ip.v15i1.224

Solomon, F. J. (2011). *Examining the feasibility of informal settlement flood early warning systems: focus on the urban flood-risk experience of Kosovo and Masiphumelele residents, Cape Town, South Africa.* Unpublished master's thesis. Cape Town, South Africa: University of Cape Town.

Statistics South Africa. (2015). *Population Living in Poverty Statistics South Africa.* doi:10.15493/BEA.DATA.10000093

Syahnur, S., & Diantimala, Y. (2021). Surviving and growing up with illegal status: The Analysis of Socio-Economic Household, Potential Conflict, the Environmental Damage, and Vulnerability of Local Community to Disaster. *Journal of Sustainable Mining*. . doi:10.46873/2300-3960.1061

Tajima, R. Y. O., Gore, T. O. M., & Fischer, T. B. (2014). Policy integration of environmental assessment and disaster management. *Journal of Environmental Assessment Policy and Management, 16*(03), 1450028. doi:10.1142/S1464333214500288

Tempelhoff, J., Van Niekerk, D., Van Eeden, E., Gouws, I., Botha, K., & Wurige, R. (2009). The December 2004-January 2005 floods in the Garden Route region of the Southern Cape, South Africa. Jàmbá. *Journal of Disaster Risk Studies, 2*(2), 93–112.

Ton, K. T., Gaillard, J. C., Adamson, C. E., Akgungor, C., & Ho, H. T. (2019). Expanding the capabilities of people with disabilities in disaster risk reduction. *International Journal of Disaster Risk Reduction, 34*, 11–17. doi:10.1016/j.ijdrr.2018.11.002

Trade and Investment KwaZulu-Natal. (2020). *Mining and beneficiation.* Retrieved from https://www.tikzn.co.za/open.php?cat=key_sectorsandpage=mining_and_beneficiation

UNISDR. (2005). *Hyogo Framework for Action 2005–2015: building the resilience of nations and communities to disasters.* World Conference on Disaster Reduction (A/CONF. 206/6). Geneva: The United Nations International Strategy for Disaster Reduction.

United Nations. (2019). *State of the world's indigenous peoples.* Retrieved from https://www.un.org/development/desa/indigenouspeoples/publications/state-of-the-worlds-indigenous-peoples.html

United Nations Office for Disaster Risk Reduction. (2019). *Leave no one behind: A call to action for gender equality and women's empowerment in disaster risk reduction.* Retrieved from https://www.undrr.org/publication/leave-no-one-behind-call-action-gender-equality-and-womens-empowerment-disaster-risk

Van der Berg, A. (2015). Public-private partnerships in local disaster management: a panacea to all local disaster management ills? *Potchefstroom Electronic Law Journal / Potchefstroomse Elektroniese Regsblad*, *18*(4), 994-1033.

Van Niekerk, D. (2014). A critical analysis of the South African Disaster Management Act and policy framework. *Disasters*, *38*(4), 858–877. doi:10.1111/disa.12081 PMID:25196341

Van Niekerk, D. (2015). Disaster risk governance in Africa - A retrospective assessment of progress in against the Hyogo Framework for Action (2000-2012). *Disaster Prevention and Management*, *24*(3), 397–416. doi:10.1108/DPM-08-2014-0168

Van Niekerk, D., Coetzee, C., & Nemakonde, L. (2020). Implementing the Sendai Framework in Africa : Progress Against the Targets (2015 – 2018). *International Journal of Disaster Risk Science, 11*(October), 179–189. . doi:10.1007/s13753-020-00266-x

Van Niekerk, D., Wentink, G. J., & Shoroma, L. B. (2018). Natural hazards governance in South Africa. Oxford Research Encyclopedia of Natural Hazard Science.

Van Niekerk, D., & Wisner, B. (2014). Integrating disaster risk management and development planning: Experiences from Africa. Routledge.

Van Riet, G. (2009). Disaster risk assessment in South Africa: Some current challenges. *South African Review of Sociology*, *40*(2), 194–208. doi:10.1080/21528586.2009.10425108

Verbano, C., & Venturini, K. (2011). Development paths of risk management: Approaches, methods, and fields of application. *Journal of Risk Research*, *14*(5), 519–550. doi:10.1080/13669877.2010.541562

Vogel, L., O'Carroll, D., Edge, T., & Robinson, C. (2016). Release of Escherichia coli from Foreshore Sand and Pore Water during Intensified Wave Conditions at a Recreational Beach. *Environmental Science & Technology*, *50*(11), 5676–5684. doi:10.1021/acs.est.6b00707 PMID:27120087

Wafi, Z. K., Malek, M. F. A., Alnajjar, S. H., & Ahmad, R. B. (2015). Early warning system for Disaster management in rural areas. In *2015 International Symposium on Technology Management and Emerging Technologies (ISTMET)* (pp. 369-372). IEEE. 10.1109/ISTMET.2015.7359061

Wentink, G., & Niekerk, D. (2017). The capacity of personnel in disaster risk management in South African municipalities. *The Journal for Transdisciplinary Research in Southern Africa*, *13*(1), 10. doi:10.4102/td.v13i1.427

World Health Organization. (2017). *Promoting the health of refugees and migrants: Draft global action plan, 2019-2023*. Retrieved from https://www.who.int/migrants/publications/global-action-plan/en/

Zou, H., He, L., Gao, F., Zhang, M., Chen, S., Wu, D., Liu, Y., He, L., Bai, H., & Ying, G. (2021). Antibiotic resistance genes in surface water and groundwater from mining-affected environments. *The Science of the Total Environment*, *772*, 145516. doi:10.1016/j.scitotenv.2021.145516 PMID:33571766

Chapter 12
Resilience in Compound Disasters:
A Case Study of KwaZulu-Natal Province in South Africa

Obianuju E. Okeke-Uzodike
Durban University of Technology, South Africa

Tyolome S. J. Kinga-Upaa
Veritas University, Nigeria

Eze Chika
Veritas University, Abuja, Nigeria

ABSTRACT

In the wake of a corona-virus outbreak, reports of additional natural disasters were recorded in various cities across the globe. Resilience at such challenging times has gained attention as one of the key approaches to disaster management. This chapter reflects on the experiences of compound disasters in the KwaZulu-Natal province of South Africa. This study applies a case study research design and data collection and analysis involving a three-phase methodological protocol. The study used KwaZulu-Natal provincial level of government as a case focusing on disaster management and resilience. The main lesson identified relates to weaknesses in the provincial disaster management framework and processes, which shows concerns in the areas of training and development and stakeholder engagement. A disaster management framework is suggested for building resilience in the system. The study adds a valuable theoretical contribution to the growing body of knowledge that seeks an understanding of managing compound disaster and building relentless resilience in disruptive times.

DOI: 10.4018/979-8-3693-2721-0.ch012

INTRODUCTION

Globally, the prevalence of disasters has increased significantly in recent times. The world has witnessed various forms of such events ranging from natural disasters to a wide array of man-made disasters – climate and environmental changes or health, environment, economic, technological, market, and political etc. More recently, for instance, the global health crisis triggered by coronavirus disease (Covid-19) outbreak shook the world so massively that it created alarm and panic. The impact was felt across all spheres of life, sectors, economies, societies, and organizations. While many governments across the world grappled with the containment of the virus, some countries simultaneously experienced severe forms of other natural disasters such as: Cyclone Harold which hit the Pacific countries; flooding in Manitoba province and Ottawa City in Canada; Typhoon Hagibis in Japan and Bangladesh; very destructive torrential rains in KwaZulu-Natal province, South Africa (Mishra et al., 2021; Ishiwatari et al., 2020). It is important to point out that some countries were simultaneously hit by more than one natural disasters. For instance, while Covid-19 pandemic imposed severe natural disasters in many communities across South Africa Mishra et al., (2021), other compounding natural disasters, sometimes emerged simultaneously with adverse consequences. The high complexity associated with the resulting compounding of the disasters, add new problems to further create and deepen long-term devastation, shock, and pose coordination and recovery challenges (Liu et al., 2022; Gissing et al., 2022; Munawar et al. 2021; Wu et al., 2022). For the province of KwaZulu-Natal, the net result has been the emergence of significant socio-economic and environmental vulnerabilities. Thus, building adaptive capacity for continuity of lives and economic activities remains the most critical concern for all stakeholders, which highlights the importance of resilience in compound disaster management.

In this chapter, the compound disaster impacting the province of KwaZulu-Natal is a combination of Covid-19 pandemic (human/public disaster) and severe torrential rains leading to flooding and landslides (natural disasters). Vast literature document approaches to building resilience focusing mostly at the community-based level. From the human resource management (HRM) perspectives, it is without a doubt that human resources play crucial roles during uncertainties and dramatic and problematic environmental changes to reduce the irreparable damage. Studies focusing on the influences of HRM in the context of a stable environment have been extensively researched, but most importantly, these studies focus on aftershock management. There is a paucity in literature on HRM practices. Especially, the paucity generally applies within the context of environmental uncertainties associated with compound disasters. Edvardsson & Durst, (2021), emphasized and recommended the need to focus on crucial topics of this nature when they found minimal research in their review study on types of crises and HRM. Similarly, Al-Maqashi et al. (2021), asserted the importance of examining the link between disaster management and HRM practices; and the role of HRM in the efficacy of managing disasters is a novel and essential knowledge area. In agreement with these authors, this chapter bridges the gap by examining how human resources management practices can be integrated in a disaster management framework to foster resilience. Specifically, this chapter explores the integration of training and development (T&D) and stakeholder engagement (SE) across the phases of a disaster management framework with an overall aim of reducing the anticipated aftermath effect, thus building or deepening resilience.

The content of this chapter is presented as follows: the chapter first presents literature review on the related constructs, which are disaster/compound disaster, training and development, stakeholder engagement, and resilience. An overview of KwaZulu-Natal and disaster occurrence then follows. The methodology adopted for the study preceded and the discussion sheds light on an integrative disaster management framework to foster resilience in disruptive times. The chapter then presents the concluding thoughts and puts forward the recommendations for policy and practice.

Review of related literature

According to Snyder, (2019), literature reviews provide foundation for all types of research, and serve as basis for knowledge development. They create guidelines for policy and practice, provide evidence of an effect and, if well conducted, have the capacity to engender new ideas and directions. Therefore, this section provides a review of literature on the study-related concepts: disaster and compound disaster; resilience; training and development; and stakeholder engagement.

Disaster and Compound Disaster

The world today is witnessing natural or manmade environmental disruptions and life-threatening changes. These events, often referred to as 'disaster', are accompanied by immeasurable damages, fatalities, and socio-economic and political losses. Disasters are highly complex, and scholars oftentimes lack a general consensus on their acceptable definition. Amarasinghe (2020) attributed the lack of acceptable definition to this to three factors: i) shift in the scientific invention; ii) nature of problems occurrence; and iii) the improved cognition among people. Going by Section 30 of the South African Disaster Management Act No. 57 of 2002, the concept "Disaster" means a progressive or sudden widespread, or localized, natural, or human-caused occurrence which:

(a) causes or threatens to cause (i) death, injury, or disease; (ii) damage to property, infrastructure, or the environment; or (iii) disruption of the life of a community; and
(b) is of a magnitude that exceeds the ability of those affected by the disaster to cope with its effects using only their own resources" (RSA, 2002).

The co-occurrence of disastrous events is not new in human history. Disasters sometimes occur simultaneously and manifest through different mechanisms involving hazardous events which may be caused by humans, nature, technology, or could be health related. For example, since the outbreak of Covid-19 pandemic, which is considered a public health disaster, many countries witnessed additional disasters either naturally or human induced actions, leading to a complex situation, mostly referred to by scholars as compound disaster. Accordingly, a compound disaster arises when the adverse consequences resulting from different disaster agents occur simultaneously (Mishra et al., 2021). It also entails two or more shock events overlapping, such as multiple weather extremes that occur concurrently (Gissing et al., 2022; AghaKouchak et al., 2020; Zscheischler et al., 2020). The coupling effects could be natural, man-made or public health hazards and the complexity of such disasters causes compound effects at catastrophic scale, extensive loss of lives, and economic damages. This potentially deepens the existing vulnerabilities, and according to Gissing et al., (2022), severely weakens the resilience of economic activities and the society at large.

KwaZulu-Natal Province and Compound Disaster

KwaZulu-Natal (KZN) province is on the southeastern part of South Africa. KwaZulu–Natal's provincial capital and second largest city is Pietermaritzburg. The largest city in KZN Province, Durban is one of the largest commercial cities in South Africa. Durban, as depicted in Figure 1. With an estimated population of 11.54million, translating to 19% of the country's population, makes the province second to the largest in the country (Stats SA, 2022). The province has three distinct areas made up of the coastal lowland region, bounded by Indian Ocean, the central grassland with savannah and the mountainous region.

The outbreak of Coronavirus disease which affected the global world did not spare South Africa and, particularly, KwaZulu-Natal. The first confirmed case of Covid-19 infection in South Africa was reported on the 5th of March 2020 in the Hilton area of KwaZulu-Natal province (National Department of Health, 2020).

Figure 1. A map of KwaZulu-Natal province of South Africa
Source: KZN Online, 2023

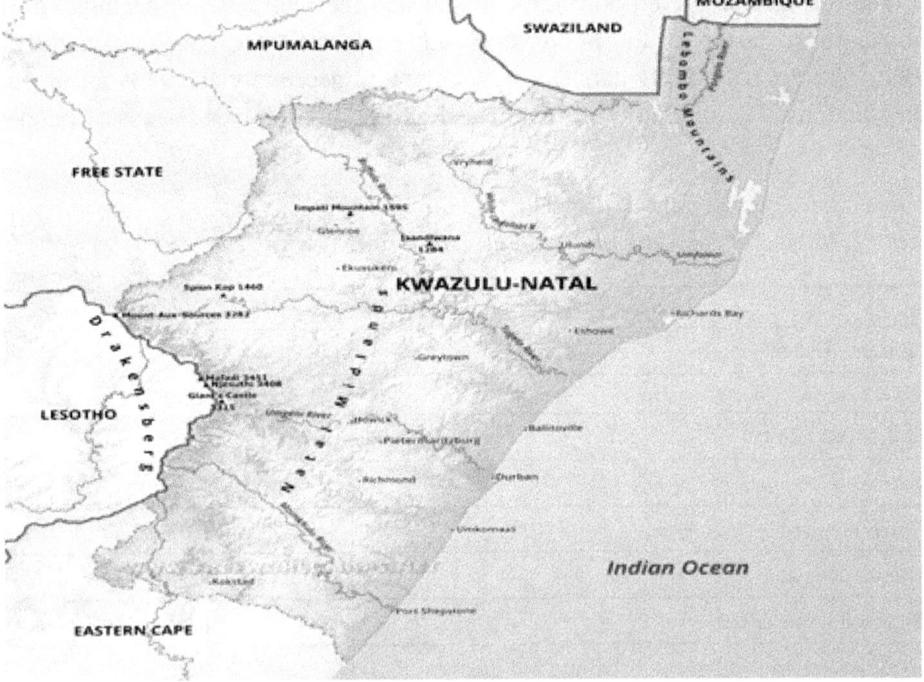

As was getting common at the time elsewhere in the world, the first reported case was followed by confirmed cases of the disease in other provinces. Subsequently, government embarked on a national drive to curtail the rapid spread of the virus. The magnitude and the severity of the Covid-19 disease outbreak led government to the declaration of National State of Disaster on the 15th of March 2020 (CoGTA 2020). As though the nation did not have enough burden laboring considerably to contain the coronavirus pandemic, the province of KwaZulu-Natal experienced additional disasters within a short space of time. On the 11-13 April 2022, a slow severe flooding and landslides (caused by torrential rain) led to the death of 448 people, over 40,000 missing, displaced over 40,000 people, completely washed away more than 12,000 houses, and recorded financial losses of R245 million to business (ReliefWeb, 2022). The disaster caused critical damages to the public infrastructure such as roads, bridges, water, and electricity. Barely a month later, over the period of 21-22 May 2022, the province was struck again by one of the deadliest storms accompanied by flooding which left over 400 people dead and 40,000 displaced (Floodlist, 2022, IFRC, 2022). An insight into the estimated financial cost of the damages is as shown in Table I while Figures II and III paint grim images of the extent of the damages.

These events triggered both the national and the international declaration of the state of disaster for the province (South African Government, 2022, IFRC, 2022). The severity of the damages poses greater uncertainty and calls for urgent interventions. Though, an immediate relief addressing the basic needs of people

Table 1. Impact of flooding on KwaZulu-Natal province

Impact	Financial Loss
Damages to rail infrastructure	R995.4 m
Water and sanitation infrastructure	R1b
SMME's affected 110	R20m
Damages to the manufacturing sector	R431m
Warehousing and logistics sector	R33.5m
Construction sector	R18.1m
Retail and wholesale sector	R46.9m
Agricultural sector	R12.6m

Source: IOL (2022)

Figure 2. Image of KZN flooding
Source: Floodlist, (2022) & IFRC, (2022)

Figure 3. Image of KZN landslide
Source: Floodlist, (2022) & IFRC, (2022)

followed, the devastating impact of the compound disaster exposed the huge socio-economic challenges facing the province of KwaZulu-Natal.

Resilience in Compound Disaster Management

Of late, it seems disasters no longer occur as isolated events but, rather, two or more extreme events occurring simultaneously and generating multiple of devastating damages and losses to society. Resilience has attracted many discussions lately as a system-based concept in addressing compounding acute events. The concept has been studied across disciplines with the same goal of reducing human exposures to disaster consequences and developing measures to reduce the impact but lacks a consensus on an acceptable definition. This chapter adopts Laurien & Mechler, (2020:1) definition of resilience as "The ability of a system, community or society to pursue its social, ecological and economic development and growth objectives, while managing its disaster risk over time in a mutually reinforcing way." The impact and recovery in concurrent disastrous events are amplified and more challenging. The key to managing such a situation is building greater system resilience to hasten the recovery process. With compound disasters increasingly becoming a new normal, researchers, policymakers and practitioners are calling for the need to address it from a development-centric manner to tackle the multiple threats it poses (NASEM, 2022). Hence, as a multidimensional, multi-scalar and multi-faceted notion (Zaman & Raihan, 2023), therefore, there is a need to apply resilience within the nature and scope of a study.

Compound disaster has posed more challenges for the field of human resource practices and its effectiveness in disaster management has become a growing area of research. Since the dawn of increasing global crises, HR practices have become an important dimension managing resilience in compound disaster. Developing the ability to manage a disaster necessitates specific intervention strategies that allow for the recognition of events, response, and recovery from the disastrous situation. Focusing on the KZN government at provincial level, this chapter elaborates on the integration of human resources training and development and stakeholder engagement across a disaster management framework to foster the resilience of a system. Training and development are apparatus that help people to contribute and be effective in their roles (Siddiqui, 2017). Thompson, (2022), notes that investing in capacities helps in building resilience against compounding disasters. Jethwaney, (2023), demonstrated how stakeholder training helps in building capacity for effective disaster management. Similarly, stakeholder engagement is a vital part of effective disaster management, often considered the relationship-building principles in managing disaster. Communication is a key aspect in the engagement processes. Engagement is influenced by the pre-disaster relationships and reflects along the cycle of disaster framework as collaboration, integration, participation and deeper involvement of all parties concerned. In disaster management framework, activities relating to engagement are information sharing, planning, decision-making, and ensuring that informational resources reflect the needs and preferences for the disaster (Panneer et al., 2021; Bajracharya & Hastings, 2020).

RESEARCH METHODOLOGY

The study adopts a case study research design defined by Yin (2003: 13) as "an empirical inquiry that investigates a contemporary phenomenon within its real-life context, especially when the boundaries between phenomenon and context are not clearly evident". The choice of case study design helps the researchers to answer the 'how' and 'why' questions (Yin, 2015), has a reputation as an effective methodology for investigating and understanding complex issues in real world settings (Harrison et al., 2017). A case study provides a real-life contemporary bounded system (Creswell, 2013), and enables holistic in-depth understanding of a specific phenomenon, event, or situation within a specified context (De Massis & Kotlar, 2014: Rosenberg & Yates, 2007) and the need for investigation (Taherdoost, 2021). Furthermore, a case study data helps to integrate both the objective and perceptual data (Priya, 2021). For methodological rigour, literature notes that a typical case study method demonstrates a clear procedure and provides structure; hence, the researchers adopted and modified Teegavarapu & Summer (2008) and Rosenberg & Yates (2007) schematic model. The model consists of 3 broad phases (with several steps): define and design; prepare, collect and analyze; analyze and conclude. The steps include: (i) problem definition/research question, (ii) identify theories and map themes, (iii) identify case, context and phenomenon, (iv) select case, (v) data collection method, (vi) analysis strategy, (vii) refine data, (viii) data reduction, and (ix) Conclusion. The modified case study schematic for this study is presented in Figure IV.

Figure 4. Case study schematic
Source: Adapted and modified from Teegavarapu & Summer (2008) and Rosenberg & Yates (2007)

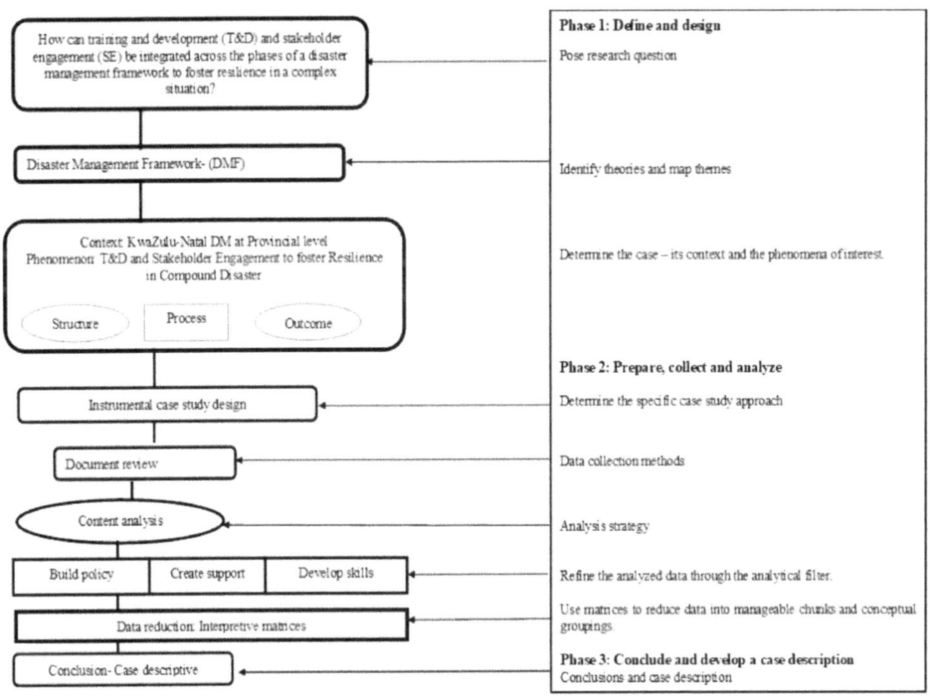

Phase 1- Define and Design: This phase consists of three steps namely, problem definition or posing of the research question; identification of theories and mapping of the themes and the identification of case, context, and phenomenon. The compound disaster (Covid-19) and the extreme weather conditions (torrential rain/flooding) that struck KwaZulu-Natal province of South Africa had a devastating impact on the economy, businesses, public infrastructure, and the livelihoods. The severity of the impact was attributed mostly to its overall management. The AGSA report (PMG, 2023) noted the following findings in the KZN municipalities:

i) Ethekwini Metropolitan Municipality: The municipality is found to account for sixty-eight per cent (68%) of the overall disaster related damages in the province of KwaZulu-Natal. The report also noted that the municipality had slow implementation of commitments, an outdated disaster management plan, lacks disaster readiness, effective oversight, and lacks coordination to support service delivery.
ii) Kwadukuza local municipality disaster management plan has not been reviewed overtime. In addition, the municipality has poor workmanship (human resources), inadequate disaster relief funding and infrastructural challenges such as no road markings or signs resulting in ineffective responses.
iii) Ilembe district municipality disaster management plan was found not effective in addressing the key deliverables, has inadequate funding for disaster relief and budgeting for repairs and maintenance.
iv) Ugu district municipality had disaster management challenges relating to inadequate scheduling, lack of intergovernmental coordination on procurement and inadequate monitoring and recording of information.
v) Zululand district municipality recorded no challenges in managing the disasters effect.
vi) OR Tambo district municipality had a slow response to disaster and was found with challenges relating to non-compliance with procurement processes, and the municipality do not submit information for auditing processes (PMG, 2023).

Overall, the African Centre for Disaster Studies (City Press, 2022), notes that government inadequacy in responding to the disaster manifested in ineffective planning, staff shortage, inadequate skills/capacities, lack of guidance, poor communication, and slow response which contributed immensely to the devastating effect of the compound disaster on the KZN province. For these reasons, Schwarz et al., (2023), assert that compound disaster as a growing research field poses the need for new methods, systems, and approaches. The context of KwaZulu-Natal compound disaster requires an alternative response approach than the norm to foster resilience, hence the question, "How can training and development and stakeholder engagement be integrated in a disaster management framework to foster resilience in a complex situation". Using KZN provincial level as a case study, Rosenberg & Yates (2007) noted the importance of identifying the underlying theories and themes in research design. The chapter focuses on the general concept of a disaster management framework (DMF) which consists of the phases of prevention/mitigation, preparedness, response, and recovery. Given the focus of the study, the underlying themes are compound disaster, HR training and development, stakeholder engagement and resilience.

Phase 2: Prepare, Collect, and Analyze

After setting an overview and the boundaries of the case, the second phase entails determining the research approach to adopt to facilitate the data collection. Furthermore, the strategy for data reduction and analysis is also discussed. The study is considered an instrumental case study design which allows for the use of a particular case (Crowe et al., 2011), to gain a broader understanding of the related issues or phenomena of interest (Rosenburg & Yates, 2007), and in this context, KZN provincial level. Compound disaster occurrence relating to health and natural causes is rare and unique and considered of extreme circumstance in the context of KZN. The province has one of the highest poverty levels (60.7%) among others, and more than 52% of its population live below the lower bound poverty line (Stats SA 2019; 2020). The impact of the disaster exacerbated on the living conditions in the province thus, making the case 'special' and the need for the study.

Data was collected by reviewing related existing literature. For compiling the paper sample, structured keywords or phrases were applied which are "Disaster *OR/AND* Compound Disaster', "Resilience *AND* Compound Disaster". These keywords/phrases were alternated and combined with "KwaZulu-Natal *AND/OR* South Africa". The researchers placed limits to the search, (i) a time dimension from 2020 to 2023 which is taken as a unit of analysis for this study (the outbreak of Covid-19 in South Africa was in 2020 and the most devastating weather extreme conditions in KZN was in 2022); (ii) interdisciplinary field (the study considered publications in business management, public administration, and geography); and (iii) open access. Applying these measures ensured that the search engines provided publications that address the topic under study. The initial search via the institution's reputable database, namely: SCOPUS, Web of Science (WOS), Wiley online Library Journals, and Google Scholar provided no result within the limits set out above. The researchers applied advanced search, and data reduced by reading through the title, abstract and the entire publication where necessary. The search engine produced one (1) publication from ScienceDirect, three (3) publications from SA *e*Publications database and six (6) from Google, making a total of ten (10) publications analyzed for the purpose of the study. Within this piece of research, these publications inductively attributed to the main focus and objective of the review covering compound disaster occurrences (covid-19 and extreme weather conditions) in the context of KZN province. This limited number of documents justifies the need for more research in this gray area. The two co-authors applied data cross-checking to corroborate and augment evidence from other sources as suggested by Kotlar & De Massis, (2013) and the retrieved papers were categorized under 'peer-reviewed journal articles', 'government reports/publications', and 'media (newspaper articles, professional bodies)' as shown in Table 2.

Table 2. Characteristics of the sampled studies

Source of the Reviewed Papers	Title	Type of Publication	Database
Mudefi, 2023	Disaster management 'deeds' in the context of April 2022 KwaZulu-Natal floods: A scoping review.	Journal article	ScienceDirect
Quintana et al., 2022	Learning from COVID and climatic events to build a resilient health system: Western Cape Department of Health.	Journal article	SA ePublications
Blecher, et al., 2021	The South African government's COVID-19 response: protecting lives and livelihoods.	Journal article	SA ePublications
Mncwabe, 2021	The Effectiveness of Government Institutions in Responding to the Epidemic of NCDs	Journal article	SA ePublications
Parliamentary Monitoring Group, 2023	AGSA Report on Flood Relief Funds	Government Report	Google
City Press, 2022	Half of KZN's flood deaths could have been avoided.	Media-Newspaper	Google
International Federation of Red Cross, 2022	South Africa: KwaZulu-Natal Floods - Emergency Appeal No. MDRZA012	Media-Newspaper	Google
IOL, 2022	Effects of the KwaZulu-Natal flood in numbers	Media-Newspaper	Google
ReliefWeb, 2022	South Africa: Floods and Landslides	Media-Newspaper	Google
Floodlist, 2022	South Africa – Evacuations After More Flooding in KwaZulu-Natal	Media-Newspaper	Google
Total			10

Source: Compiled by the authors, 2024

Phase 3: Developing a Case Description

A typical traditional disaster management framework (DMF) model consists of four cyclical phases namely, prevention/mitigation, preparedness, response, and recovery. These phases are grouped under pre-disaster (prevention and preparedness), and post-disaster (response and recovery) respectively as shown in shown in Figure V.

Each phase of the DMF is associated with multiple action plans and procedures with the length overlapping each other, depending on the severity of the disaster. The effect therein differs among organizations and the society at large. At each phase, all governments' levels and other stakeholders have a clear delineation of roles and responsibilities for an effective management of the disaster.

The pre-disaster stage of the framework focuses on measures taken to reduce possible human and property losses due to potential hazard in anticipation of disasters occurring. As such it helps to facilitate realistic predictions of what would be affected (Patel et al., 2023). The prevention or mitigation stage aims to eliminate or minimize future occurrence of disasters in the long term. The stage entails incorporating appropriate measures which determine its effectiveness and may include identifying risks and potential hazards, analysis, and appraisal, strengthening structures, spatial planning, resource allocation and creating public awareness (Norris et al., 2015; Khan et al., 2008). Similarly, disaster preparedness refers to actions taken in advance of a disaster happening, to ensure adequate response to its impacts, and relief and recovery from its consequences (Coppola, 2020). The stage deals with disaster

Figure 5. A typical disaster management framework
Source: Le Cozannet et al., (2020)

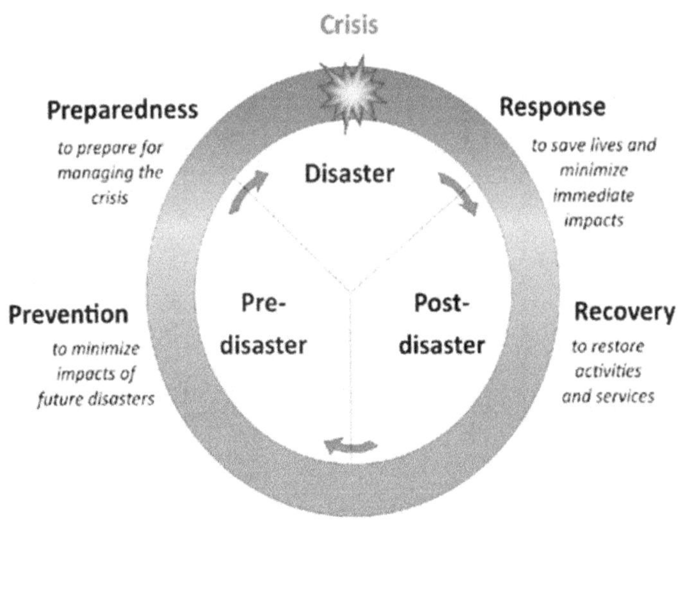

planning, disaster training/exercise, and warning systems, therefore considered vital in mitigating the adverse effects of disaster should it happen. The goal of these activities is to achieve a satisfactory level of readiness with the associated efforts according to Chan & Ho (2018), ranging from individual-level activities, household actions, community efforts, to implementation of government strategies.

Similarly, the post-disaster stage involves the response and the recovery related activities in managing disasters. The response stage in the DMF aims to provide immediate assistance to minimize the impact and support lives. The focus is to provide emergency relief to maintain life while search, rescue and other humanitarian needs continues. Recovery phase in the disaster management model focuses on returning to normal. It involves a lot of cyclical activities aimed at restoring the lives of the population, rebuilding the infrastructures, and monitoring the process. Activities at this level enhance prevention, increase preparedness, and minimize vulnerability. Ritchie & Jiang, (2019), added that effective disaster management strategies minimize potential hazards and accelerate the system's recovery process.

The Case of KwaZulu-Natal Province

The Disaster Management Act (DMA) No. 57 of 2002 is the primary legislation for disaster management in South Africa. The framework focuses clearly on prevention/mitigation, emergency preparedness, rapid effective response, and post disaster recovery. In managing disasters, the framework comprises four (4) key performance areas (KPAs) and three (3) enablers as shown in Figure VI.

Figure 6. Disaster management legislative framework
Source: CoGTA, 2023

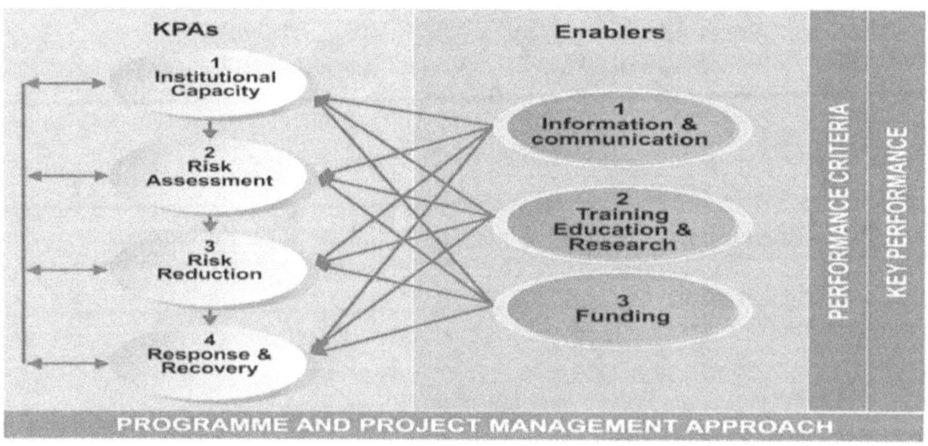

According to CoGTA (2023), the first key performance area which is 'institutional capacity' focuses on establishing arrangements for cooperation at various levels of government. Accordingly, the Chapter 4, Sections 28/29, and subsections 1/2 of the DMA as amended (CoGTA, 2016), requires that each province implements a disaster management framework and establishes Provincial Disaster Management Centres (PDMC) to coordinate its activities at provincial, metropolitan and municipal spheres of government in uniform with other organs of the state, provincial statutory functionaries, and non-governmental organisations. This entails involving all stakeholders in strengthening the capabilities across all spheres of government. Similarly, the second KPA addresses monitoring and assessment of the associated risks to minimize the likelihood of harm or hazards resulting from the disaster. The focus of the third KPA is on planning and implementation of approaches, programmes and projects set out to reduce disaster risks by integrating the principles of prevention and mitigation. The last KPA focuses on the priorities relating to effectiveness in disaster management. The framework as per Figure VI shows that each of the three enablers (information and communication, training education and research, and funding) feeds into each key performance area to ensure an effective management of a disaster (CoGTA, 2023). These three enablers addresses data or information sharing and communication, creating public awareness through training, education and research and the provision of the resources through funding to manage the entirety of the system.

The KZN provincial disaster management framework is aligned to the national DM framework requirements at pre-and post-phases in a disaster management cycle. A reflection on the compound disaster occurrence in KZN shows the extent the principles of the national disaster management framework were applied in practice. Focusing on the objective of the chapter, a review of related literature in line with the KPAs and the enablers is presented in Table 3.

In managing disaster for building resilience, the South African national DM model emphasizes the importance of conducting workshops (T&D) and the need for coordination among stakeholders (SE) across the levels of prevention or mitigation, preparedness, response, and recovery. In addition, the legislative national framework noted training, education and research, and information sharing and communication as vital enablers in measuring the KPAs in disaster management processes. In view of this, data emerging from Table 3 clearly shows the various challenges in the disaster management processes at the KZN provincial level. Data revealed challenges relating to T&D and SE emanating from the key performance areas. Data also indicated that the three enablers, information and communication, training, education and research, and funding are not sufficiently supporting the KPAs. Overall, the key resilience elements (in the context of this chapter), which are training and development (T&D) and stakeholder engagement (SE) emerged

Table 3. A review of compound disaster related publications at KZN provincial government level

Sources	Findings on DM Process at KZN	Interpretation	Elements of Resilience
PMG, 2023 IFRC, 2022 City Press, 2022 Floodlist, 2022 ReliefWeb, 2022 IOL, 2022	Slow in implementation of commitments, outdated/redundant disaster management plan, lack of disaster readiness, oversight, and lack of coordination among stakeholders. Poor workmanship (human resources), inadequate disaster relief funding, poor infrastructure resulting in ineffective or slow response. Misalignment of DM plan with the key deliverables, inadequate funding for disaster relief and budgeting for repairs and maintenance. Inadequate scheduling, lack of intergovernmental coordination and non-compliance on procurement process, inadequate monitoring and information capturing, non-submission of data for auditing processes, ineffective planning, staff shortage, inadequate skills/capacities, lack of guidance, and poor communication.	Challenges with KPAs 1, 2, 3 and 4.	T&D and SE
Mudefi, 2022	Gaps in disaster management processes, lack of disaster preparedness, mitigation, and over subscription to disaster response and recovery, capacity, and coordination among key stakeholders	Challenges with KPAs 1, 3 and 4	T&D and SE
Blecher et al., 2021	Economic effects on livelihoods, Budget/financial challenges	KPA 4	SE
Mncwabe, 2021	Insufficient human resources and funding challenges,	KPA 1 and 3	T&D
Quintana, 2022	Challenges relates to structural and relational aspects of governance; stewardship, leadership, and management; partnerships and intersectoral collaboration; system capacities; and learning oriented culture.	KPAs 1, 2, 3, and 4	T&D, SE

Source: Compiled by the authors, 2024

NB: Each KPA is supported by the three enablers'-information and communication, training education and research, and funding.

as concerned areas in DM processes at the KwaZulu-Natal provincial government level. This finding gave insight to the devastating effect of the compound disaster on the livelihoods in the case of the KZN provincial government level.

Existing literature is full of models, frameworks, and procedures in managing and coping with disasters, yet the frequency and the impact is still on the increase and becoming more complicated. Kruczkiewicza et al., (2021) argued that this is often due to siloed, fragmented, and inadequate approaches applied in managing disasters. Scholars have argued that the traditional disaster management models are rarely updated or evaluated and often burdened with protocols and responsibilities for satisfying top management (Sawalha, 2020; Alexander, 2015). This is also the case for the KZN DM process as revealed by the data. Though, the traditional disaster models are beneficial in certain cases and circumstances, Alrehaili, et al., (2022), argued that it is not generally applicable to all types of disasters, thus why Sawalha, (2020) asserts the need for contemporary robust approaches to better understand and manage disasters to build resilience. Hoa, et al., (2020) noted a growing demand for a context-specific model in compound disaster management. Shah et al., (2023), concurs and noted that when disaster strikes concerns for resilience call for the need for an integrated conventional disaster management model. A viewpoint that promotes an integration of elements that add significant value for performance and resilience across the DM lifecycle. In view of this, the chapter thus proposes a conceptual framework integrating training and development, and stakeholder engagement across a traditional disaster management model as shown in Figure VII.

Resilience in compound disaster is a complex array of interacting and interconnected elements that foster the system to bounce back. The role of human resource training and development and stakeholder engagement in building a resilient system cannot be overlooked. Training and development at pre-and post-disaster stage entails creating awareness campaigns, and training on safety and security to instill knowledge of the stakeholders about disaster. Related activities include education,

Figure 7. A unified framework for compound disaster management

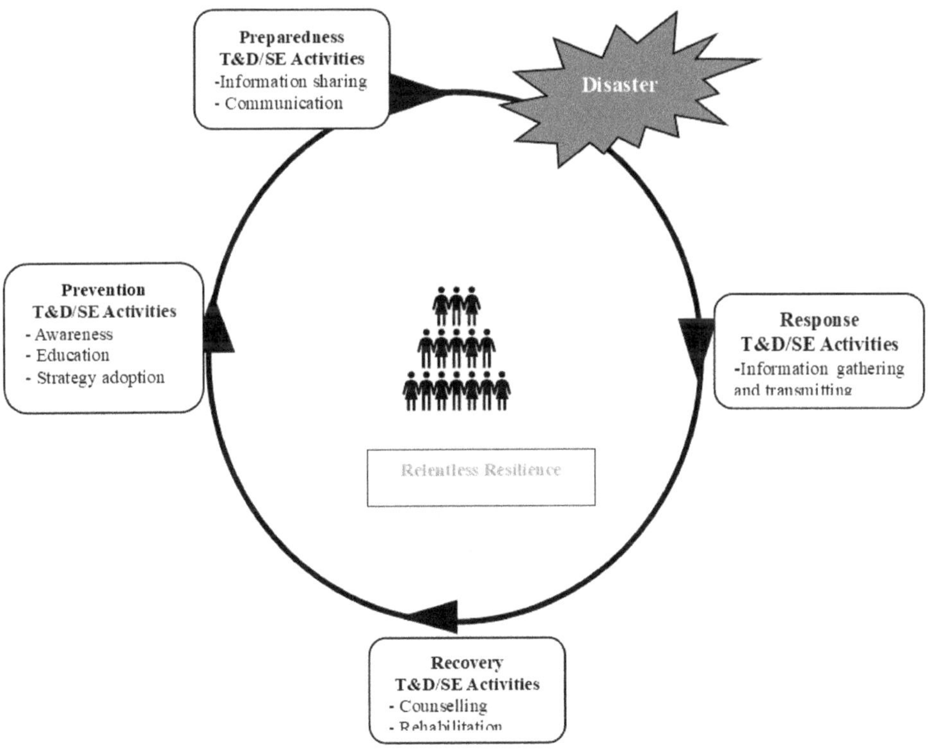

drills, tabletop exercises, and full-scale exercises of what stakeholders would do in an event of emergency. Hamid et al., (2021), asserted that the level of disaster literacy in formal education remains minimal. The authors demonstrated in their study how integration of disaster literacy into school subjects provided people with knowledge about the region's local culture and practices used in mitigating earthquakes and tsunamis. In addition, Sofyana et al., (2022), in their study demonstrated how systemic interprofessional education assists in responding to complex problems resulting from disaster. This is on the premise that trained people can protect themselves and others better, and effective training prevents or reduces the after effect of disasters (Torani et al., 2019). Sofyana et al., (2024), concurs and showed in their study that integrated training model effectively increases the capacity in managing disaster. On a general note, disaster training helps individuals to learn and create the abilities to carry out tasks adequately. Hence, in building resilience, the learning process within the DM framework helps to understand how actions can be taken prior, during and after disaster. Hence, disaster management illustrates a cyclic process used by all stakeholders in reducing the aftermath effect and enhancing the ability to bounce back. Creating appropriate solutions for challenges linking to compound disaster requires a better approach than the norm. Appropriate actions at each point in the cycle are important in reducing the level of vulnerability.

CONCLUSION

Disasters are inevitable events disrupting the functioning of the communities or society. The frequency of disaster occurrence and its severity have been on the increase and often, do not occur as isolated events but rather pile up. Disasters feature great uncertainties and when experienced concurrently, the disruptions are compounded and devastating in terms of economic, social, psychological, and otherwise thereby leaving no time for communities to bounce back. This chapter presents a review and critical assessment of compound disaster occurrence at the KwaZulu-Natal province of South Africa. Findings revealed that the devastating effect of the compound disaster in KZN is attributed to an array of weaknesses in the conventional disaster management process employed which is typically considered administrative in nature. These weaknesses among others majorly relate to lack of capacity, inadequate skills/knowledge, and lack of coordination among the stakeholders. Jayasekara et al., (2022), notes that lack of coordination among stakeholders creates a communication gap which hampers the disaster preparedness and response planning mechanisms as in the case of KZN. These findings highlight the importance of human resource training and development and stakeholder engagement in managing disaster. Disaster management is an on-going process with specific activities, principles and strategies deployed across its phases. In an event of occurrence, disaster management entails a process of strategic use of resources to reduce the anticipated harm across the stages of prevention, preparedness, response, and recovery. In times of disaster, government becomes the key agent for humanitarian relief and the restoration of public services. Hence, building resilience entails well-coordinated actions across levels of government and other stakeholders. In the case of a compound disaster, a more contemporary approach than the norm becomes optimal. In view of this, the chapter promotes a value-based approach which incorporates human resource training and development and continuous engagement of the stakeholders across the phases of prevention/mitigation, preparedness, responses, and recovery to foster the resilience ability of the system and the society at large. The findings of the study draw on the implications for policy and the enablement of a more integrated framework in managing compound disasters. For the future marked by compounding and cascading events, the study adds a valuable theoretical contribution to the growing body of knowledge that seeks an understanding of building not only resilience but relentless resilience in disruptive times.

REFERENCES

AghaKouchak, A., Chiang, F., Huning, L. S., Love, C. A., Mallakpour, I., & Mazdiyasni, O. (2020). Climate extremes and compound hazards in a warming world. Annual Review of Earth and Planetary Sciences, 48, 519–548. doi:10.1146/annurev-earth-071719-055228

Al-Maqashi, M. A. H., Rayalwar, A. P., & Al-Maqashi, S. A. H. (2021). The Role of Human Resources Management in Organisational Crisis Management. *International Journal of Scientific Research in Engineering and Management*, *06*(01), 1–6.

Alexander, D. (2015). *Disaster and emergency planning for preparedness, response and recovery*. Oxford Research Encyclopaedia of Natural Hazard Science. doi:10.1093/acrefore/9780199389407.013.12

Alrehaili, N., Almutairi, Y., Alghamdi, H., & Almuthaybiri, M. (2022). A Structural Review on Disaster Management Models and Their Contributions. *International Journal of Disaster Management*, *5*(2), 93–108. doi:10.24815/ijdm.v5i2.27087

Amarasinghe, A. G. (2020). Concept of Disaster is in Disarray: Need a Consensus among Experts. *International Journal of Geography and Geology*, *9*(2), 84–92. doi:10.18488/journal.10.2020.92.84.92

Bajracharya, B., & Hastings, P. (2020). Stakeholder engagement for disaster management in master-planned communities. *Australian Journal of Emergency Management, 35*(3), 41–47.

Blecher, M. S., Davén, J., Meyer-Rath, G., Silal, S. P., Makrelov, K., & Van Niekerk, M. (2021). The South African government's COVID-19 response: Protecting lives and livelihoods. *South African Health Review, 2021*(1), 1–17.

Chan, E. Y. Y., & Ho, J. Y. (2018). Urban community disaster and emergency health risk perceptions and preparedness. In R. Shaw, K. Shiwaku, & T. Izumi (Eds.), *Science and Technology in Disaster Risk Reduction in Asia: Potentials and Challenges* (pp. 95–110). Academic Press. doi:10.1016/B978-0-12-812711-7.00007-9

City Press. (2022, November 13). *Half of KZN's flood deaths could have been avoided.* City Press newspaper. https://www.news24.com/citypress/news/half-of-kzns-flood-deaths-could-have-been-avoided-20221112

Coppola, D. P. (2020). Preparedness. In Introduction to International Disaster Management (pp. 331-392). doi:10.1016/B978-0-12-817368-8.00005-1

Creswell, J. W. (2013). *Qualitative inquiry and research design: Choosing among five approaches*. Academic Press.

Crowe, S., Cresswell, K., Robertson, A., Huby, G., Avery, A., & Sheikh, A. (2011). The case study approach. *BMC Medical Research Methodology, 11*(1), 100. doi:10.1186/1471-2288-11-100 PMID:21707982

De Massis, A., & Kotlar, J. (2014). The case study method in family business research: Guidelines for qualitative scholarship. *Journal of Family Business Strategy, 5*(1), 15–29. doi:10.1016/j.jfbs.2014.01.007

Department: Cooperative Governance. (2016). Disaster Management Amendment Act No 16 of 2015. National Gazettes, No. 39943 of 22 April 2016.

Department: Cooperative Governance and Traditional Affairs (CoGTA). (2023). *Overview of Disaster Management System in South Africa*. International Dialogue series on Disaster Management Systems South Africa, 17 April 2023.

Department of Co-Operative Governance and Traditional Affairs (CoGTA). 2020. *Disaster Management Act: Declaration of a National State of Disaster: COVID-19 (coronavirus)*. Disaster Management Act, 2002. NO. 313, 15 March 2020. https://www.gov.za/sites/default/files/gcis_document/202003/43096gon313.pdf

Edvardsson, I. R., & Durst, S. (2021). Human Resource Management in Crisis Situations: A Systematic Literature Review. *Sustainability (Basel), 13*(22), 12406. doi:10.3390/su132212406

Floodlist. (2022, May 23). South Africa – Evacuations after more flooding in KwaZulu-Natal. https://floodlist.com/africa/south-africa-floodi-kwazulu-natal-may-2022

Gissing, A., Timms, M., Browning, S., Crompton, R., & McAneney, J. (2022). Compound natural disasters in Australia: A historical analysis. *Environmental Hazards, 21*(2), 159–173. doi:10.1080/17477891.2021.1932405

Hamid, N., Trihatmoko, E., Herlina, M., & Aroyandini, E. N. (2021). Developing a Model for Disaster Education to Improve Students' Disaster Mitigation Literacy. *Journal of Disaster Research, 16*(8), 1243–1256. doi:10.20965/jdr.2021.p1243

Hao, F., Xiao, Q., & Chon, K. (2020). COVID-19 and China's Hotel Industry: Impacts, a Disaster Management Framework, and Post-Pandemic Agenda. *International Journal of Hospitality Management, 90*, 102636. doi:10.1016/j.ijhm.2020.102636 PMID:32834356

Harrison, H., Birks, M., Franklin, R., & Mills, J. (2017). Case Study Research: Foundations and Methodological Orientations. *Forum Qualitative Social Research, 18*(1), 19. http://nbn-resolving.de/urn:nbn:de:0114-fqs1701195

International Federation of Red Cross (IFRC). (2022). *South Africa: KwaZulu-Natal Floods - Emergency Appeal No. MDRZA012 - Operational Strategy*. IFRC.

IOL. (2022, April 26). *Effects of the KwaZulu-Natal flood in numbers*. https://www.iol.co.za/news/south-africa/kwazulu-natal/effects-of-the-kwazulu-natal-flood-in-numbers-32c45aae-9611-4d8a-b137-25fc9843b8f0

Ishiwatari, M., Koike, T., Hiroki, K., Toda, T., & Katsube, T. (2020). Managing disasters amid Covid-19 pandemic: Approaches of response to flood disasters. *Progress in Disaster Science, 6*, 100096. doi:10.1016/j.pdisas.2020.100096 PMID:34171012

Jayasekara, R., Siriwardana, C., Dilanthi Amaratunga, D., & Richard Haigh, R. (2022). Evaluating the network of stakeholders in Multi-Hazard Early Warning Systems for multiple hazards amidst biological outbreaks: Sri Lanka as a case in point. *Progress in Disaster Science, 14*, 100228. doi:10.1016/j.pdisas.2022.100228 PMID:35402893

Jethwaney, J. (2023). Capacity building through stakeholder training in Media and Communication for effective disaster management. In A. Singh (Ed.), *International Handbook of Disaster Research* (pp. 1371–1388). Springer. doi:10.1007/978-981-19-8388-7_86

Khan, H., Vasilescu, L. G., & Khan, A. (2008). Disaster management cycle – a theoretical approach. *Management and Marketing Journal*, *6*(1), 43–50.

Kotlar, J., & De Massis, A. (2013). Goal setting in family firms: Goal diversity, social interactions, and collective commitment to family–centered goals. *Entrepreneurship Theory and Practice*, *37*(6), 1263–1288. doi:10.1111/etap.12065

Kruczkiewicz, A., Klopp, J., Fisher, J., Mason, S., McClain, S., Sheekh, N. M., Moss, R., Parks, R. M., & Braneon, C. (2021). Opinion: Compound risks and complex emergencies require new approaches to preparedness. *Proceedings of the National Academy of Sciences of the United States of America*, *118*(19), e2106795118. doi:10.1073/pnas.2106795118 PMID:33952695

Laurien, F., & Mechler, R. (2020, May 11). *Resilience thinking and action in the context of compound risks*. https://floodresilience.net/blogs/resilience-thinking-and-action-in-the-context-of-compound-risks/

Le Cozannet, G., Kervyn, M., Russo, S., Speranza, C. I., Ferrier, P., & Foumelis, M. (2020). Space-based earth observations for disaster risk management. *Surveys in Geophysics*, *41*(6), 1209–1235. doi:10.1007/s10712-020-09586-5

Liu, J. D., Dong, C. Q., An, S., & Mai, Q. (2022). Dynamic evolution analysis of the emergency collaboration network for compound disasters: A case study involving a public health emergency and an accident disaster during Covid-19. *Health Care*, *10*, 500. PMID:35326978

Mishra, A., Bruno, E., & Zilberman, D. (2021). Compound natural and human disasters: Managing drought and COVID-19 to sustain global agriculture and food sectors. *The Science of the Total Environment*, *754*, 142210. doi:10.1016/j.scitotenv.2020.142210 PMID:32920415

Mncwabe, M. (2021). The effectiveness of government institutions in responding to the epidemic of NCDs. *Economic and Social Rights in South Africa*, *22*(4), 11-16. https://hdl.handle.net/10520/ejc-esrrev-v22-n4-a3

Mudefi, E. (2023). Disaster management 'deeds' in the context of April 2022 KwaZulu-Natal floods: A scoping review. *International Journal of Disaster Risk Reduction*, *98*, 104122. doi:10.1016/j.ijdrr.2023.104122

Munawar, H. S., Khan, S. I., Anum, N., Qadir, Z., Kouzani, A. Z., & Parvez Mahmud, M. A. (2021). Post-flood risk management and resilience building practices: A case study. *Applied Sciences (Basel, Switzerland)*, *11*(11), 4823. doi:10.3390/app11114823

National Academies of Sciences, Engineering, and Medicine. (2022). *Resilience for compounding and cascading events.* Washington, DC: The National Academies Press. doi:10.17226/26659

National Department of Health. (2020). *Corona virus (COVID-19) outbreak.* https://www.health.gov.za/index.php/outbreaks/145-corona-virus-outbreak/465-corona-virus-outbreak

Norris, A. C., Martinez, S., Labaka, L., Madanian, S., Gonzalez, J. J., & Parry, D. (2015). Disaster E-Health: A new paradigm for collaborative healthcare in disasters. *Proceedings of the ISCRAM 2015 Conference*.

OnlineK. Z. N. (2023). http://www.kznonline.gov.za/images

Panneer, S., Kantamaneni, K., Pushparaj, R. R. B., Shekhar, S., Bhat, L., & Rice, L. (2021). Multistakeholder participation in disaster management-The case of the COVID-19 pandemic. *Health Care*, *9*(2), 203. doi:10.3390/healthcare9020203 PMID:33668669

Parliamentary Monitoring Group. (2023, June 13). *AGSA report on flood relief funds*. Ad Hoc Joint Committee on Flood Disaster Relief and Recovery. https://pmg.org.za/committee-meeting/37213/

Patel, R. K., Pamidimukkala, A., Kermanshachi, S., & Etminani-Ghasrodashti, R. (2023). Disaster preparedness and awareness among university students: A structural equation analysis. *International Journal of Environmental Research and Public Health*, *20*(5), 4447. doi:10.3390/ijerph20054447 PMID:36901454

Priya, A. (2021). Case study methodology of qualitative research: Key attributes and navigating the conundrums in its application. *Sociological Bulletin*, *70*(1), 94–110. doi:10.1177/0038022920970318

Quintana, A., Angeletti-du Toit, L., Gilson, L., Smith, W., & Vallabhjee, K. (2022). Learning from COVID and climatic events to build a resilient health system: Western Cape Department of Health. *South African Health Review*, *25*, 1–9. doi:10.10520/ejc-healthr-v2022-n1-a11

ReliefWeb. (2022, April 24). *South Africa: Floods and Landslides*. https://reliefweb.int/disaster/fl-2022-000201-zaf

Republic of South Africa. (2002). *Disaster Management Act No. 57 of 2002*. Government Printers.

Ritchie, B. W., & Jiang, Y. (2019). A review of research on tourism risk, crisis and disaster management: Launching the annals of tourism research curated collection on tourism risk, crisis and disaster management. *Annals of Tourism Research, 79*, 102812. doi:10.1016/j.annals.2019.102812

Rosenberg, J. P., & Yates, P. M. (2007). Schematic representation of case study research designs. *Journal of Advanced Nursing, 60*(4), 447–452. doi:10.1111/j.1365-2648.2007.04385.x PMID:17822427

Sawalha, I. H. (2020). A contemporary perspective on the disaster management cycle. *Foresight, 22*(4), 469–482. doi:10.1108/FS-11-2019-0097

Schwarz, J., Kascel, V., Azmat, M., & Kummer, S. (2023). Collation of best practices for preparedness: Lessons from disasters in Pakistan and Japan. *Journal of Humanitarian Logistics and Supply Chain Management, 13*(3), 311–330. doi:10.1108/JHLSCM-07-2022-0086

Shah, A. A., Ullah, A., Khan, N. A., Shah, M. H., Ahmed, R., Hassan, S. T., Maur, T., & Xu, C. (2023). Identifying obstacles encountered at different stages of the disaster management cycle (DMC) and its implications for rural flooding in Pakistan. *Frontiers in Environmental Science, 11*, 1088126. doi:10.3389/fenvs.2023.1088126

Siddiqui, S. T. (2017). Can employee training and development increase organizational resilience against economic crises? *International Journal of Managerial Studies and Research, 5*(4), 78–82. doi:10.20431/2349-0349.0504011

Snyder, H. (2019). Literature review as a research methodology: An overview and guidelines. *Journal of Business Research, 104*, 333–339. doi:10.1016/j.jbusres.2019.07.039

Sofyana, H., Ibrahim, K., Afriandi, I., & Herawati, E. (2024). The implementation of disaster preparedness training integration model based on Public Health Nursing (ILATGANA-PHN) to increase community capacity in natural disaster-prone areas. *BMC Nursing, 23*(105), 105. Advance online publication. doi:10.1186/s12912-024-01755-w PMID:38326888

Sofyana, H., Ibrahim, K., Afriandi, I., Herawati, E., & Wahito Nugroho, H. S. (2022). The need for a preparedness training model on disaster risk reduction based on culturally sensitive Public Health Nursing (PHN). *International Journal of Environmental Research and Public Health, 19*(24), 16467. doi:10.3390/ijerph192416467 PMID:36554351

South African Government. (2022, May 23). *KwaZulu-Natal provides an update on latest storms and floods.* Statement by the KwaZulu-Natal Provincial Government and eThekwini Municipality following the recent storms in KwaZulu-Natal held on 22 May 2022. https://www.gov.za/speeches/kwazulu-natal-provides-update-latest-storms-and-floods-23-may-2022-0000?gclid=Cj0KCQjwkOqZBhDNARIsAACsbfIwmoOAUZZrrKT_P-G74NL2FsUsH0qU2qyemn7rQgiOY0UkrjGP2rUaAomwEALw_wcB

Statistics South Africa. (2019). *Five facts about Poverty in South Africa.* https://www.statssa.gov.za/?p=12075

Statistics South Africa. (2020). *Poverty in Perspective Statistics South Africa* https://www.parliament.gov.za

Statistics South Africa. (2022). *Mid-year population estimates 2022.* Statistical Release P0302. Pretoria. https://www.statssa.gov.za/publications/P0302/P03022022.pdf

Taherdoost, H. (2021). Data collection methods and tools for research; a step-by-step guide to choose data collection technique for academic and business research projects. *International Journal of Academic Research in Management, 10*(1), 10–38.

Teegavarapu, S., & Summer, J. D. (2008). Case study method for design research. *Proceedings of IDETC/DTM ASME 2008 International Design Engineering Technical Conferences & Computers and Information in Engineering Conference.*

Thompson, D. D. P. (2022). Compounding challenges for disaster resilience in small island developing states. *Disaster Prevention and Resilience,* (1), 4. doi:10.20517/dpr.2021.04

Torani, S., Majd, P. M., Maroufi, S. S., Dowlati, M., & Sheikhi, R. A. (2019). The importance of education on disasters and emergencies: A review article. *Journal of Education and Health Promotion, 8*(1), 85. doi:10.4103/jehp.jehp_262_18 PMID:31143802

Yin, R. K. (2003). *Case study research: Design and methods* (3rd ed., Vol. 5). Sage.

Zaman, M. O., & Raihan, M. M. H. (2023). Community resilience to natural disasters: A systemic review of contemporary methods and theories. *Natural Hazards Review*, *3*(3), 583–594. doi:10.1016/j.nhres.2023.05.003

Zscheischler, J., Martius, O., Westra, S., Bevacqua, E., Raymond, C., Horton, R. M., van den Hurk, B., AghaKouchak, A., Jézéquel, A., Mahecha, M. D., Maraun, D., Ramos, A. M., Ridder, N. N., Thiery, W., & Vignotto, E. (2020). A typology of compound weather and climate events. *Nature Reviews. Earth & Environment*, *1*(7), 333–347. doi:10.1038/s43017-020-0060-z

Compilation of References

21st Conference of the Parties to the United Nations Framework Convention on Climate Change. (2015). *Understanding Human Rights and Climate Change*. www.ohchr.org/Documents/Issues/ClimateChange/COP21.pdf

Aboualola, M., Abualsaud, K., Khattab, T., Zorba, N., & Hassanein, H. S. (2023). *Edge Technologies for Disaster Management: A Survey of Social Media and Artificial Intelligence Integration*. IEEE. doi:10.1109/ACCESS.2023.3293035

Abraham, M. (1988). The Bhopal Tragedy—A Community Action Perspective. *Community Development Journal: An International Forum*, *23*(4), 229–234. https://www.jstor.org/stable/44258578. doi:10.1093/cdj/23.4.229

Achillopoulou, D. V., Mitoulis, S. A., Argyroudis, S. A., & Wang, Y. (2020). Monitoring of transport infrastructure exposed to multiple hazards: A roadmap for building resilience. *The Science of the Total Environment*, *746*, 141001. doi:10.1016/j.scitotenv.2020.141001 PMID:32795756

Adams, R. M., Evans, C., Wolkin, A., Thomas, T., & Peek, L. (2022). Social vulnerability and disasters: Development and evaluation of a CONVERGE training module for researchers and practitioners. Disaster Prevention and Management. *International Journal (Toronto, Ont.)*, *31*(6), 13–29.

Adams, R. M., Prelip, M. L., Glik, D. C., Donatello, I., & Eisenman, D. P. (2018). Facilitating partnerships with community-and faith-based organizations for disaster preparedness and response: Results of a national survey of public health departments. *Disaster Medicine and Public Health Preparedness*, *12*(1), 57–66. doi:10.1017/dmp.2017.36 PMID:28735594

Adger, W. N. (2000). Social and ecological resilience: Are they related? *Progress in Human Geography*, *24*(3), 347–364. doi:10.1191/030913200701540465

Afkhamiaghda, M., & Elwakil, E. (2023). Challenges review of decision making in post-disaster construction. *International Journal of Construction Management*, *23*(14), 2409–2418. doi:10.1080/15623599.2022.2061751

AghaKouchak, A., Chiang, F., Huning, L. S., Love, C. A., Mallakpour, I., & Mazdiyasni, O. (2020). Climate extremes and compound hazards in a warming world. Annual Review of Earth and Planetary Sciences, 48, 519–548. doi:10.1146/annurev-earth-071719-055228

Aghda, S. M. F., Bagheri, V., & Razifard, M. (2018). Landslide susceptibility mapping using fuzzy logic system and its influences on mainlines in Lashgarak Region, Tehran, Iran. *Geotechnical and Geological Engineering*, *36*, 915–937. doi:10.1007/s10706-017-0365-y

Agrawal, A., & Gibson, C. C. (1999). Enchantment and disenchantment: The role of community in natural resource conservation. *World Development*, *27*(4), 629–649. doi:10.1016/S0305-750X(98)00161-2

Ahmed, A. A. A., Mahalakshmi, A., ArulRajan, K., Alanya-Beltran, J., & Naved, M. (2023). Integrated artificial intelligence effect on crisis management and lean production: Structural equation modelling frame work. *International Journal of System Assurance Engineering and Management*, *14*(1), 220–227. doi:10.1007/s13198-022-01679-1

Ahmed, S., & Dey, K. (2020). Resilience modeling concepts in transportation systems: A comprehensive review based on mode, and modeling techniques. *Journal of Infrastructure Preservation and Resilience*, *1*(1), 1–20. doi:10.1186/s43065-020-00008-9

Ainuddin, S., & Routray, J. K. (2012). Earthquake hazards and community resilience in Baluchistan. *Natural Hazards*, *63*(2), 909–937. doi:10.1007/s11069-012-0201-x

Ajayi, V. O. (2023). A Review on Primary Sources of Data and Secondary Sources of Data. *European Journal of Education and Pedagogy*, *2*(3), 1–7.

Akaike, H. (1973). Information theory and an extension of the maximum likelihood principle. In B.N. Petrov, and P. Ciaki (Eds.), *2nd International Symposium on Information Theory* (pp. 267– 81). Budapest: Akademiai Kiad6.

Akhai, S., Mala, S., & Jerin, A. A. (2020). Apprehending air conditioning systems in context to COVID-19 and human health: A brief communication. *International Journal of Healthcare Education & Medical Informatics, 7*(1&2), 28-30.

Akhai, S., Singh, V. P., & John, S. (2016). Investigating Indoor Air Quality for the Split-Type Air Conditioners in an Office Environment and Its Effect on Human Performance. *Journal of Mechanical Civil Engineering*, *13*(6), 113–118.

Akhai, S., Thareja, P., & Singh, V. P. (2017). Assessment of Indoor Environment Health Sustenance in Air Conditioned Class Rooms. *Advanced Research in Civil and Environmental Engineering*, *4*(1&2), 1–9. doi:10.24321/2393.8307.201701

Albattat, A. R. S., & Som, A. P. M. (2013). Employee Dissatisfaction and Turnover Crises in the Malaysian Hospitality Industry. *International Journal of Business and Management*, *8*(5), 62–71.

AlBattat, A. R. S., & Som, A. P. M. (2014). Emergency planning and disaster recovery in Malaysian hospitality industry. *Procedia: Social and Behavioral Sciences*, *144*, 45–53. doi:10.1016/j.sbspro.2014.07.272

Albattat, A. R., & Mat Som, A. P. (2019). *Disaster planning and preparedness in the hotel industry*. Emerald. doi:10.1108/9781787699373

Albattat, A., & Som, A. P. M. (2015). Emergency Preparedness of the Hotel Industry: The Case of Jordan. *Advances in Environmental Biology*, *9*(3), 19–22.

Aldakkhelallah, A., & Simic, M. (2021). Autonomous vehicles in intelligent transportation systems. In *Human Centred Intelligent Systems: Proceedings of KES-HCIS 2021 Conference* (pp. 185-198). Springer Singapore.

Aldrich, D. P., & Meyer, M. A. (2015). Social capital and community resilience. *The American Behavioral Scientist*, *59*(2), 254–269. doi:10.1177/0002764214550299

Aldunce, P., Beilin, R., Howden, M., & Handmer, J. (2015). Resilience for disaster risk management in a changing climate: Practitioners' frames and practices. *Global Environmental Change*, *30*, 1–11. doi:10.1016/j.gloenvcha.2014.10.010

Alexander, D. (2015). *Disaster and emergency planning for preparedness, response and recovery*. Oxford Research Encyclopaedia of Natural Hazard Science. doi:10.1093/acrefore/9780199389407.013.12

Alexander, D. E. (2013). Resilience and disaster risk reduction: An etymological journey. *Natural Hazards and Earth System Sciences*, *1*, 1257–1284.

Al-Fadly, A. (2020). Impact of COVID-19 on SMEs and employment. *Entrepreneurship and Sustainability Issues*, *8*(2), 629–648. doi:10.9770/jesi.2020.8.2(38)

Allam, A. S., Bassioni, H. A., Kamel, W., & Ayoub, M. (2020). Estimating the standardized regression coefficients of design variables in daylighting and energy performance of buildings in the face of multicollinearity. *Solar Energy*, *211*, 1184–1193. doi:10.1016/j.solener.2020.10.043

Allison, P. (2012). Logistic regression using SAS: Theory and application (2nd ed.). SAS Institute.

Al-Maqashi, M. A. H., Rayalwar, A. P., & Al-Maqashi, S. A. H. (2021). The Role of Human Resources Management in Organisational Crisis Management. *International Journal of Scientific Research in Engineering and Management*, *06*(01), 1–6.

Almeida, F., & Santos, J. D. (2020). The effects of COVID-19 on job security and unemployment in Portugal. *The International Journal of Sociology and Social Policy*, *40*(9), 995–1003. doi:10.1108/IJSSP-07-2020-0291

Alrehaili, N. R., Almutairi, Y. N., Alghamdi, H. M., & Almuthaybiri, M. S. (2022). A Structural review on disaster management models and their contributions. *International Journal of Disaster Management*, *5*(2), 93–108. doi:10.24815/ijdm.v5i2.27087

Althuwaynee, O. F., Pradhan, B., Park, H.-J., & Lee, J. H. (2014). A novel ensemble decision tree-based CHi-squared Automatic Interaction Detection (CHAID) and multivariate logistic regression models in landslide susceptibility mapping. *Landslides*, *11*(6), 1063–1078. doi:10.1007/s10346-014-0466-0

Amani, D., & Ismail, I. J. (2022). Investigating the predicting role of COVID-19 preventive measures on building brand legitimacy in the hospitality industry in Tanzania: Mediation effect of perceived brand ethicality. *Future Business Journal, 13*(1), 1–14. doi:10.1186/s43093-022-00128-6

Amarasinghe, A. G. (2020). Concept of Disaster is in Disarray: Need a Consensus among Experts. *International Journal of Geography and Geology, 9*(2), 84–92. doi:10.18488/journal.10.2020.92.84.92

Amaratunga, R. D. G., Siriwardena, M. L., Malalgoda, C. I., Pathirage, C. P., & Thayaparan, M. (2011). *Lifelong learning needs for disaster management education in the built environment.* https://salford-repository.worktribe.com/output/1449597/lifelong-learning-needs-for-disaster-management-education-in-the-built-environment

Andrew, M. (2018). Revisiting community-based disaster risk management. In *Reducing Disaster Risks* (pp. 42–52). Routledge.

Ann, S., & Blum, S. C. (2020). Motivating senior employees in the hospitality industry. *International Journal of Contemporary Hospitality Management, 32*(1), 324–346. doi:10.1108/IJCHM-08-2018-0685

Anton, S. G., & Nucu, A. E. A. (2020). The effect of financial development on renewable energy consumption. A panel data approach. *Renewable Energy, 147*, 330–338. doi:10.1016/j.renene.2019.09.005

Anyanwu, J. C., & Salami, A. O. (2021). The impact of COVID-19 on African economies: An introduction. *African Development Review, 33*(S1, Suppl 1), S1. doi:10.1111/1467-8268.12531 PMID:34149237

Arabameri, A., Pradhan, B., & Rezaei, K. (2019). Gully erosion zonation mapping using integrated geographically weighted regression with certainty factor and random forest models in GIS. *Journal of Environmental Management, 232*(September), 928–942. doi:10.1016/j.jenvman.2018.11.110

Arif, S. M., & Gupta, S. (2022). Assessment of A Higher Water Tank Taking into Account the Capacity and Foundation Wall in Earthquake Regions. In *2022 11th International Conference on System Modeling & Advancement in Research Trends (SMART)* (pp. 708-712). IEEE. 10.1109/SMART55829.2022.10047220

Arifin, S., Wicaksono, S. S., Sumarto, S., Martitah, M., & Sulistianingsih, D. (2021). Disaster resilient village-based approach to disaster risk reduction policy in Indonesia: A regulatory analysis. *Jamba (Potchefstroom, South Africa), 13*(1), a1021. doi:10.4102/jamba.v13i1.1021 PMID:34230845

Arimoto, M. (2023). *Preliminary Cross-Cultural Research on Resilience Based on the Sendai Framework for Disaster Risk Reduction: Toward extending questionnaire survey by combining ethnography after flood\ndamage in a local town of Japan* (Doctoral dissertation, Tohoku University).

Arndt, C., Davies, R., Gabriel, S., Harris, L., Makrelov, K., Modise, B., . . . Anderson, L. (2020). Impact of Covid-19 on the South African economy. *Southern Africa-Towards Inclusive Economic Development Working Paper, 111.*

Arora, P. (2013). *Leveraging Human Resource for Effective Disaster Management.* Academic Press.

Asghar, S., Alahakoon, D., & Churilov, L. (2006). A comprehensive conceptual model for disaster management. *Journal of Humanitarian Assistance, 1360*(222), 1–15.

Atalay, I. (1982). Türkiye Jeomorfolojisine Giriş (Genişletilmiş 2. Baskı), *Ege Üniversitesi Edebiyat Fakültesi Yayınları, 9*(1982)

Atkinson, D. (2021). Preparing for the worst: South African municipalities' readiness to manage disasters related to potential shale gas mining. *International Journal of Disaster Risk Reduction, 65*, 102537. doi:10.1016/j.ijdrr.2021.102537

Aydınoğlu, A. Ç., & Altürk, G. (2021). Heyelan Duyarlılık Haritalarının İstatistik ve Makine Öğrenmesi Teknikleri Kullanılarak Üretilmesi: Taşlıdere Havzası Örneği (Rize). *Coğrafya Dergisi. The Journal of Geography, 0*(43), 159–176. doi:10.26650/JGEOG2021-814561

Bajracharya, B., & Hastings, P. (2020). Stakeholder engagement for disaster management in master-planned communities. *Australian Journal of Emergency Management, 35*(3), 41–47.

Baltazar, C. S., & Rossetto, E. V. (2020). Mozambique Field Epidemiology and Laboratory Training Program as responders workforce during Idai and Kenneth cyclones: A commentary. *The Pan African Medical Journal, 36*(1). Advance online publication. doi:10.11604/pamj.2020.36.264.21087 PMID:33088393

Banerji, P., & Singh, N. (2013). Comparative analysis of disaster management between Japan & India. *IOSR Journal of Business and Management, 13*(6), 62–74. doi:10.9790/487X-1366274

Banica, A, Kourtit, K. & Nijkamp, P. (2020) Natural disaster as a development opportunity: a spatial economic resilience interpretation. *Review of Regional Research, 40*(3).

Bannour, W., Maalel, A., & Ben Ghezala, H. H. (2023). Emergency management case-based reasoning systems: A survey of recent developments. *Journal of Experimental & Theoretical Artificial Intelligence, 35*(1), 35–58. doi:10.1080/0952813X.2021.1952654

Barka, A. (1984). Erzincan havzasının bazı nanotektonik özellikleri, 1. Ulusal Deprem Sempozyumu, 223-247.

Barnes, M. L., Wang, P., Cinner, J. E., Graham, N. A., Guerrero, A. M., Jasny, L., Lau, J., Stutcliffe, S. R., & Zamborain-Mason, J. (2020). *An Invitation to Reflexive sociology.* University of Chicago Press.

Baum, T., & Hai, N. T. T. (2020). Hospitality, tourism, human rights and the impact of COVID-19. *International Journal of Contemporary Hospitality Management, 32*(7), 2397–2407. doi:10.1108/IJCHM-03-2020-0242

Baum, T., Kralj, A., Robinson, R. N., & Solnet, D. J. (2016). Tourism workforce research: A review, taxonomy and agenda. *Annals of Tourism Research*, *60*, 1–22. doi:10.1016/j.annals.2016.04.003

Baum, T., Mooney, S. K., Robinson, R. N., & Solnet, D. (2020). COVID-19's impact on the hospitality workforce–new crisis or amplification of the norm? *International Journal of Contemporary Hospitality Management*, *32*(9), 2813–2829. doi:10.1108/IJCHM-04-2020-0314

Baytiyeh, H., & Naja, M. K. (2014). Revolutionizing engineering education in the Middle East region to promote earthquake-disaster mitigation. *European Journal of Engineering Education*, *39*(5), 573–583. doi:10.1080/03043797.2014.895705

Bean, H. (2021). Researching Integrated Disaster Risk Management from an Interpretive/Critical Perspective. *Journal of Integrated Disaster Risk Management*, *11*(2), 31150. doi:10.5595/001c.31150

Bega, S. (2022, March). Southern Africa's environment migrants are the forgotten refugees. Mail and Guardian. *Environment*, *19*. https://mg.co.za/environment/2022-03-19-southernafricas-environment-migrants-are-the-forgotten-refugees/

Beg, K., Bhadra, B. K., Sharma, J. R., Punia, M. P., & Chaurey, R. (2013). Seismic Data Analysis for Disaster Management in Jodhpur District of Rajasthan Using GIS, GPS and Remote Sensing Techniques. *Jour. Ind. Geol. Cong*, *5*, 1.

Bell, A., Fairbrother, M., & Jones, K. (2019). Fixed and random effects models: Making an informed choice. *Quality & Quantity*, *53*(2), 1051–1074. doi:10.1007/s11135-018-0802-x

Belle, J. A. (2016). *The Integration of Disaster Risk Reduction and Climate Change Adaptation Strategies into Wetlands Management in the Eastern Free State, South Africa*. Unpublished Doctoral Dissertation. Bloemfontein, South Africa. University of the Free State.

Below, R., Wirtz, A., & Guha-Sapir, D. (2009). *Disaster category classification and peril terminology for operational purposes*. No. UCL-Université Catholique de Louvain.

Bennett, D. (2020). Five years later: Assessing the implementation of the four priorities of the Sendai framework for inclusion of people with disabilities. *International Journal of Disaster Risk Science*, *11*(2), 155–166. doi:10.1007/s13753-020-00267-w

Berger, A. A. (2000). The meanings of culture: Culture: Its many meanings. *M/C Journal*, *3*(2). Available at: https://journal.mediaculture.org.au/mcjournal/article/view/1833?__hstc=87814215.82af9c9a98fa600b1bb630f9cde2cb5f.1540944000065.1540944000066.1540944000067.1&__hssc=87814215.1.1540944000068&__hsfp=2025384311

Bhardwaj, J. R., Chawla, R., & Sharma, R. K. (2007). *Chemical disaster management: Current status and perspectives*. https://nopr.niscpr.res.in/handle/123456789/1140

Birkmann, J., Sorg, L., & Welle, T. (2017). Disaster vulnerability. The Palgrave handbook of unconventional risk transfer, 329-356.

Birkmann, J., Cardona, O. D., Carreño, M. L., Barbat, A. H., Pelling, M., Schneiderbauer, S., & Welle, T. (2013). Framing vulnerability, risk and societal responses: The MOVE framework. *Natural Hazards*, *67*(2), 193–211. doi:10.1007/s11069-013-0558-5

Blaikie, P., Cannon, T., Davis, I., & Wisner, B. (2003). The challenge of disasters and our approach. *Environmentalism: Critical Concepts*, *4*, 233.

Blaikie, P., Cannon, T., Davis, I., & Wisner, B. (2014). *At risk: natural hazards, people's vulnerability and disasters*. Routledge. doi:10.4324/9780203714775

Blanchard, B. W., & Branch, R. (2003). *Outlines of competencies to develop successful 21st century hazard or disaster or emergency or hazard risk managers*. Presentation at the *Natural Hazards Workshop*, Boulder, Colorado.

Blecher, M. S., Davén, J., Meyer-Rath, G., Silal, S. P., Makrelov, K., & Van Niekerk, M. (2021). The South African government's COVID-19 response: Protecting lives and livelihoods. *South African Health Review*, *2021*(1), 1–17.

Bonfanti, A., Vigolo, V., & Yfantidou, G. (2021, April 1). The impact of the Covid-19 pandemic on customer experience design: The hotel managers' perspective. *International Journal of Hospitality Management*, *94*, 10287. doi:10.1016/j.ijhm.2021.102871 PMID:34866744

Booyens, I., Rogerson, C. M., Rogerson, J. M., & Baum, T. (2022). Covid-19 crisis management responses of small tourism firms in South Africa. *Tourism Review International*, *26*(1), 121–137. doi:10.3727/154427221X16245632411872

Borowska-Stefańska, M., Goniewicz, K., Grama, V., Hornak, M., Masierek, E., Morar, C., ... Wiśniewski, S. (2023). Evaluating approaches to wartime mass evacuation management in eastern NATO territories: a literature review. *Safety &Defense, 1.*

Botha, D, Van Niekerk, D, Wentink, G, Coetzee, C, Forbes, K, Maartens, Y, Annandale, E, Tshona, T, & Raju, E. (2011). Disaster risk management: Status assessment at municipalities in South Africa. *The African Centre for Disaster Studies, North-West University.*

BreuerM.deHaanE. (2023). Using and interpreting fixed effects models. *Available at* SSRN. doi:10.2139/ssrn.4539828

Brown, K. (2014). Global environmental change I: A social turn for resilience? *Progress in Human Geography*, *38*(1), 107–117. doi:10.1177/0309132513498837

Brunsdon, C., Fotheringham, A. S., & Charlton, M. E. (1996). Geographically Weighted Regression: A Method for Exploring Spatial Nonstationarity. *Geographical Analysis*, *28*(4), 281–298. doi:10.1111/j.1538-4632.1996.tb00936.x

Bull, M. T., Sanhueza, F., Oyarzo, C., & Cea, P. (2017). *Towards A Culture Of Prevention: Including Disaster Management In Engineering Undergraduate Programs*. https://www.wcee.nicee.org/wcee/article/16WCEE/WCEE2017-2893.pdf

Busayo, E. T., Kalumba, A. M., Afuye, G. A., Ekundayo, O. Y., & Orimoloye, I. R. (2020). Assessment of the Sendai framework for disaster risk reduction studies since 2015. *International Journal of Disaster Risk Reduction, 50*, 101906. doi:10.1016/j.ijdrr.2020.101906

Butler, L., Yigitcanlar, T., & Paz, A. (2020). Smart urban mobility innovations: A comprehensive review and evaluation. *IEEE Access : Practical Innovations, Open Solutions, 8*, 196034–196049. doi:10.1109/ACCESS.2020.3034596

Cadwell, P. (2021). Translation and interpreting in disaster situations. The Routledge handbook of translation and health, 253-268.

Callaway, B., & Karami, S. (2023). Treatment effects in interactive fixed effects models with a small number of time periods. *Journal of Econometrics, 233*(1), 184–208. doi:10.1016/j.jeconom.2022.02.001

Can, T., Duman, T. Y., Nefeslioglu, H. A., Durmaz, S., Gokceoglu, C., & Sonmez, H. (2005). Earthflows in a small catchment from Eastern Black Sea Region Turkey: conditional Environmental factors and susceptibility assessments. In *Proceedings of the International Symposium on Latest Natural Disasters: New Challenges for Engineering Geology, Geotechnics and Civil Protection* (pp. 82-83). Academic Press.

Cao, L. (2023). AI and data science for smart emergency, crisis and disaster resilience. *International Journal of Data Science and Analytics, 15*(3), 231–246. doi:10.1007/s41060-023-00393-w PMID:37035277

Cao, Y., Xu, C., Aziz, N. M., & Kamaruzzaman, S. N. (2023). BIM–GIS Integrated Utilization in Urban Disaster Management: The Contributions, Challenges, and Future Directions. *Remote Sensing (Basel), 15*(5), 1331. doi:10.3390/rs15051331

Celse, J., & Kensen, M. (2023). Can positional concerns be a threat to disaster management? Assessing the prevalence of positional concerns among socially vulnerable populations in Trinidad & Tobago. *Environmental Hazards, 22*(2), 152–176. doi:10.1080/17477891.2022.2096556

Centre for Research on the Epidemiology of Disasters. (2023). *Disaster Year in Review: Cred Crunch.* https://reliefweb.int/report/world/cred-crunch-newsletter-issue-no-70-april-2023-disasters-year-review-2022

Chabana, D. (2020). *Family influence on youth disaster risk and vulnerability perceptions in a South African context* (Doctoral dissertation, North-West University (South Africa)).

Chalkias, C., Kalogirou, S., & Ferentinou, M. (2014). Landslide susceptibility, Peloponnese Peninsula in South Greece. *Journal of Maps, 10*(2), 211–222. doi:10.1080/17445647.2014.884022

Chalkias, C., Polykretis, C., Karymbalis, E., Soldati, M., Ghinoi, A., & Ferentinou, M. (2020). Exploring spatial non-stationarity in the relationships between landslide susceptibility and conditioning factors: A local modeling approach using geographically weighted regression. *Bulletin of Engineering Geology and the Environment, 79*(6), 2799–2814. doi:10.1007/s10064-020-01733-x

Chan, I. C. C., Ma, J., Ye, H., & Law, R. (2021, January 1). A Comparison of Hotel Guest Experience Before and During Pandemic: Evidence from Online Reviews. *Springer eBooks*, 549-556.

Chan, E. Y. Y., & Ho, J. Y. (2018). Urban community disaster and emergency health risk perceptions and preparedness. In R. Shaw, K. Shiwaku, & T. Izumi (Eds.), *Science and Technology in Disaster Risk Reduction in Asia: Potentials and Challenges* (pp. 95–110). Academic Press. doi:10.1016/B978-0-12-812711-7.00007-9

Chatiza, K. (2020). *Cyclone Idai in Zimbabwe: An analysis of policy implications for post-disaster institutional development to strengthen disaster risk management*. https://doi.org// doi:10.21201

Cheer, J. M. (2020). Human flourishing, tourism transformation and COVID-19: A conceptual touchstone. *Tourism Geographies*, *22*(3), 1–11. doi:10.1080/14616688.2020.1765016

Chen, W., Zhao, X., Shahabi, H., Shirzadi, A., Khosravi, K., Chai, H., Zhang, S., Zhang, L., Ma, J., Chen, Y., Wang, X., Bin Ahmad, B., & Li, R. (2019). Spatial prediction of landslide susceptibility by combining evidential belief function, logistic regression and logistic model tree. *Geocarto International*, *34*(11), 1177–1201. doi:10.1080/10106049.2019.1588393

Chipangura, P., van Niekerk, D., & Van der Waldt, G. (2017). Disaster Risk Policy Problem Framing: Insights from Societal Perceptions in Zimbabwe. *International Journal of Disaster Risk Reduction*, *22*, 317–324. doi:10.1016/j.ijdrr.2017.02.012

Chowdhuri, I., Pal, S. C., Arabameri, A., Ngo, P. T. T., Chakrabortty, R., Malik, S., Das, B., & Roy, P. (2020). Ensemble approach to develop landslide susceptibility map in landslide dominated Sikkim Himalayan region, India. *Environmental Earth Sciences*, *79*(20), 1–28. doi:10.1007/s12665-020-09227-5

City Press. (2022, November 13). *Half of KZN's flood deaths could have been avoided*. City Press newspaper. https://www.news24.com/citypress/news/half-of-kzns-flood-deaths-could-have-been-avoided-20221112

Clark, D. B. (1973). The Concept of Community: A Re-Examination. *The Sociological Review*, *21*(3), 397–416. doi:10.1111/j.1467-954X.1973.tb00230.x

Coaffee, J., & Clarke, J. (2015). On securing the generational challenge of urban resilience. *The Town Planning Review*, *86*(3), 249–255. doi:10.3828/tpr.2015.16

Coaffee, J., & Lee, P. (2016). *Urban resilience: planning for risk, crisis and uncertainty*. Palgrave. doi:10.1007/978-1-137-28884-4

CoGTA – KZN. (2022). *April 2022 inclement weather conditions report. Report No: 1.22 of May 2022*. Provincial Disaster Management Centre.

CoGTA. (2023). *Municipalities received their first tranche of the Local Government Equitable Share*. Available at: https://www.cogta.gov.za/index.php/2023/07/13/municipalities-received-their-first-tranche-of-the-local-government-equitable-share/

Cole, D. H. (2017). Laws, norms, and the Institutional Analysis and Development framework. *Journal of Institutional Economics*, *13*(4), 829–847. doi:10.1017/S1744137417000030

Comfort, L. K. (2005). Risk, security, and disaster management. *Annual Review of Political Science*, *8*(1), 335–356. doi:10.1146/annurev.polisci.8.081404.075608

Coppola, D. P. (2020). Preparedness. In Introduction to International Disaster Management (pp. 331-392). doi:10.1016/B978-0-12-817368-8.00005-1

Creswell, J. W. (2013). *Qualitative inquiry and research design: Choosing among five approaches*. Academic Press.

Creswell, J. (2014). *Research Design: Qualitative, Quantitative and Mixed Methods Approaches* (4th ed.). SAGE Publications. doi:10.7748/nr.12.1.82.s2

Croissant, Y., & Millo, G. (2019). *Panel data econometrics with R*. Wiley.

Crowe, S., Cresswell, K., Robertson, A., Huby, G., Avery, A., and Sheikh, A. (2017). The case study approach. . doi:10.1016/B978-008044910-4.00408-9

Crowe, S., Cresswell, K., Robertson, A., Huby, G., Avery, A., & Sheikh, A. (2011). The case study approach. *BMC Medical Research Methodology*, *11*(1), 100. doi:10.1186/1471-2288-11-100 PMID:21707982

Crozier, M. J., & Glade, T. (2005). Landslide hazard and risk: issues, concepts and approach. In T. Glade, M. Anderson, & M. J. Crozier (Eds.), *Landslide hazard and risk*. Wiley. doi:10.1002/9780470012659.ch1

Cumming, G. S., & Allen, C. R. (2017). Protected areas as social-ecological systems: Perspectives from resilience and social-ecological systems theory. *Ecological Applications*, *27*(6), 1709–1717. doi:10.1002/eap.1584 PMID:28618079

Cutter, S. L., Emrich, C. T., Webb, J. J., & Morath, D. (2009). Social vulnerability to climate variability hazards: A review of the literature. *Final Report to Oxfam America, 5*, 1-44.

Cutter, S. L. (2010). *Social science perspectives on hazards and vulnerability science*. Springer Netherlands.

Cutter, S. L., Ash, K. D., & Emrich, C. T. (2014). The geographies of community disaster resilience. *Global Environmental Change*, *29*, 65–77. doi:10.1016/j.gloenvcha.2014.08.005

Cutter, S. L., Boruff, B. J., & Shirley, W. L. (2012). Social vulnerability to environmental hazards. In *Hazards vulnerability and environmental justice* (pp. 143–160). Routledge. doi:10.4324/9781849771542

Cvetković, V. M., Tanasić, J., Ocal, A., Kešetović, Ž., Nikolić, N., & Dragašević, A. (2021). Capacity development of local self-governments for disaster risk management. *International Journal of Environmental Research and Public Health*, *18*(19), 10406. doi:10.3390/ijerph181910406 PMID:34639706

Dağ, S., & Bulut, F. (2012). Coğrafi Bilgi Sistemleri Tabanlı Heyelan Duyarlılık Haritalarının Hazırlanmasına Bir Örnek: Çayeli (Rize, KD Türkiye). *Jeoloji Muhendisligi Dergisi*, *36*(1), 35–62.

Das, P. (2012). Disaster management in India: Policy review and institutional structure. *Asiapac Journal of Socical Sciences*, *4*, 37–52.

Davidson, J., Jacobson, C., Lyth, A., Dedekorkut-Howes, A., Baldwin, C., Ellison, J., Holbrook, N. J., Howes, M. J., Serrao-Neumann, S., Singh-Peterson, L., & Smith, T. F. (2016). Interrogating resilience: Toward a typology to improve its operationalization. *Ecology and Society*, *21*(2), 27. doi:10.5751/ES-08450-210227

Davids, R., Rouget, M., Boon, R., & Roberts, D. (2018). Spatial analyses of threats to ecosystem service hotspots in Greater Durban, South Africa. *PeerJ*, *6*, e5723. doi:10.7717/peerj.5723 PMID:30386691

Davoudi, S., Brooks, E., & Mehmood, A. (2013). Evolutionary resilience and strategies for climate adaptation. *Planning Practice and Research*, *28*(3), 307–322. doi:10.1080/02697459.2013.787695

Davoudi, S., & Porter, L. (2012). Applying the resilience perspective to planning: Critical thoughts from theory and practice. *Planning Theory & Practice*, *13*(2), 299. doi:10.1080/14649357.2012.677124

De Massis, A., & Kotlar, J. (2014). The case study method in family business research: Guidelines for qualitative scholarship. *Journal of Family Business Strategy*, *5*(1), 15–29. doi:10.1016/j.jfbs.2014.01.007

Del Lama Soares, F. (2023). Climate Change and Disaster Risk Management Policy Integration in Pacific Island Countries: Trajectories and Trends. In *Climate Change Strategies: Handling the Challenges of Adapting to a Changing Climate* (pp. 389–410). Springer Nature Switzerland. doi:10.1007/978-3-031-28728-2_18

Department of Co-Operative Governance and Traditional Affairs (CoGTA). 2020. *Disaster Management Act: Declaration of a National State of Disaster: COVID-19 (coronavirus).* Disaster Management Act, 2002. NO. 313, 15 March 2020. https://www.gov.za/sites/default/files/gcis_document/202003/43096gon313.pdf

Department: Cooperative Governance and Traditional Affairs (CoGTA). (2023). *Overview of Disaster Management System in South Africa*. International Dialogue series on Disaster Management Systems South Africa, 17 April 2023.

Department: Cooperative Governance. (2016). Disaster Management Amendment Act No 16 of 2015. National Gazettes, No. 39943 of 22 April 2016.

Deprez, S., & Labattut, E. (2020). *Lessons Learnt from CARE's Shelter Responses to Cyclone Idai in Malawi.*

Deria, A., Ghannad, P., & Lee, Y. C. (2020). Evaluating implications of flood vulnerability factors with respect to income levels for building long-term disaster resilience of low-income communities. *International Journal of Disaster Risk Reduction*, *48*, 101608. doi:10.1016/j.ijdrr.2020.101608

Disaster Education in Japan: Preparing for Natural Disasters to Protect Kids' Lives. (n.d.). *Web Japan.* Available at: https://web-japan.org/kidsweb/ja/cool/20/202011_disaster-prevention-education_ja.html

Disaster risk reduction in India: Status report 2020. (n.d.). Available at: https://www.preventionweb.net/publication/disaster-risk-reduction-india-status-report-2020-0

Djalali, A., Ingrassia, P. L., & Ragazzoni, L. (2016). *Role of Hospitals in a Disaster*. ELSEVIER. doi:10.1016/B978-0-323-28665-7.00006-6

Djalante, R., Nurhidayah, L., Van Minh, H., Phuong, N. T. N., Mahendradhata, Y., Trias, A., Lassa, J., & Miller, M. A. (2020). COVID-19 and ASEAN responses: Comparative policy analysis. *Progress in Disaster Science*, *8*, 100129. doi:10.1016/j.pdisas.2020.100129 PMID:34173447

Dreze, J., & Sen, A. (2002). *India: Development and participation*. OUP Oxford.

Druker, M. J. (1974). Development and structure of public attitudes toward the multiple levels of American government. The University of Wisconsin-Madison.

DuBard, C. A., & Massing, M. W. (2007). Trends in emergency Medicaid expenditures for recent and undocumented immigrants. *Journal of the American Medical Association*, *297*(10), 1085–1092. doi:10.1001/jama.297.10.1085 PMID:17356029

Du, S., Wang, Q., & Guo, L. (2014). Spatially varying relationships between land-cover change and driving factors at multiple sampling scales. *Journal of Environmental Management*, *137*, 101–110. doi:10.1016/j.jenvman.2014.01.037 PMID:24603033

Duze, M. P. (2016). *A framework for disaster risk reduction for local government: A case study of eThekwini Municipality*. Unpublished master's thesis. University of KwaZulu-Natal, Westville Campus.

Duze, M., & Reddy, P. S. (2020). Disaster Risk Management at the Local Level: The Case of Ethekwini City Council in South Africa. Advances in African Economic, Social, and Political Development. In P. S. Reddy & H. Wissink (Eds.), *Reflections on African Cities in Transition*. Springer., doi:10.1007/978-3-030-46115-7_12

Ebba, E (2022). *Managing disasters in Zimbabwe. A qualitative analysis on Disaster Risk Reduction in Zimbabwe*. Uppsala University.

Ebrahimnejad, S., Villeneuve, M., & Tavakkoli-Moghaddam, R. (2023). An optimization model for evacuating people with disability in extreme disaster conditions: A case study. *Scientia Iranica*, *30*(4), 1498–1517.

Economic and Social Commission for Asia and the Pacific (ESCAP). (2006). Enhancing regional cooperation in infrastructure development, including that related to disaster management. Author.

Edvardsson, I. R., & Durst, S. (2021). Human Resource Management in Crisis Situations: A Systematic Literature Review. *Sustainability (Basel)*, *13*(22), 12406. doi:10.3390/su132212406

Elmqvist, T., Andersson, E., McPhearson, T., Bai, X., Bettencourt, E., Brondizo, J., Colding, G., Daily, G., Folke, C., Grimm, N., Haase, D., Ospina, D., Parnell, S., Polasky, S., Seto K.C. and Van Der Leeuw, S. (2021) Urbanization in and for the Anthropocene. *Nature Partner Journals/ Urban Sustainability, 1*(6), 1-6. . doi:10.1038/s42949-021-00018-w

Emrich, C. T., & Cutter, S. L. (2011). Social vulnerability to climate-sensitive hazards in the southern United States. *Weather, Climate, and Society*, *3*(3), 193–208. doi:10.1175/2011WCAS1092.1

Engelbrecht, F., Le Roux, A., & Mbiyozo, A. N. (2022). *With every second city and town projected to suffer increased flooding, South Africa must be better prepared.* Daily Maverick. https://www.dailymaverick.co.za/article/2022-05-12-poor-climate-adaptation-outdated-infrastructure-served-as-catalysts-for-kzn-floods/

Environmental Protection Agency (EPA). (2000). *Abandoned mine site characterization and cleanup handbook.* USEPA.

Erener, A., & Düzgün, H. S. B. (2010). Improvement of statistical landslide susceptibility mapping by using spatial and global regression methods in the case of More and Romsdal (Norway). *Landslides*, *7*(1), 55–68. doi:10.1007/s10346-009-0188-x

Erinç, S. (1944). *Glazialmorphologische Untersuchungen im Nordostanatolischen Randgebirge.* PhD Dissertation Series, Geographical Institute of the University of Istanbul.

Erinç, S. (1965). *Yağış müessiriyeti üzerine bir deneme ve yeni bir indis. İstanbul Üniversitesi. Edebiyat Fakültesi.* Coğrafya Enstitüsü Yayınları.

Eyler, J. (2002). Reflection: Linking service and learning – Linking students and communities. *The Journal of Social Issues*, *58*(3), 517–534. doi:10.1111/1540-4560.00274

Faivre, N., Sgobbi, A., Happaerts, S., Raynal, J., & Schmidt, L. (2018). Translating the Sendai Framework into action: The EU approach to ecosystem-based disaster risk reduction. *International Journal of Disaster Risk Reduction*, *32*, 4–10. doi:10.1016/j.ijdrr.2017.12.015

Falk, I., & Kilpatrick, S. (2000). What Is Social Capital? A Study of Rural Communities. *Sociologia Ruralis*, *40*(1), 87–110. doi:10.1111/1467-9523.00133

Faraji Sabokbar, H., Shadman Roodposhti, M., & Tazik, E. (2014). Landslide susceptibility mapping using geographically-weighted principal component analysis. *Geomorphology*, *226*, 15–24. doi:10.1016/j.geomorph.2014.07.026

Fatti, C. E., & Patel, Z. (2013). Perceptions and responses to urban flood risk: Implications for climate governance in the South. *Applied Geography (Sevenoaks, England)*, *36*, 13–22. doi:10.1016/j.apgeog.2012.06.011

Felbermayr, G., & Gröschl, J. (2014). Naturally negative: The growth effects of natural disasters. *Journal of Development Economics*, *111*, 92–106. doi:10.1016/j.jdeveco.2014.07.004

Ferreira Nogueira, D. (2019). *Mobile-Based Early WarningSystems in Mozambique: An exploratory study on the viability to integrate Cell Broadcast into disaster mitigation routines*. Academic Press.

Feuillet, T., Coquin, J., Mercier, D., Cossart, E., Decaulne, A., Jónsson, H. P., & Sæmundssonthorsteinn, T. (2014). Focusing on the spatial non-stationarity of landslide predisposing factors in northern Iceland: Do paraglacial factors vary over space? *Progress in Physical Geography*, *38*(3), 354–377. doi:10.1177/0309133314528944

Few, R., Scott, Z., Wooster, K., Avila, M. F., & Tarazona, M. (2016). Strengthening capacities for disaster risk management II: Lessons for effective support. *International Journal of Disaster Risk Reduction*, *20*, 154–162. doi:10.1016/j.ijdrr.2016.02.005

Field, C. B. (Ed.). (2012). *Managing the risks of extreme events and disasters to advance climate change adaptation: special report of the intergovernmental panel on climate change*. Cambridge University Press. doi:10.1017/CBO9781139177245

Findlay, A. M., & Li, F. L. N. (2004). Methodological issues in researching migration. *The Professional Geographer*, *51*(1), 50–59. doi:10.1111/0033-0124.00144

Floodlist. (2022, May 23). South Africa – Evacuations after more flooding in KwaZulu-Natal. https://floodlist.com/africa/south-africa-floodi-kwazulu-natal-may-2022

Folke, C. (2006). Resilience, The emergence of a perspective for social-ecological systems analyses. *Global Environmental Change*, *16*(3), 253–267. doi:10.1016/j.gloenvcha.2006.04.002

Fotheringham, A. S., Brunsdon, C., & Charlton, M. (2003). *Geographically weighted regression: the analysis of spatially varying relationships* (1st ed.). John Wiley & Sons.

Fotheringham, A. S., Charlton, M. E., & Brunsdon, C. (2001). Spatial Variations in School Performance: A Local Analysis Using Geographically Weighted Regression. *Geographical and Environmental Modelling*, *5*(1), 43–66. doi:10.1080/13615930120032617

Fotheringham, A. S., & Oshan, T. M. (2016). Geographically weighted regression and multicollinearity: Dispelling the myth. *Journal of Geographical Systems*, *18*(4), 303–329. doi:10.1007/s10109-016-0239-5

Freckleton, D., Heaslip, K., Louisell, W., & Collura, J. (2012). Evaluation of resiliency of transportation networks after disasters. *Transportation Research Record: Journal of the Transportation Research Board*, *2284*(1), 109–116. doi:10.3141/2284-13

Fuchs, S., Röthlisberger, V., Thaler, T., Zischg, A., & Keiler, M. (2017). Natural hazard management from a coevolutionary perspective: Exposure and policy response in the European Alps. *Annals of the Association of American Geographers*, *107*(2), 382–392. doi:10.1080/24694452.2016.1235494 PMID:28267154

Furco, A., & Billig, S. H. (Eds.). (2002). *Service learning: The essence of the pedagogy*. IAP.

Galera-Zarco, C., & Floros, G. (2023). A deep learning approach to improve built asset operations and disaster management in critical events: An integrative simulation model for quicker decision making. *Annals of Operations Research*, 1–40. doi:10.1007/s10479-023-05247-z

Ganin, A. A., Mersky, A. C., Jin, A. S., Kitsak, M., Keisler, J. M., & Linkov, I. (2019). Resilience in intelligent transportation systems (ITS). *Transportation Research Part C, Emerging Technologies*, *100*, 318–329. doi:10.1016/j.trc.2019.01.014

Garrido-Moreno, A., García-Morales, J., & Martín-Rojas, R. (2021). Going beyond the curve: Strategic measures to recover hotel activity in times of COVID-19. *International Journal of Hospitality Management*, *96*, 102928. doi:10.1016/j.ijhm.2021.102928 PMID:36540887

GDCDE. (2016). *Rize Province Güneysu Micro-catchment Landslide Susceptibility and Hazard Map Production Project*. General Directorate of Combating with Desertification and Erosion, Ankara, Turkey. Retrived From https://webdosya.csb.gov.tr/db/cem/icerikler/taslidire-heyelan-20211108102301.pdf

Gholamhosseinian, A., & Seitz, J. (2022). A comprehensive survey on cooperative intersection management for heterogeneous connected vehicles. *IEEE Access : Practical Innovations, Open Solutions*, *10*, 7937–7972. doi:10.1109/ACCESS.2022.3142450

Giddy, J. K., & Rogerson, J. M. (2021). Nature-based tourism enterprise adaptive responses to COVID-19 in South Africa. *Geo Journal of Tourism and Geosites*, *36*(2spl), 36. doi:10.30892/gtg.362spl18-700

Gil-Rivas, V., & Kilmer, R. P. (2016). Building community capacity and fostering disaster resilience. *Journal of Clinical Psychology*, *72*(12), 1318–1332. doi:10.1002/jclp.22281 PMID:26990644

Gissing, A., Timms, M., Browning, S., Crompton, R., & McAneney, J. (2022). Compound natural disasters in Australia: A historical analysis. *Environmental Hazards*, *21*(2), 159–173. doi:10.1080/17477891.2021.1932405

Glago, F. J. (2021). Flood disaster hazards; causes, impacts and management: a state-of-the-art review. *Natural hazards-impacts, adjustments and resilience*, 29-37.

glm. (2022). *glm: Fitting Generalized Linear Models*. Retrieved from https://www.rdocumentation.org/packages/stats/versions/3.6.2/topics/glm

Gogo, J. (2020). Formulate Disaster Management Policy. *The Herald*. Available online at: https://www.herald.co.zw/formulate-disaster-management-policy/

Gossling, S., Scott, D., & Hall, C. M. (2021). Pandemics, tourism and global change: A rapid assessment of COVID-19. *Journal of Sustainable Tourism*, *29*(1), 1–20. doi:10.1080/09669582.2020.1758708

Goto, K., Goff, J., & Paris, R. (2021). Ten years since the 2011 Tohoku-oki tsunami-Progress in paleotsunami research. *Earth-Science Reviews*, *216*, 103598. doi:10.1016/j.earscirev.2021.103598

Government of India. (n.d.). *UNDP Disaster Risk Management Programme 2002-2009: Evaluation and Review of Lessons Learnt.* Author.

Gray, B., Eaton, J., Christy, J., Duncan, J., Hanna, F., & Kasi, S. (2021). A proactive approach: Examples for integrating disaster risk reduction and mental health and psychosocial support programming. *International Journal of Disaster Risk Reduction, 54*, 1–9. doi:10.1016/j.ijdrr.2021.102051 PMID:33732610

Greer, S. (2017). *The Concept of Community: Readings with Interpretations*. Routledge. doi:10.4324/9781315131443

Gross, T. (2013). "We the People"-What is the Legal Meaning of "People"? *Revue europeenne de droit public= European review of public law= Eur. Zeitschrift des offtl.Rechts= Rivista europea di diritto pubblico, 25*(4), 1791–1822.

Guerdan, B. R. (2009). Disaster preparedness and disaster management. *The American Journal of Chinese Medicine, 6*, 32–40.

Gülhan, S. (2021). An overview of disaster resilience. *Turkish Journal of Health Science and Life, 4*(3), 106–115.

Gupta, A. K. (2018). Disaster governance and legal systems in India. *Disaster risk governance in India and cross cutting issues*, 39-60.

Haak-Saheem, W. (2020) Talent management in Covid -19 crisis – how Dubai manages and sustains its global talent pool. *Asian Business and Management, 19*(3), 298-301. https://doi.org/doi:10.1057/s41291-020-

Hamid, N., Trihatmoko, E., Herlina, M., & Aroyandini, E. N. (2021). Developing a Model for Disaster Education to Improve Students' Disaster Mitigation Literacy. *Journal of Disaster Research, 16*(8), 1243–1256. doi:10.20965/jdr.2021.p1243

Hao, F., Xiao, Q., & Chon, K. (2020). COVID-19 and China's Hotel Industry: Impacts, a Disaster Management Framework, and Post-Pandemic Agenda. *International Journal of Hospitality Management, 90*, 102636. doi:10.1016/j.ijhm.2020.102636 PMID:32834356

Haque, M. (2019). Indigenous knowledge and practices in disaster management: Experiences of the coastal people of Bangladesh. Disaster Risk Reduction: Community Resilience and Responses, 59–72.

Harrison, H., Birks, M., Franklin, R., & Mills, J. (2017). Case Study Research: Foundations and Methodological Orientations. *Forum Qualitative Social Research, 18*(1), 19. http://nbn-resolving.de/urn:nbn:de:0114-fqs1701195

Heaslip, K., Louisell, W., &Collura, J. (2009). *A methodology to evaluate transportation resiliency for regional networks* (No. 09-3194).

Hellerstein, J. K., & Neumark, D. (2020). Social Capital, Networks, and Economic Wellbeing. IZA DP No. 13413, Bonn, Germany.

Henningsen, A., & Henningsen, G. (2019). Analysis of panel data using R. In *Panel data econometrics* (pp. 345–396). Academic Press. doi:10.1016/B978-0-12-814367-4.00012-5

Hill, T. D., Davis, A. P., Roos, J. M., & French, M. T. (2020). Limitations of fixed-effects models for panel data. *Sociological Perspectives*, *63*(3), 357–369. doi:10.1177/0731121419863785

Hollnagel, E. (Ed.). (2013). *Resilience engineering in practice: A guidebook*. Ashgate Publishing, Ltd.

Hong, H., Pradhan, B., Sameen, M. I., Chen, W., & Xu, C. (2017). Spatial prediction of rotational landslide using geographically weighted regression, logistic regression, and support vector machine models in Xing Guo area (China). *Geomatics, Natural Hazards & Risk*, *8*(2), 1997–2022. doi:10.1080/19475705.2017.1403974

Hong, H., Pradhan, B., Sameen, M. I., Kalantar, B., Zhu, A., & Chen, W. (2018). Improving the accuracy of landslide susceptibility model using a novel region-partitioning approach. *Landslides*, *15*(4), 753–772. doi:10.1007/s10346-017-0906-8

Hönke, J., & Kranz, N. (2013). Cleaning up Their Act, or More? Mining Companies and Environmental Protection in South Africa. In T. A. Börzel & C. R. Thauer (Eds.), *Business and Governance in South Africa. Governance and Limited Statehood Series*. Palgrave Macmillan. doi:10.1057/9781137317810_10

Houston, L. J. H. (2024). Climate action and the pursuit of environmental sustainability in Africa: An analysis of Africa's involvement in COP26 discussions. In *Human Rights and the Environment in Africa* (pp. 375–395). Routledge.

Hsiao, C. (2022). Analysis of panel data (No. 64). Cambridge University Press.

Huabin, W., Gangjun, L., Weiya, X., & Gonghui, W. (2005). GIS-based landslide hazard assessment: An overview. *Progress in Physical Geography*, *29*(4), 548–567. doi:10.1191/0309133305pp462ra

Humaida, N., Saputra, M. H., & Hadiyan, Y. (2023). Urban gardening for mitigating heat island effect. *IOP Conference Series. Earth and Environmental Science*, *1133*(1), 012048. doi:10.1088/1755-1315/1133/1/012048

Human Rights Campaign. (2021). *Violence against the transgender and gender non-conforming community in 2020*. Retrieved from https://www.hrc.org/resources/violence-against-the-transgender-and-gender-non-conforming-community-in-2020

Huq, M. E. (2021). Resilience for Disaster Management: Opportunities and Challenges. In Climate Vulnerability and Resilience in the Global South. Cham: Springer International Publishing (Climate Change Management). doi:10.1007/978-3-030-77259-8_22

Hyasat, A. (2022). The impact of the covid-19 pandemic upon tourism & hospitality employees in Jordan. *Geo Journal of Tourism and Geosites*, *43*(2), 1099–1107. doi:10.30892/gtg.43331-925

IISD (International Institute for Sustainable Development). (2023). The Reporting Service for Environment and Development Negotiations. *Earth Negotiations Bulletin*, *12*(842).

Iliopoulou, C., Konstantinidou, M. A., Kepaptsoglou, K. L., & Stathopoulos, A. (2020). ITS technologies for decision making during evacuation operations: A review. *Journal of Transportation Engineering. Part A, Systems*, *146*(4), 04020010. doi:10.1061/JTEPBS.0000329

Imai, K., & Kim, I. S. (2021). On the use of two-way fixed effects regression models for causal inference with panel data. *Political Analysis*, *29*(3), 405–415. doi:10.1017/pan.2020.33

International Federation of Red Cross (IFRC). (2022). *South Africa: KwaZulu-Natal Floods - Emergency Appeal No. MDRZA012 - Operational Strategy*. IFRC.

Interparliamentary Panel on Climate Change (IPCC). (n.d.). *The IPCC's Sixth Assessment Report Impacts, adaptation options and investment areas for a climate-resilient southern Africa*. https://cdkn.org/sites/default/files/2022-%2003/IPCC%20Regional%20Factsheet_Southern%20Africa_Web.pdf

IOL. (2022, April 26). *Effects of the KwaZulu-Natal flood in numbers*. https://www.iol.co.za/news/south-africa/kwazulu-natal/effects-of-the-kwazulu-natal-flood-in-numbers-32c45aae-9611-4d8a-b137-25fc9843b8f0

Ishiwatari, M., Koike, T., Hiroki, K., Toda, T., & Katsube, T. (2020). Managing disasters amid Covid-19 pandemic: Approaches of response to flood disasters. *Progress in Disaster Science*, *6*, 100096. doi:10.1016/j.pdisas.2020.100096 PMID:34171012

Ishiwatari, M., Ranghieri, F., Taniguchi, K., & Mimura, S. (2021). Learning from megadisasters in Japan: Sharing lessons with the world. *Journal of Disaster Research*, *16*(6), 942–946. doi:10.20965/jdr.2021.p0942

Islam, M. A., Rashid, S. I., Hossain, N. U. I., Fleming, R., & Sokolov, A. (2023). An integrated convolutional neural network and sorting algorithm for image classification for efficient flood disaster management. *Decision Analytics Journal*, *7*, 100225. doi:10.1016/j.dajour.2023.100225

Islam, R., Jones, S., & Hudnall, M. (2023). Transportation system functions during hurricane response: A systematic review of modes and methods. *Transportation Research Interdisciplinary Perspectives*, *18*, 100786. doi:10.1016/j.trip.2023.100786

Islam, R., Kamaruddin, R., Ahmad, S. A., Jan, S., & Anuar, A. R. (2016). A review on mechanism of flood disaster management in Asia. *International Review of Management and Marketing*, *6*(1), 29–52.

Ismaeel, S. S., Midi, H., & Sani, M. (2021). Robust multicollinearity diagnostic measure for fixed effect panel data model. *Malaysian Journal of Fundamental and Applied Sciences*, *17*(5), 636–646. doi:10.11113/mjfas.v17n5.2391

Ito, E. (2021). *The Dark Realism of Miyazawa Kenji: Social Activism in the Science-Fiction Children's Story, The Life of Gusukō Budori, and its Precursor Work*. Dissertations and Theses. doi:10.15760/etd.7725

Izumi, T., Shaw, R., Djalante, R., Ishiwatari, M., & Komino, T. (2019). Disaster risk reduction and innovations. *Progress in Disaster Science*, *2*, 100033. doi:10.1016/j.pdisas.2019.100033

JasmineC. A. (2019) Impacts of Covid-19 on company and efforts to support organization. doi:10.2139/ssrn.3590726

Jawad, S., & Liu, J. (2020). Electrical vehicle charging services planning and operation with interdependent power networks and transportation networks: A review of the current scenario and future trends. *Energies*, *13*(13), 3371. doi:10.3390/en13133371

Jayasekara, R., Siriwardana, C., Dilanthi Amaratunga, D., & Richard Haigh, R. (2022). Evaluating the network of stakeholders in Multi-Hazard Early Warning Systems for multiple hazards amidst biological outbreaks: Sri Lanka as a case in point. *Progress in Disaster Science*, *14*, 100228. doi:10.1016/j.pdisas.2022.100228 PMID:35402893

Jethwaney, J. (2023). Capacity building through stakeholder training in Media and Communication for effective disaster management. In A. Singh (Ed.), *International Handbook of Disaster Research* (pp. 1371–1388). Springer. doi:10.1007/978-981-19-8388-7_86

Jewkes, R., & Murcott, A. (1996). Meanings of community. *Social Science & Medicine*, *43*(4), 555–563. doi:10.1016/0277-9536(95)00439-4 PMID:8844956

Jha, A. K., Miner, T. W., & Stanton-Geddes, Z. (Eds.). (2013). *Building urban resilience: principles, tools, and practice*. World Bank Publications. doi:10.1596/978-0-8213-8865-5

Jones, P., & Comfort, D. (2020). The COVID-19 crisis, tourism and sustainable development. *Athens Journal of Tourism*, *7*(2), 75–86. doi:10.30958/ajt.7-2-1

Kalantar, B., Ueda, N., Lay, U. S., Al-Najjar, H. A. H., & Halin, A. A. (2019). Conditioning Factors Determination for Landslide Susceptibility Mapping Using Support Vector Machine Learning. *International Geoscience and Remote Sensing Symposium (IGARSS)*, 9626–9629. 10.1109/IGARSS.2019.8898340

Kalantar, B., Ueda, N., Saeidi, V., Ahmadi, K., Halin, A. A., & Shabani, F. (2020). Landslide Susceptibility Mapping: Machine and Ensemble Learning Based on Remote Sensing Big Data. *Remote Sensing (Basel)*, *12*(11), 1737. doi:10.3390/rs12111737

Kalnins, A. (2018). Multicollinearity: How common factors cause Type 1 errors in multivariate regression. *Strategic Management Journal*, *39*(8), 2362–2385. doi:10.1002/smj.2783

Kamal, A., Songwathana, P., & Sia, W. S. (2012). Knowledge and skills of emergency care during a disaster for community health volunteers: A literature review. *Nurse Media Journal of Nursing*, *2*(2), 371–381.

Kamara, J. K., Wali, N., Agho, K., & Renzaho, A. M. N. (2018). Resilience to climate induced disasters and its overall impact on well-being in Southern Africa: A mixed methods systematic review. *International Journal of Environmental Research and Public Health*, *15*(11), 2375. doi:10.3390/ijerph15112375 PMID:30131077

Kamete, A. (2004). Home industries and the formal city in Harare, Zimbabwe, Reconsidering informality: *Perspectives From Urban Africa*, 120–138.

Karpouza, M., Chousianitis, K., Bathrellos, G. D., Skilodimou, H. D., Kaviris, G., & Antonarakou, A. (2021). Hazard zonation mapping of earthquake-induced secondary effects using spatial multi-criteria analysis. *Natural Hazards, 109*(1), 637–669. doi:10.1007/s11069-021-04852-0

Kathleen Geale, S. (2012). The ethics of disaster management. *Disaster Prevention and Management: An International Journal, 21*(4), 445-462. doi:10.1108/09653561211256152

Katz, J. E., Rice, R. E., Acord, S., Dasgupta, K., & David, K. (2004). 'Personal Mediated Communication and the Concept of Community In Theory and Practice. *Annals of the International Communication Association, 28*(1), 315–371. doi:10.1080/23808985.2004.11679039

Kavzoglu, T., Sahin, E. K., & Colkesen, I. (2014). Landslide susceptibility mapping using GIS-based multi-criteria decision analysis, support vector machines, and logistic regression. *Landslides, 11*(3), 425–439. doi:10.1007/s10346-013-0391-7

Ketin, İ. (1966). Tectonic units of Anatolia (Asia minor). *Bulletin of the Mineral Research and Exploration, 66*(66), 20–37. https://dergipark.org.tr/en/pub/bulletinofmre/issue/3892/51961

Khalid, K., Usman, M., & Mehdi, M. A. (2021). The determinants of environmental quality in the SAARC region: A spatial heterogeneous panel data approach. *Environmental Science and Pollution Research International, 28*(6), 6422–6436. doi:10.1007/s11356-020-10896-9 PMID:32997249

Khan, A., Bibi, S., Lyu, J., Latif, A., & Lorenzo, A. (2020). COVID-19 and sectoral employment trends: Assessing resilience in the US leisure and hospitality industry. *Current Issues in Tourism, 24*(7), 952–969. doi:10.1080/13683500.2020.1850653

Khan, H., Vasilescu, L. G., & Khan, A. (2008). Disaster management cycle – a theoretical approach. *Management and Marketing Journal, 6*(1), 43–50.

Khorram-Manesh, A., Ashkenazi, M., Djalali, A., Ingrassia, P. L., Friedl, T., Von Armin, G., Lupesco, O., Kaptan, K., Arculeo, C., Hreckovsi, B., Komadina, R., Fisher, P., Voigt, S., James, J., & Gursky, E. (2015). Education in disaster management and emergencies: Defining a new European course. *Disaster Medicine and Public Health Preparedness, 9*(3), 245–255. doi:10.1017/dmp.2015.9 PMID:25781077

Kim, J. J., Han, H., & Ariza-Montes, A. (2021). The impact of hotel attributes, well-being perception, and attitudes on brand loyalty: Examining the moderating role of the COVID-19 pandemic. *Journal of Retailing and Consumer Services, 62*, 102634. doi:10.1016/j.jretconser.2021.102634

Kim, J., Kim, J., & Wang, Y. (2021). Uncertainty risks and strategic reaction of restaurant firms amid COVID-19: An assessment of antecedents, consequences and intermediate. *International Journal of Hospitality Management, 95*, 102894.

Kim, S. J., & Bostwick, W. (2020). Social vulnerability and racial inequality in COVID-19 deaths in Chicago. *Health Education & Behavior, 47*(4), 509–513. doi:10.1177/1090198120929677 PMID:32436405

Kitagawa, K. (2016). Disaster preparedness, adaptive politics and lifelong learning: A case of Japan. *International Journal of Lifelong Education, 35*(6), 629–647. doi:10.1080/02601370.2016.1231230

Kitagawa, K. (2021). Conceptualising 'disaster education'. *Education Sciences, 11*(5), 233. doi:10.3390/educsci11050233

Koçyiğit, A. (1985). The Karayazı faul. *Bulletion of Geological Society of Turkey, 28*(1), 67–72.

Koivisto, J. E., & Nohrstedt, D. (2017). A policymaking perspective on disaster risk reduction in Mozambique. *Environmental Hazards, 16*(3), 210–227. doi:10.1080/17477891.2016.1218820

Kong, T. M., de Villiers, A. C., Ntloana, M. B., Pollard, S., & Vogel, C. (2020). Implementing capacity development for disaster risk reduction as a social learning system. *International Journal of Disaster Risk Reduction, 50*, 101740. doi:10.1016/j.ijdrr.2020.101740

Kornejady, A., Ownegh, M., Rahmati, O., & Bahremand, A. (2018). Landslide susceptibility assessment using three bivariate models considering the new topo-hydrological factor: HAND. *Geocarto International, 33*(11), 1155–1185. doi:10.1080/10106049.2017.1334832

Koshimura, S., Hayashi, S., & Gokon, H. (2014). The impact of the 2011 Tohoku earthquake tsunami disaster and implications to the reconstruction. *Soil and Foundation, 54*(4), 560–572. doi:10.1016/j.sandf.2014.06.002

Kotlar, J., & De Massis, A. (2013). Goal setting in family firms: Goal diversity, social interactions, and collective commitment to family–centered goals. *Entrepreneurship Theory and Practice, 37*(6), 1263–1288. doi:10.1111/etap.12065

Kropko, J., & Kubinec, R. (2020). Interpretation and identification of within-unit and cross-sectional variation in panel data models. *PLoS One, 15*(4), e0231349. doi:10.1371/journal.pone.0231349 PMID:32315338

Kruczkiewicz, A., Klopp, J., Fisher, J., Mason, S., McClain, S., Sheekh, N. M., Moss, R., Parks, R. M., & Braneon, C. (2021). Opinion: Compound risks and complex emergencies require new approaches to preparedness. *Proceedings of the National Academy of Sciences of the United States of America, 118*(19), e2106795118. doi:10.1073/pnas.2106795118 PMID:33952695

Kubo, T., Chimed-Ochir, O., Cossa, M., Ussene, I., Toyokuni, Y., Yumiya, Y., & Salio, F. (2022). First Activation of the WHO Emergency Medical Team Minimum Data Set in the 2019 Response to Tropical Cyclone Idai in Mozambique. *Prehospital and Disaster Medicine, 37*(6), 727–734. doi:10.1017/S1049023X22001406 PMID:36325992

Kulatunga, U. (2010). Impact of culture towards disaster risk reduction. *International Journal of Strategic Property Management, 14*(4), 304–313. doi:10.3846/ijspm.2010.23

Kullig, J. (2000). Community resiliency: The potential for community health nursing theory development. *Public Health Nursing (Boston, Mass.), 17*(5), 374–385. doi:10.1046/j.1525-1446.2000.00374.x PMID:11013000

Kumar, P., & Akhai, S. (2022). Effective energy management in smart buildings using VRV/VRF systems. *Additive Manufacturing in Industry 4.0: Methods, Techniques, Modeling, and Nano Aspects*, 27.

Kunguma, O. (2020). *South African disaster management framework: Assessing the status and dynamics of establishing information management and communication systems in Provinces*. Unpublished Doctoral Dissertation. University of the Free State, South Africa: Bloemfontein.

Kunguma, O. (2022). *Opinion: KZN floods expose significant socio-economic and environmental vulnerabilities*. Disaster Management Training and Education Centre for Africa: University of University of the Free State, South Africa: Bloemfontein. https://www.ufs.ac.za/templates/news-archive/campus-news/2022/april/opinion-kzn-floods-expose-significant-socio-economic-and-environmental-vulnerabilities

Kunguma, O. (2022). *COVID-19 disaster, an opportunity for the South African disaster management legislation betterment?* Academia Letters., doi:10.20935/AL4645

Kunguma, O., Ncube, A., & Mokhele, M. (2021). COVID-19 disaster response: South African disaster managers' faith in mandating legislation tested? *Jamba (Potchefstroom, South Africa)*, *13*(1). Advance online publication. doi:10.4102/jamba.v13i1.1099 PMID:34691364

Kurmann, A., Lalé, E., & Ta, L. (2020). The Impact of COVID-19 on US Employment and Hours: Real-Time Estimates with Homebase Data. *EurekAlert*. www.eurekalert.org/pub_releases/2020-05/du-rds050520.php

Kuru, K., & Khan, W. (2020). A framework for the synergistic integration of fully autonomous ground vehicles with smart city. *IEEE Access : Practical Innovations, Open Solutions*, *9*, 923–948. doi:10.1109/ACCESS.2020.3046999

Kyne, D., & Aldrich, D.P. (2019). Capturing bonding, bridging and linking social capital through publicly available data. *Risks, Hazards and Crisis in Public Policy*, 1-26.

Lagos, D. G., Poulaki, P., & Lambrou, P. (2021). COVID-19 and Its Impact on Tourism Industry. *Advances in Experimental Medicine and Biology*, *1318*, 815–824. doi:10.1007/978-3-030-63761-3_45 PMID:33973213

Lai, I. K. W., & Wong, J. W. C. (2020). Comparing crisis management practices in the hotel industry between initial and pandemic stages of COVID-19. *International Journal of Contemporary Hospitality Management*, *32*(10), 3135–3156. doi:10.1108/IJCHM-04-2020-0325

Landry, S. M., Koeser, A. K., Kane, B., Hilbert, D. R., McLean, D. C., Andreu, M., & Staudhammer, C. L. (2021). Urban forest response to Hurricane Irma: The role of landscape characteristics and sociodemographic context. *Urban Forestry & Urban Greening*, *61*, 127093. doi:10.1016/j.ufug.2021.127093

Larson, M. (n.d.). *Reading Kenji Miyazawa after 3.11: Region, Utopia, and Modernity*. Available at: https://researchmap.jp/19760415/published_papers/33470174/attachment_file.pdf

Lassa, J. A., Surjan, A., Caballero-Anthony, M., & Fisher, R. (2019). Measuring political will: An index of commitment to disaster risk reduction. *International Journal of Disaster Risk Reduction*, *34*, 64–74. doi:10.1016/j.ijdrr.2018.11.006

Laurien, F., & Mechler, R. (2020, May 11). *Resilience thinking and action in the context of compound risks*. https://floodresilience.net/blogs/resilience-thinking-and-action-in-the-context-of-compound-risks/

Lavery, M. R., Acharya, P., Sivo, S. A., & Xu, L. (2019). Number of predictors and multicollinearity: What are their effects on error and bias in regression? *Communications in Statistics. Simulation and Computation*, *48*(1), 27–38. doi:10.1080/03610918.2017.1371750

Lawrence, S., & Suresh, E. S. M. (2012). The need for integration of disaster management into engineering curriculum. *International Journal of Modern Engineering Research*, *2*(6), 4524-4526.

Le Cozannet, G., Kervyn, M., Russo, S., Speranza, C. I., Ferrier, P., & Foumelis, M. (2020). Space-based earth observations for disaster risk management. *Surveys in Geophysics*, *41*(6), 1209–1235. doi:10.1007/s10712-020-09586-5

Lee, M., Mesicek, L., Bae, K., & Ko, H. (2023). AI advisor platform for disaster response based on big data. *Concurrency and Computation*, *35*(16), e6215. doi:10.1002/cpe.6215

Lehmann, S. (2018). Implementing the urban nexus approach for improved resources-efficiency of developing cities in Southeast-Asia, City. *Cultura e Scuola*, *13*, 46–56. doi:10.1016/j.ccs.2017.10.003

Lekgau, R. J., & Tichaawa, T. M. (2021). Adaptive strategies employed by the MICE sector in response to COVID-19. *Geo Journal of Tourism and Geosites*, *38*(4), 1203–1210. doi:10.30892/gtg.38427-761

Leroy, S. A. G., Gracheva, R., & Medvedev, A. (2022). Natural hazards and disasters around the Caspian Sea. *Natural Hazards*, *114*(3), 2435–2478. doi:10.1007/s11069-022-05522-5

Lewis, F., Browne, M., & Houdet, J. (2021). COVID-19–What is the impact for tourism as a vehicle of SMME development and incentivizing environmental management in South Africa. *African Journal of Hospitality, Tourism and Leisure*, *10*(3), 987–998. doi:10.46222/ajhtl.19770720-144

Li, C., Li, F., Wu, Z., & Cheng, J. (2017). Exploring spatially varying and scale-dependent relationships between soil contamination and landscape patterns using geographically weighted regression. *Applied Geography (Sevenoaks, England)*, *82*, 101–114. doi:10.1016/j.apgeog.2017.03.007

Lindner, T., Puck, J., & Verbeke, A. (2020). Misconceptions about multicollinearity in international business research: Identification, consequences, and remedies. *Journal of International Business Studies*, *51*(3), 283–298. doi:10.1057/s41267-019-00257-1

Lin, Q., Wang, Y., Glade, T., Zhang, J., & Zhang, Y. (2020). Assessing the spatiotemporal impact of climate change on event rainfall characteristics influencing landslide occurances based on multiple GCM projections in China. *Climatic Change, 162*(2), 761–779. doi:10.1007/s10584-020-02750-1

Liu, J. D., Dong, C. Q., An, S., & Mai, Q. (2022). Dynamic evolution analysis of the emergency collaboration network for compound disasters: A case study involving a public health emergency and an accident disaster during Covid-19. *Health Care, 10*, 500. PMID:35326978

Li, Y., Liu, X., Han, Z., & Dou, J. (2020). Spatial proximity-based geographically weighted regression model for landslide susceptibility assessment: A case study of Qingchuan area, China. *Applied Sciences (Basel, Switzerland), 10*(3), 1107. Advance online publication. doi:10.3390/app10031107

Lo, A., Cheung, C., & Law, R. (2006). The survival of hotels during disaster: A case study of Hong Kong in 2003. *Asia Pacific Journal of Tourism Research, 11*(1), 65–79. doi:10.1080/10941660500500733

Logie, C. H., & Turan, J. M. (2020). How do we balance tensions between COVID-19 public health responses and stigma mitigation? Learning from HIV research. *AIDS and Behavior, 24*(7), 2003–2006. doi:10.1007/s10461-020-02856-8 PMID:32266502

Ltifi, M., & Hichri, A. (2022). The effects of corporate governance on the customer's recommendations: A study of the banking sector at the time of COVID-19. *Journal of Knowledge Management, 26*(1), 165–191. doi:10.1108/JKM-06-2020-0471

Lu, X., Liao, W., Fang, D., Lin, K., Tian, Y., Zhang, C., ... Zhao, P. (2020). Quantification of disaster resilience in civil engineering: A review. *Journal of Safety Science and Resilience, 1*(1), 19-30.

Lundgren, M., & Strandh, V. (2022). Navigating a double burden–Floods and social vulnerability in local communities in rural Mozambique. *International Journal of Disaster Risk Reduction, 77*, 103023. doi:10.1016/j.ijdrr.2022.103023

Madubula, N. P. (2021). *A comparative analysis of the vulnerability of informal and formal households to disaster risks: The case of Bekkersdal* (Doctoral dissertation).

Mahrez, Z., Sabir, E., Badidi, E., Saad, W., & Sadik, M. (2021). Smart urban mobility: When mobility systems meet smart data. *IEEE Transactions on Intelligent Transportation Systems, 23*(7), 6222–6239. doi:10.1109/TITS.2021.3084907

Malilay, J., Heumann, M., Perrotta, D., Wolkin, A. F., Schnall, A. H., Podgornik, M. N., Cruz, M. A., Horney, J. A., Zane, D., Roisman, R., Greenspan, J. R., Thoroughman, D., Anderson, H. A., Wells, E. V., & Simms, E. F. (2014). The role of applied epidemiology methods in the disaster management cycle. *American Journal of Public Health, 104*(11), 2092–2102. doi:10.2105/AJPH.2014.302010 PMID:25211748

Mamabolo, M., & Sebola, M. (2021). *The role and adequacy of disaster management unit within the South African Municipalities*. The Business and Management Review. doi:10.24052/BMR/V12NU02/ART-09

Mandal, M., Ghosh, D., & Ghosh, B. (2016). Environmental Impact of Sand Mining: A Case Study along the Lower Reaches of Ajay River, West Bengal, India. *Journal of Environment and Sociobiology*, *13*, 99–108.

Mandal, S., & Maiti, R. (2015). *Semi-quantitative approaches for landslide assessment and prediction*. Springer. doi:10.1007/978-981-287-146-6

Mandip, G. (2012). Green HRM: People management commitment to environmental sustainability. *Research Journal of Recent Sciences, ISSN*, *2277*, 2502.

Manjunatha, A., Chengappa, P., & Chandrakanth, M. (2006). Externalities Due To Sand Mining and Distillery Effluent in Water Streams of India. *Journal of Global Economics*, *2*, 135–151. doi:10.1956/JGE.V2I2.209

Mannan, S., Haque, D. M. E., & Sarker, N. C. D. (2021). A study on national DRR policy in alignment with the SFDRR: Identifying the scopes of improvement for Bangladesh. *Progress in Disaster Science*, *12*, 100206. doi:10.1016/j.pdisas.2021.100206 PMID:34632365

Manyena, B. (2009). *Disaster resilience in development and humanitarian interventions* [PhD Thesis]. Northumbria.

Manyena, B. (2014). Disaster resilience: A question of 'multiple faces' and 'multiple spaces? *International Journal of Disaster Risk Reduction*, *8*, 1–9. doi:10.1016/j.ijdrr.2013.12.010

Manyena, B. (2016). After Sendai: Is Africa Bouncing Back or Bouncing Forward from Disasters? *International Journal of Disaster Risk Science*, *7*(1), 41–53. doi:10.1007/s13753-016-0084-7

Manyena, B., O'Brien, G., O'keefe, P., & Rose, J. (2011). Disaster resilience: A bounce back or bounce forward ability? *Local Environment*, *16*(5), 417–424. doi:10.1080/13549839.2011.583049

Manyena, S. B. (2006). Rural local authorities and disaster resilience in Zimbabwe. *Disaster Prevention and Management*, *15*(5), 810–820. doi:10.1108/09653560610712757

Manyena, S. B. (2006). The concept of resilience revisited. *Disasters*, *30*(4), 434–450. doi:10.1111/j.0361-3666.2006.00331.x PMID:17100752

Martins, S.F, Cunha, C.J., & Fernando, S. (2018). Secondary Data in Research – Uses and Opportunities. *Iberoamerican Journal of Strategic Management, 17*(4), 1-4.

Maru, Y. T., Smith, M. S., Sparrow, A., Pinho, P. F., & Dube, O. P. (2014). A linked vulnerability and resilience framework for adaptation pathways in remote, disadvantaged communities. *Global Environmental Change*, *28*, 337–350. doi:10.1016/j.gloenvcha.2013.12.007

Maslin, M. A., Lang, J., & Harvey, F. (2023). A short history of the successes and failures of the international climate change negotiations. *UCL Open Environment*, *5*, 5. doi:10.14324/111.444/ucloe.000059 PMID:37475957

Mathada, H., & Kori, E. (2012). An Evaluation of Sand and Gravel Extraction in Nzhelele River Valley, Limpopo Province of South Africa. *3rd International Conference on Biology, Environment, and Chemistry IPCBEE*, 46.

Mavhura, E. (2016). Disaster legislation: A critical review of the Civil Protection Act of Zimbabwe. *Natural Hazards*, *80*(1), 605–621. doi:10.1007/s11069-015-1986-1

Mavhura, E., Collins, A., & Bongo, P. P. (2017). Flood vulnerability and relocation readiness in Zimbabwe. *Disaster Prevention and Management*, *26*(1), 41–54. doi:10.1108/DPM-05-2016-0101

Mayer, K. (2020). *Financial health is workers' biggest wellness concern.* https://hrexecutive.com/financial-health-is-employees-biggest-covid-19-wellness-concern/

Maylie, D. (2020). *South Africa's tourism industry next stop recovery.* Available at https://www.ifc.org/wps/wcm/connect/region__ext_content/ifc_external_corporate_site/subsaharan+africa/news/south+africas+tourism+industry+next+stop+recovery

Meena, H. R., & Hridya, S. (2022). Preparedness of the Marginalized People in Disaster Management During Extreme Climate Events. *Analytical Approaches for Vulnerability and Adaptation to Climate Change in Agriculture,* 170.

Meerow, S. J. P., & Newell, J. P. (2016). Urban resilience for whom, what, when, where, and why. *Urban Geography*, *40*(3), 309–329. doi:10.1080/02723638.2016.1206395

Meerow, S., Newell, J. P., & Stults, M. (2016). Defining urban resilience: A review. *Landscape and Urban Planning*, *147*, 38–49. doi:10.1016/j.landurbplan.2015.11.011

Mehta, M. (2007). *Gender matters: Lessons for disaster risk reduction in South Asia*. The International Centre for Integrated Mountain Development (ICIMOD), Kathmandu, Nepal. Retrieved from https://www.preventionweb.net/files/2406_GenderandDisasters.pdf

Memon, R. A., Leung, D. Y. C., & Chunho, L. (2008). A review on the generation, determination and mitigation of urban heat island. *Journal of Environmental Sciences (China)*, *20*(1), 120–128. doi:10.1016/S1001-0742(08)60019-4 PMID:18572534

Méndez, M., Flores-Haro, G., & Zucker, L. (2020). The (in) visible victims of disaster: Understanding the vulnerability of undocumented Latino/a and indigenous immigrants. *Geoforum*, *116*, 50–62. doi:10.1016/j.geoforum.2020.07.007 PMID:32834081

Meneguette, R. I., De Grande, R., & Loureiro, A. A. (2018). *Intelligent transport system in smart cities*. Springer International Publishing. doi:10.1007/978-3-319-93332-0

Merilainen, E. (2019). The dual discourse of urban resilience: Robust city and self-organised neighbourhoods. *Disasters*, *44*(1), 125–151. doi:10.1111/disa.12367 PMID:31231816

Merkle, T., Tajeddini, K., Mensah Herv, D., Amoako-Atta, E., Hossain, M., Illés, C. B., & Dunay, A. (2022). Impact of COVID-19 Pandemic on Hotel Employees in the Greater Accra Region of Ghana. *Sustainability (Basel), 14*(5), 2509. doi:10.3390/su14052509

Merriam, S. B. (2010). Qualitative case studies. In P. Peterson, E. Baker, & B. McGaw (Eds.), *International encyclopedia of education* (3rd ed., pp. 456–462). Elsevier Ltd. doi:10.1016/B978-0-08-044894-7.01532-3

Meyer, I. H. (2015). Resilience in the study of minority stress and health of sexual and gender minorities. *Psychology of Sexual Orientation and Gender Diversity, 2*(3), 209–213. doi:10.1037/sgd0000132

Migration Policy Institute. (2021). *South Africa as an Immigration Destination: History, Recent Trends, and Current Challenges.* https://www.migrationpolicy.org/article/south-africa-immigration-destination-history

Miller, F., Osbahr, H., Boyd, E., Thomalla, F., Bharwani, S., Ziervogel, G., & Nelson, D. (2010). Resilience and vulnerability: Complementary or conflicting concepts? *Ecology and Society, 15*(3), 1–11. doi:10.5751/ES-03378-150311

Miller, H. E., Engemann, K. J., & Yage, R. R. (2006). Disaster planning and management. *Communications of the IIMA, 6*(2), 4.

Mishra. (n.d.). *What is Disaster? Alternative Approaches to Disaster Analysis.* Odisha State Open University E-pathshala. https://osou.ac.in/eresources/Disaster.pdf

Mishra, A., Bruno, E., & Zilberman, D. (2021). Compound natural and human disasters: Managing drought and COVID-19 to sustain global agriculture and food sectors. *The Science of the Total Environment, 754*, 142210. doi:10.1016/j.scitotenv.2020.142210 PMID:32920415

Mizutori, M. (2020). Reflections on the Sendai Framework for disaster risk reduction: Five years since its adoption. *International Journal of Disaster Risk Science, 11*(2), 147–151. doi:10.1007/s13753-020-00261-2

Mkhwanazi, N., & Mpanza, S. (2020). Examining the nature of disaster risk reduction strategies in local municipalities: a case study of Ndwedwe Local Municipality. *International Conference on Public Administration and Development Alternatives (IPADA).*

Mncwabe, M. (2021). The effectiveness of government institutions in responding to the epidemic of NCDs. *Economic and Social Rights in South Africa, 22*(4), 11-16. https://hdl.handle.net/10520/ejc-esrrev-v22-n4-a3

Mngeni, A., Betek, C., Musampa, C. M., & Nakin, M. D. V. (2017). The physical and environmental impacts of sand mining. *Transactions of the Royal Society of South Africa, 72*(1), 1–5. doi:10.1080/0035919X.2016.1209701

Mohanty, M. P., Mudgil, S., & Karmakar, S. (2020). Flood management in India: A focussed review on the current status and future challenges. *International Journal of Disaster Risk Reduction, 49*, 101660. doi:10.1016/j.ijdrr.2020.101660

Moosa, B. (2021). *A policy framework towards fostering urban resilience: the case of KwaZulu-Natal, South Africa* (Doctoral dissertation, University of Johannesburg).

Moraa, R. O., & Kipngetich, R. (2021). Impact of COVID-19 on employee retention in the hospitality industry. *The Strategic Journal of Business and Change Management*, *8*(1), 81–90. doi:10.61426/sjbcm.v8i1.1895

Moseley, L. (2004). Educational needs for disaster management. *Australian Journal of Emergency Management*, *19*(4), 28–31.

Moşteanu, N. R., Faccia, A., & Cavaliere, L. P. L. (2020). Disaster Management, Digitalization, and Financial Resources: Key factors to keep the organization ongoing. *Proceedings of the 2020 4th International Conference on Cloud and Big Data Computing*, 118-122. 10.1145/3416921.3416930

Mpanje, D., Gibbons, P., & McDermott, R. (2018). Social capital in vulnerable urban settings: An analytical framework. *Journal of International Humanitarian Action*, *3*(4), 4. Advance online publication. doi:10.1186/s41018-018-0032-9

Mudefi, E. (2023). Disaster management 'deeds' in the context of April 2022 KwaZulu-Natal floods: A scoping review. *International Journal of Disaster Risk Reduction*, *98*, 104122. doi:10.1016/j.ijdrr.2023.104122

Mugabe, V. A., Gudo, E. S., Inlamea, O. F., Kitron, U., & Ribeiro, G. S. (2021). Natural disasters, population displacement and health emergencies: Multiple public health threats in Mozambique. *BMJ Global Health*, *6*(9), e006778. doi:10.1136/bmjgh-2021-006778 PMID:34489329

Munawar, H. S., Khan, S. I., Anum, N., Qadir, Z., Kouzani, A. Z., & Parvez Mahmud, M. A. (2021). Post-flood risk management and resilience building practices: A case study. *Applied Sciences (Basel, Switzerland)*, *11*(11), 4823. doi:10.3390/app11114823

Munsaka, E., Mudavanhu, C., Sakala, L., Manjeru, P., & Matsvange, D. (2021). When Disaster Risk Management Systems Fail: The Case of Cyclone Idai in Chimanimani District, Zimbabwe. *International Journal of Disaster Risk Science*, *5*(7), 1–9. doi:10.1007/s13753-021-00370-6

Munsamy, L. (2018). *Development of an integrated disaster risk management model for the municipalities in the Free State Province*. Unpublished Doctoral Dissertation. Bloemfontein, South Africa: Central University of Technology, Free State.

Murwirapachena, G., & Kabange, M. M. (2023). Can commercialisation address consumer debt in local government? A case study of South African metropolitan municipalities. *Canadian Journal of Development Studies. Canadian Journal of Development Studies*, 1–25.

Mutsau, S., & Billiat, E. (2015). Leveraging Schools Systems As A Locus For Disaster Risk Reduction In Zimbabwe. *Journal of Education and Practice*, *6*(29), 163–169.

Mwale, F. D., Adeloye, A. J., & Beevers, L. (2015). Quantifying vulnerability of rural communities to flooding in SSA: A contemporary disaster management perspective applied to the Lower Shire Valley, Malawi. *International Journal of Disaster Risk Reduction*, *12*, 172–187. doi:10.1016/j.ijdrr.2015.01.003

Nahayo, L., Li, L., Habiyaremye, G., Richard, M., Mukanyandwi, V., Hakorimana, E., & Mupenzi, C. (2018). Extent of disaster courses delivery for the risk reduction in Rwanda. *International Journal of Disaster Risk Reduction*, *27*, 127–132. doi:10.1016/j.ijdrr.2017.09.046

Nakaya, T. (2022). *GWR 4.0, Geographically Weighted Modelling Software*. Retrieved from https://gwr.maynoothuniversity.ie/gwr4-software/

Napierała, T., Leśniewska-Napierała, K., & Burski, R. (2020). Impact of Geographic Distribution of COVID-19 Cases on Hotels' Performances: Case of Polish Cities. *Sustainability (Basel)*, *12*(11), 4697–4697. doi:10.3390/su12114697

Nath, S. K., Roy, D., & Thingbaijam, K. K. S. (2008). Disaster mitigation and management for West Bengal, India–An appraisal. *Current Science*, 858–864.

National Academies of Sciences, Engineering, and Medicine. (2022). *Resilience for compounding and cascading events.* Washington, DC: The National Academies Press. doi:10.17226/26659

National Department of Health. (2020). *Corona virus (COVID-19) outbreak.* https://www.health.gov.za/index.php/outbreaks/145-corona-virus-outbreak/465-corona-virus-outbreak

National Disaster Management Authority. (2022). *National Policy on Mitigation and Rehabilitation Measures for people displaced by coastal and river erosion.* https://ndma.gov.in/sites/default/files/PDF/National_Policy_on_MRM_CRE.pdf

NDMD (National Disaster Management Division). (2004). *'Disaster Management in India'- A Status Report*. Ministry of Home Affairs, Government of India.

Neal, J., & Neal, Z. (2019). Implementation capital: Merging frameworks of implementation outcomes and social capital to support the use of evidence-based practices. *Implementation Science : IS*, *14*(16), 16. Advance online publication. doi:10.1186/s13012-019-0860-z PMID:30764850

Nefeslioglu, H. A., & Gokceoglu, C. (2011). Probabilistic risk assessment in medium scale for rainfall-induced earthflows: Catakli catchment area (Cayeli, Rize, Turkey). *Mathematical Problems in Engineering*, *2011*, 1–21. Advance online publication. doi:10.1155/2011/280431

Nefeslioglu, H. A., Gokceoglu, C., & Sonmez, H. (2008). An assessment on the use of logistic regression and artificial neural networks with different sampling strategies for the preparation of landslide susceptibility maps. *Engineering Geology*, *97*(3–4), 171–191. doi:10.1016/j.enggeo.2008.01.004

News24. (2022). *Flood-ravaged KwaZulu-Natal still waiting on R1bn disaster funds*. News24. https://www.news24.com/news24/southafrica/news/flood-ravaged-kwazulu-natal-still-waiting-on-r1bn-disaster-funds-20220511

Ngcamu, B. S. (2011). *Disaster risk management in local government: a case study of Foreman and Kennedy Road informal settlements, eThekwini Municipality, KwaZulu-Natal*. Unpublished Doctoral dissertation. Durban University of Technology: KwaZulu-Natal.

Ngcamu, B. (2022). *Climate change and disaster preparedness issues in Eastern Cape and Kwazulu-Natal*. Town and Regional Planning. doi:10.18820/2415-0495/trp81i1.5

Nhamo, G., Dube, K., & Chikodzi, D. (2020). Impacts and implications of COVID-19 on the global hotel industry and Airbnb. In *Counting the Cost of COVID-19 on the global tourism industry* (pp. 183–204). Springer. doi:10.1007/978-3-030-56231-1_8

Nicola, M., Alsafi, Z., Sohrabi, C., Kerwan, A., Al-Jabir, A., Iosifidis, C., Agha, M., & Agha, R. (2020). The socio-economic implications of the coronavirus pandemic (COVID-19): A review. *International Journal of Surgery*, *78*, 185–193. doi:10.1016/j.ijsu.2020.04.018 PMID:32305533

Niino, M. (2012, June). Caught between Solidarity and Hesitation: The 3/11 Earthquake and Theatre in Japan. *Critical Stages/Scènes*, (6), https://www.critical-stages.org/6/caught-between-solidarity-and-hesitation-the-311-earthquake-and-theatre-in-japan/

Nilsson, P. (2014). Natural amenities in urban space-A geographically weighted regression approach. *Landscape and Urban Planning*, *121*, 45–54. doi:10.1016/j.landurbplan.2013.08.017

Nipa, T. J., Kermanshachi, S., Patel, R., & Tafazzoli, M. (2020, September). Disaster preparedness education: Construction curriculum requirements to increase students' preparedness in pre-and post-disaster activities. In *Associated Schools of Construction (ASC) International Conference* (pp. 142-131). Academic Press.

Nipa, T. J., & Kermanshachi, S. (2021). Dimensions of Resilience Measurement in Critical Transportation Infrastructure. In *International Conference on Transportation and Development 2021* (pp. 302-312). 10.1061/9780784483541.028

Nkombi, Z., & Wentink, G. J. (2022). The role of public participation in disaster risk reduction initiatives: The case of Katlehong township. *Jàmbá: Journal of Disaster Risk Studies*, *14*(1), a1203. https://doi.org/https://doi.org/10.4102/jamba.v14i1.1203

Noguchi, N., Inoue, S., Shimanoe, C., Shibayama, K., Matsunaga, H., Tanaka, S., Ishibashi, A., & Shinchi, K. (2016). What kinds of skills are necessary for physicians involved in International disaster response? *Prehospital and Disaster Medicine*, *31*(4), 397–406. doi:10.1017/S1049023X16000418 PMID:27221114

Nojavan, M., Salehi, E., & Omidvar, B. (2018). Conceptual change of disaster management models: A thematic analysis. *Jamba (Potchefstroom, South Africa)*, *10*(1), 1–11. doi:10.4102/jamba.v10i1.451 PMID:29955258

Norasakkunkit, V., Uchida, Y., & Toivonen, T. (2012). Caught between culture, society, and globalization: Youth marginalization in post-industrial Japan. *Social and Personality Psychology Compass*, *6*(5), 361–378. doi:10.1111/j.1751-9004.2012.00436.x

Norris, A. C., Martinez, S., Labaka, L., Madanian, S., Gonzalez, J. J., & Parry, D. (2015). Disaster E-Health: A new paradigm for collaborative healthcare in disasters. *Proceedings of the ISCRAM 2015 Conference.*

Norris, F. H., Stevens, S. P., Pfefferbaum, B., Wyche, K. F., & Pfefferbaum, R. L. (2008). Community resilience as a metaphor, theory, set of capacities, and strategy for disaster readiness. *American Journal of Community Psychology, 41*(1-2), 127–150. doi:10.1007/s10464-007-9156-6 PMID:18157631

Noy, I., & Yonson, R. (2018). Economic vulnerability and resilience to natural hazards: A survey of concepts and measurements. *Sustainability (Basel), 10*(8), 2850. doi:10.3390/su10082850

Nugmanova, A., Arndt, W. H., Hossain, M. A., & Kim, J. R. (2019). Effectiveness of ring roads in reducing traffic congestion in cities for long run: Big Almaty ring road case study. *Sustainability (Basel), 11*(18), 4973. doi:10.3390/su11184973

Nyahunda, L., Tirivangasi, H. M., & Mabila, T. E. (2022). Challenges faced by humanitarian organisations in rendering services in the aftermath of Cyclone Idai in Chimanimani, Zimbabwe. *Cogent Social Sciences, 8*(1), 1–11. doi:10.1080/23311886.2022.2030451

O' Keefe, P., Westgate, K., & Wisner, B. (1976). Taking the naturalness out of natural disasters. *Nature, 260*(5552), 566–567. doi:10.1038/260566a0

O'Brien, G., O'keefe, P., Gadema, Z., & Swords, J. (2010). Approaching disaster management through social learning. *Disaster Prevention and Management, 19*(4), 498–508. doi:10.1108/09653561011070402

Obergassel, W., Arens, C., Beuermann, C., Elsner, C., Hermwille, L., Kreibich, N., ... Spitzner, M. (2022). Last Call for 1.5 Degrees. *Carbon & Climate Law Review, 16*(4), 225–242. doi:10.21552/cclr/2022/4/4

OECD. (2015). *Disaster Risk Financing: A global survey of practices and challenges*. OECD Publishing.

Ohaeri, M., Ogbonna, D., Gobo, A., & Ngah, S. (2021). Environmental Impacts of Sand Mining in Some Coastal Communities in Port Harcourt Metropolis Nigeria. *Journal of Applied Life Sciences International*. . doi:10.9734/jalsi/2021/v24i1030265

Onigbinde, L. (2018). *The impacts of natural disasters on educational attainment: Cross-country evidence from macro data*. https://repository.usfca.edu/thes/1078/

OnlineK. Z. N. (2023). http://www.kznonline.gov.za/images

Oosthuizen, R. C. (2018). *A critical analysis of disaster risk management in local governance with reference to Sarah Baartman District Municipality*. Unpublished Doctoral Dissertation. Port Elizabeth, South Africa. Nelson Mandela University. Port Elizabeth.

Oshan, T., Li, Z., Kang, W., Wolf, L., & Fotheringham, A. (2019). MGWR: A Python Implementation of Multiscale Geographically Weighted Regression for Investigating Process Spatial Heterogeneity and Scale. *ISPRS International Journal of Geo-Information, 8*(6), 269. doi:10.3390/ijgi8060269

Ostrom, E. (2011). Background in institutional analysis and development framework. *Policy Studies Journal: the Journal of the Policy Studies Organization*, *39*(1), 7–27. doi:10.1111/j.1541-0072.2010.00394.x

Ovens, A., & Smith, W. (2006). Skill: Making sense of a complex concept. *Journal of Physical Education New Zealand*, *39*(1), 72.

Oxfam. (2019). *Cyclone Idai in Zimbabwe: An analysis of policy implications, for post-disaster institutional development to strengthen disaster risk management*. Available at: https://oxfamilibrary.openrepository.com/bitstream/handle/10546/620892/bp-impact-response-cyclone-idai-zimbabwe-071119-en.pdf;jsessionid=2110164995CF25C5A08EC8A5E792F09B?sequence=1

OziliP. K.ArunT. (2020). Spillover of COVID-19: impact on the global economy, available at SSRN: https://ssrn.com/abstract=3562570 or doi:10.2139/ssrn.3562570

Pachar, S., Dudeja, D., Batra, N., Tomar, V., Bhimavarapu, J. P., & Singh, A. K. (2023). Recent Developments in Machine Learning Predictive Analytics for Disaster Resource Allocation. *Engineering Proceedings*, *59*(1), 19. https://www.mdpi.com/2673-4591/59/1/19

Page, M. J., McKenzie, J. E., Bossuyt, P. M., Boutron, I., Hoffmann, T. C., Mulrow, C. D., Shamseer, L., Tetzlaff, J. M., Akl, E. A., Brennan, S. E., Chou, R., Glanville, J., Grimshaw, J. M., Hróbjartsson, A., Lalu, M. M., Li, T., Loder, E. W., Mayo-Wilson, E., McDonald, S., ... Moher, D. (2021). The PRISMA 2020 statement: An updated guideline for reporting systematic reviews. *British Medical Journal*, *372*(71), n71. Advance online publication. doi:10.1136/bmj.n71 PMID:33782057

Panday, S., Rushton, S., Karki, J., Balen, J., & Barnes, A. (2021). The role of social capital in disaster resilience in remote communities after the 2015 Nepal earthquake, International. *Journal of Disaster Reduction*, *55*, 102112.

Panneer, S., Kantamaneni, K., Pushparaj, R. R. B., Shekhar, S., Bhat, L., & Rice, L. (2021). Multistakeholder participation in disaster management-The case of the COVID-19 pandemic. *Health Care*, *9*(2), 203. doi:10.3390/healthcare9020203 PMID:33668669

Pan, Y., & Zhang, L. (2023). Integrating BIM and AI for smart construction management: Current status and future directions. *Archives of Computational Methods in Engineering*, *30*(2), 1081–1110. doi:10.1007/s11831-022-09830-8

Papke, L. E., & Wooldridge, J. M. (2023). A simple, robust test for choosing the level of fixed effects in linear panel data models. *Empirical Economics*, *64*(6), 2683–2701. doi:10.1007/s00181-022-02337-y

Parliamentary Monitoring Group. (2023, June 13). *AGSA report on flood relief funds*. Ad Hoc Joint Committee on Flood Disaster Relief and Recovery. https://pmg.org.za/committee-meeting/37213/

Partelow, S. (2018). A review of the social-ecological systems framework. *Ecology and Society*, *23*(4), art36. doi:10.5751/ES-10594-230436

Patel, R. K., Pamidimukkala, A., Kermanshachi, S., & Etminani-Ghasrodashti, R. (2023). Disaster preparedness and awareness among university students: A structural equation analysis. *International Journal of Environmental Research and Public Health*, *20*(5), 4447. doi:10.3390/ijerph20054447 PMID:36901454

Pathirage, C. (2014). *Knowledge factors and associated challenges for successful disaster knowledge sharing*. Prepared for the global assessment report on disaster risk reduction.

Patriksson, M. (2015). *The traffic assignment problem: models and methods*. Courier Dover Publications.

Payne, J. (2004). *The changing meaning of skill*. Available at: https://ora.ox.ac.uk/objects/uuid:32bdde50-3585-4af4-99c6-9ae28344e68d

Pearson, C. M., & Clair, J. A. (1998). Reframing crisis management. *Academy of Management Review*, *23*(1), 59–76. doi:10.2307/259099

Peduzzi, P. (2019). The disaster risk, global change, and sustainability nexus. *Sustainability (Basel)*, *11*(4), 957. doi:10.3390/su11040957

Pennar, K. (1997). *The tie that leads to prosperity: The economic value of social bonds is only beginning to be measured*. Business Weekly.

People Definition & Meaning. (n.d.). Available at: https://www.merriam-webster.com/dictionary/people

Perera, I., Meedeniya, D., Banerjee, I., & Choudhury, J. (2013, December). Educating users for disaster management: An exploratory study on using immersive training for disaster management. In *2013 IEEE International Conference in MOOC, Innovation and Technology in Education (MITE)* (pp. 245-250). IEEE. 10.1109/MITE.2013.6756343

Pérez, J. L., Gutierrez-Torre, A., Berral, J. L., & Carrera, D. (2018). A resilient and distributed near real-time traffic forecasting application for Fog computing environments. *Future Generation Computer Systems*, *87*, 198–212. doi:10.1016/j.future.2018.05.013

Perney, M. E. P., & D'angelo, G. (2023). Local Governance Support Tools for Disaster Risk Reduction and Climate Adaptation Strategies: The EU Contribution in the Case Study of the Municipality of Naples. *Sustainability (Basel)*, *15*(15), 11716. doi:10.3390/su151511716

Perry, R. W. (2018). Defining Disaster: An Evolving Concept. In Handbook of Disaster Research. Cham: Springer International Publishing. doi:10.1007/978-3-319-63254-4_1

Pitchaiah, P. (2017). Impacts of Sand Mining on Environment–A Review. *International Journal of Geoinformatics*, *4*(1), 1–6. doi:10.14445/23939206/IJGGS-V4I1P101

Pitchaiah, S. P., & Sultana, N. (2016). Impact of Sand Mining on Asian Countries - A Review. *International Journal of Geoinformatics*, *3*(5), 1–5. doi:10.14445/23939206/IJGGS-V3I5P101

Powell, T. M., & Bui, T. (2016). Supporting Social and Emotional Skills After a Disaster: Findings from a Mixed Methods Study. *School Mental Health, 8*(1), 106–119. doi:10.1007/s12310-016-9180-5

Preiser, R., Biggs, R., De Vos, A., & Folke, C. (2018). Social-ecological systems as complex adaptive systems. *Ecology and Society, 23*(4), art46. doi:10.5751/ES-10558-230446

Prentice, C., Altinay, L., & Woodside, A. G. (2021). Transformative service research and COVID-19. *Service Industries Journal, 41*(1-2), 1–8. doi:10.1080/02642069.2021.1883262

Price, R. K., & Vojinovic, Z. (2008). Urban flood disaster management. *Urban Water Journal, 5*(3), 259–276. doi:10.1080/15730620802099721

Priya, A. (2021). Case study methodology of qualitative research: Key attributes and navigating the conundrums in its application. *Sociological Bulletin, 70*(1), 94–110. doi:10.1177/0038022920970318

Pulvers, R. (2013). Miyazawa Kenji's Prophetic Green Vision: Japan's Great Writer/Poet on The 80th Anniversary of His Death. *Asia-Pacific Journal-Japan Focus, 11*(44).

Quintana, A., Angeletti-du Toit, L., Gilson, L., Smith, W., & Vallabhjee, K. (2022). Learning from COVID and climatic events to build a resilient health system: Western Cape Department of Health. *South African Health Review, 25*, 1–9. doi:10.10520/ejc-healthr-v2022-n1-a11

Raadsen, M. P., Bliemer, M. C., & Bell, M. G. (2020). Aggregation, disaggregation and decomposition methods in traffic assignment: Historical perspectives and new trends. *Transportation Research Part B: Methodological, 139*, 199–223. doi:10.1016/j.trb.2020.06.008

Radvanovsky, R. S., & McDougall, A. (2018). *Critical infrastructure: homeland security and emergency preparedness*. CRC Press.

Raharjo, E. P., Sarjana, S., & Safitri, M. (2022). Transportation infrastructure planning in supporting disaster mitigation: Case study in Mount Gamalama. *Jamba (Potchefstroom, South Africa), 14*(1), 1123. doi:10.4102/jamba.v14i1.1123 PMID:35401943

Rahmani, M., Muzwagi, A., & Pumariega, A. J. (2022). Cultural factors in disaster response among diverse children and youth around the world. *Current Psychiatry Reports, 24*(10), 481–491. doi:10.1007/s11920-022-01356-x PMID:35953637

Rane, N. (2023). Integrating Building Information Modelling (BIM) and Artificial Intelligence (AI) for Smart Construction Schedule. *Cost, Quality, and Safety Management: Challenges and Opportunities.*

Rangarirai, W., Marnani, C., & Rahmat, A. (2023). Analyzing the impact of community-based approaches on disaster preparedness to the risk of tropical cyclone induced flooding in Chimanimani District, Zimbabwe. *3rd International Conference on Disaster Management. Earth Environ. Sci., 1173.*

Rashid, Y., Rashid, A., Warraich, M., Sabir, S., & Waseem, A. (2019). Case Study Method: A Step-by-Step Guide for Business Researchers. *International Journal of Qualitative Methods, 18*. Advance online publication. doi:10.1177/1609406919862424

Rebeeh, Y. A., Pokharel, S., Abdella, G. M., & Hammuda, A. S. (2019). Disaster management in industrial areas: Perspectives, challenges and future research. *Journal of Industrial Engineering and Management, 12*(1), 133–153. doi:10.3926/jiem.2663

ReliefWeb. (2022). *South Africa: Floods and Landslides - Apr 2022*. ReliefWeb. https://reliefweb.int/disaster/fl-2022-000201-zaf

ReliefWeb. (2022, April 24). *South Africa: Floods and Landslides*. https://reliefweb.int/disaster/fl-2022-000201-zaf

ReliefWeb. (2023). *South Africa: Fire - DREF Operation Update (MDRZA014)*. Retrieved from https://reliefweb.int/report/south-africa/south-africa-fire-dref-operation-update-mdrza014

Renne, J., Wolshon, B., Pande, A., Murray-Tuite, P., & Kim, K. (2022). *Creating Resilient Transportation Systems: Policy, Planning, and Implementation*. Elsevier.

Republic of South Africa. (2000). *Municipal Systems Act, 2000 (Act 33 of 2000)*. Government.

Republic of South Africa. (2002). *Disaster Management Act No. 57 of 2002*. Government Printers.

Republic of South Africa. (2002). Disaster Management Act, 2002 (Act 57 of 2002). Pretoria: Government Printer.

Republic of South Africa. (2015). Disaster Management Act, 2015 (Act 16 of 2015). Pretoria, South Africa. Government Printers.

Reuters. (2022, April 14). *It is bad: survivors count costs of South Africa's devastating floods*. Reuters. https://www.reuters.com/business/environment/it-is-bad-survivors-count-costs-south-africas-devastating-floods-2022-04-14

Riet, G. (2009). Disaster risk assessment in South Africa: Some current challenges. *South African Review of Sociology, 40*(2), 194–208. doi:10.1080/21528586.2009.10425108

Ritchie, B. W., & Jiang, Y. (2019). A review of research on tourism risk, crisis and disaster management: Launching the annals of tourism research curated collection on tourism risk, crisis and disaster management. *Annals of Tourism Research, 79*, 102812. doi:10.1016/j.annals.2019.102812

Robinson, R. N., Martins, A., Solnet, D., & Baum, T. (2019). Sustaining precarity: Critically examining tourism and employment. *Journal of Sustainable Tourism, 27*(7), 1008–1025. doi:10.1080/09669582.2018.1538230

Rogerson, C. M., & Rogerson, J. M. (2020). COVID-19 tourism impacts in South Africa: Government and industry responses. *Geo Journal of Tourism and Geosites, 31*(3), 1083–1091. doi:10.30892/gtg.31321-544

Rogerson, J. M. (2021). Tourism business responses to South Africa's COVID-19 pandemic emergency. *Geo Journal of Tourism and Geosites*, *35*(2), 338–347. doi:10.30892/gtg.35211-657

Rolland, E., Patterson, R. A., Ward, K., & Dodin, B. (2010). Decision support for disaster management. *Operations Management Research, 3*, 68-79. https://link.springer.com/article/10.1007/s12063-010-0028-0

Rosenberg, J. P., & Yates, P. M. (2007). Schematic representation of case study research designs. *Journal of Advanced Nursing*, *60*(4), 447–452. doi:10.1111/j.1365-2648.2007.04385.x PMID:17822427

Rotarangi, S. J., & Stephenson, J. (2014). Resilience pivots: Stability and identity in a social-ecological-cultural system. *Ecology and Society*, *19*(1), 28. doi:10.5751/ES-06262-190128

Rouhanizadeh, B., Kermanshachi, S., & Dhamangaonkar, V. S. (2020, March). Reconstruction of critical and interdependent infrastructure due to catastrophic natural disasters: lessons learned. In *Construction Research Congress 2020* (pp. 895-904). Reston, VA: American Society of Civil Engineers. 10.1061/9780784482858.096

Runga, A. (2023). *An Investigation into the effectiveness of climate-related policies on disaster preparedness and response in Zimbabwe. The Case of Cyclone Idai in Chimanimani District* (Master's thesis, University of Agder).

Ryscavage, R. (2003). The Transition from Conflict to Peace. *Emergency Relief*, 284-294

Saha, A., & Chowdhury, S. (2020). Legal Framework of Disaster Management in India along with the Functioning of NDRF. *Int'l JL Mgmt. &. Human.*, *3*, 529.

Sahin, E. K., Colkesen, I., Acmali, S. S., Akgun, A., & Aydinoglu, A. C. (2020). Developing comprehensive geocomputation tools for landslide susceptibility mapping: LSM tool pack. *Computers & Geosciences*, *144*(July), 104592. doi:10.1016/j.cageo.2020.104592

Salmerón Gómez, R., García Pérez, J., López Martín, M. D. M., & García, C. G. (2016). Collinearity diagnostic applied in ridge estimation through the variance inflation factor. *Journal of Applied Statistics*, *43*(10), 1831–1849. doi:10.1080/02664763.2015.1120712

Salvucci, V., & Santos, R. (2020). Vulnerability to natural shocks: Assessing the short-term impact on consumption and poverty of the 2015 flood in Mozambique. *Ecological Economics*, *176*, 106713. doi:10.1016/j.ecolecon.2020.106713

Sandin, P. (2018). Conceptualizations of disasters in philosophy. Disasters: Core concepts and ethical theories, 13–26. doi:10.1007/978-3-319-92722-0_2

Santana, G. (2004). Crisis management and tourism. *Journal of Travel & Tourism Marketing*, *15*(4), 1–8. doi:10.1300/J073v15n04_05

Sarabia, M. M., Kägi, A., Davison, A. C., Banwell, N., Montes, C., Aebischer, C., & Hostettler, S. (2020). The challenges of impact evaluation: Attempting to measure the effectiveness of community-based disaster risk management. *International Journal of Disaster Risk Reduction*, *49*, 101732. doi:10.1016/j.ijdrr.2020.101732

Saravi, S., Kalawsky, R., Joannou, D., Rivas Casado, M., Fu, G., & Meng, F. (2019). Use of artificial intelligence to improve resilience and preparedness against adverse flood events. *Water (Basel)*, *11*(5), 973. doi:10.3390/w11050973

Şaroğlu, F., Emre, Ö., & Boray, A. (1987). Türkiye'nin Diri Fayları Ve Depremsellikleri (Active Faults and Their Seismicity of Turkey) General Directorate of Mineral Research and Exploration, Ankara-Turkey (1987). Report No: 8174.

Şaroğlu, F., & Yılmaz, Y. (1986). Geological Evolution and Basin Models During Neotectonic Episode in the Eastern Anatolia. *Bulletin of the Mineral Research and Exploration*, *107*(107), 70–93. https://dergipark.org.tr/en/pub/bulletinofmre/issue/3926/52248

Sawalha, I. H. (2020). A contemporary perspective on the disaster management cycle. *Foresight*, *22*(4), 469-482.

Sawalha, I. H. (2020). A contemporary perspective on the disaster management cycle. *Foresight*, *22*(4), 469–482. doi:10.1108/FS-11-2019-0097

Schwarz, J., Kascel, V., Azmat, M., & Kummer, S. (2023). Collation of best practices for preparedness: Lessons from disasters in Pakistan and Japan. *Journal of Humanitarian Logistics and Supply Chain Management*, *13*(3), 311–330. doi:10.1108/JHLSCM-07-2022-0086

Scott, Z., Wooster, K., Few, R., Thomson, A., & Tarazona, M. (2016). Monitoring and evaluating disaster risk management capacity. *Disaster Prevention and Management*, *25*(3), 412–422. doi:10.1108/DPM-01-2016-0002

Seddiky, M. A., Giggins, H., & Gajendran, T. (2020). International principles of disaster risk reduction informing NGOs strategies for community based DRR mainstreaming: The Bangladesh context. *International Journal of Disaster Risk Reduction*, *48*, 101580. doi:10.1016/j.ijdrr.2020.101580

Sen, A. (1993). Capability and well-being. *The Quality of Life, 30*, 270-293.

Şen, O. (2019). *İklim ve hidrolojik modeller ile Rize'de bulunan su havzalarının taşkın risk tayini*. Unpublished PhD Thesis, Istanbul: Istanbul Technical University, Institute of Science and Technology.

Sen, A. (1980). Equality of What? In S. McMurrin (Ed.), *Tanner Lectures on Human Values* (Vol. I). University of Utah Press.

Şengör, A. M. C., & Kidd, W. S. F. (1979). Post-collisional tectonics of the Turkish-Iranian plateau and a comparison with Tibet. *Tectonophysics*, *55*(3–4), 361–376. doi:10.1016/0040-1951(79)90184-7

Serdar, M. Z., Koç, M., & Al-Ghamdi, S. G. (2022). Urban transportation networks resilience: Indicators, disturbances, and assessment methods. *Sustainable Cities and Society*, *76*, 103452. doi:10.1016/j.scs.2021.103452

Seuwou, P., Banissi, E., &Ubakanma, G. (2020). The future of mobility with connected and autonomous vehicles in smart cities. *Digital twin technologies and smart cities*, 37-52.

Shah, A. A., Ullah, A., Khan, N. A., Shah, M. H., Ahmed, R., Hassan, S. T., Maur, T., & Xu, C. (2023). Identifying obstacles encountered at different stages of the disaster management cycle (DMC) and its implications for rural flooding in Pakistan. *Frontiers in Environmental Science*, *11*, 1088126. doi:10.3389/fenvs.2023.1088126

Shamsuddin, S. (2020) Resilience resistance: The challenges and implications of urban resilience implementation. *Journal of Cities*. . doi:10.1016/j.cities.2020.102763

Shareeda, A., Al-Hashimi, M., & Hamdan, A. (2021). Smart cities and electric vehicles adoption in Bahrain. *Journal of Decision Systems*, *30*(2-3), 321–343. doi:10.1080/12460125.2021.1911024

Sharifi, A., & Yamagata, Y. (2016). Urban Resilience Assessment: Multi Dimensions Criteria and Indicators. In Y. Yamagata & H. Maruyama (Eds.), *Urban Resilience* (pp. 259–276). Springer. doi:10.1007/978-3-319-39812-9_13

Shaw, K., & Maythorne, L. (2012). Managing for local resilience: Towards a strategic approach. *Public Policy and Administration*, *28*(1), 43–65. doi:10.1177/0952076711432578

Shimizu, M. (2012). Resilience in Disaster Management and Public Policy: A Case Study of the Tohoku Disaster. *Risk, Hazards & Crisis in Public Policy*, *3*(4), 40–59. doi:10.1002/rhc3.17

Shi, P., Xu, W., Ye, T., He, C., & Li, N. (2011). Developing Disaster Risk Science– Discussion on the Disaster Reduction Implementation Science. *Journal of Natural Disaster Science*, *32*(2), 79–88. doi:10.2328/jnds.32.79

Shrestha, N. (2020). Detecting multicollinearity in regression analysis. *American Journal of Applied Mathematics and Statistics*, *8*(2), 39–42. doi:10.12691/ajams-8-2-1

Siddiqui, S. T. (2017). Can employee training and development increase organizational resilience against economic crises? *International Journal of Managerial Studies and Research*, *5*(4), 78–82. doi:10.20431/2349-0349.0504011

Siercks, E. J. (2022). *Scaling the Nation: Local Literary Production and National Literature in Postwar Japan, 1946-1955*. University of California. Available at https://search.proquest.com/openview/5fe6ccd8e023e9c581ebdbb98631a7bd/1?pq-origsite=gscholar&cbl=18750&diss=y

Singh, S. R., Eghdami, M. R., & Singh, S. (2014). The concept of social vulnerability: A review from disasters perspectives. *International Journal of Interdisciplinary and Multidisciplinary Studies*, *1*(6), 71–82.

Sitorus, S. (2021). Qualitative Method (Case Study Research). *Journal of Communication Education*. . doi:10.58217/joce-ip.v15i1.224

Sivakumar, V., & Fazel-Rastgar, F. (2021, February). Heavy rainfall resulting from extreme weather disturbances in eastern coastal parts of South Africa: 11 April 2022. In *Proceedings of the Earth and Environmental Sciences International Webinar Conference* (pp. 161-186). Cham: Springer International Publishing.

Škof, N.M. (2013). Miyazawa Kenji: Interpretation of his Literature in Present Japan. *Asian Studies (Quezon City, Philippines)*, *1*(2), 91–106.

Smart, K., Ma, E., Qu, H., & Ding, L. (2021, April 1). COVID-19 impacts, coping strategies, and management reflection: A lodging industry case. *International Journal of Hospitality Management*, *94*, 102859–102859. doi:10.1016/j.ijhm.2021.102859 PMID:34785846

Snyder, H. (2019). Literature review as a research methodology: An overview and guidelines. *Journal of Business Research*, *104*, 333–339. doi:10.1016/j.jbusres.2019.07.039

Soehardi, S., Untari, D., & Raya, U. J. (2020). The effect of the COVID-19 pandemic on hotel employees, hotel occupancy rates and hotel tax income in Jakarta, Indonesia. *Systematic Reviews in Pharmacy*, *11*, 964–972.

Sofyana, H., Ibrahim, K., Afriandi, I., & Herawati, E. (2024). The implementation of disaster preparedness training integration model based on Public Health Nursing (ILATGANA-PHN) to increase community capacity in natural disaster-prone areas. *BMC Nursing*, *23*(105), 105. Advance online publication. doi:10.1186/s12912-024-01755-w PMID:38326888

Sofyana, H., Ibrahim, K., Afriandi, I., Herawati, E., & Wahito Nugroho, H. S. (2022). The need for a preparedness training model on disaster risk reduction based on culturally sensitive Public Health Nursing (PHN). *International Journal of Environmental Research and Public Health*, *19*(24), 16467. doi:10.3390/ijerph192416467 PMID:36554351

Solanki, A., Gupta, V., & Joshi, M. (2022). Application of machine learning algorithms in landslide susceptibility mapping, Kali Valley, Kumaun Himalaya, India. *Geocarto International*, *37*(27), 16846–16871. Advance online publication. doi:10.1080/10106049.2022.2120546

Solomon, F. J. (2011). *Examining the feasibility of informal settlement flood early warning systems: focus on the urban flood-risk experience of Kosovo and Masiphumelele residents, Cape Town, South Africa*. Unpublished master's thesis. Cape Town, South Africa: University of Cape Town.

South African Government. (2022, May 23). *KwaZulu-Natal provides an update on latest storms and floods*. Statement by the KwaZulu-Natal Provincial Government and eThekwini Municipality following the recent storms in KwaZulu-Natal held on 22 May 2022. https://www.gov.za/speeches/kwazulu-natal-provides-update-latest-storms-and-floods-23-may-2022-0000?gclid=Cj0KCQjwkOqZBhDNARIsAACsbfIwmoOAUZZrrKT_P-G74NL2FsUsH0qU2qyemn7rQgi-OY0UkrjGP2rUaAomwEALw_wcB

South African Tourism. (2021). *South African tourism annual report 2020/21*. https://live.southafrica.net/media/299000/south-african-tourism-annual-report-2020-21.pdf

Spiekermann, R., Kienberger, S., Norton, J., Briones, F., & Weichselgartner, J. (2015). The DisasterKnowledge Matrix – Reframing and evaluating the knowledge challenges in disaster risk reduction. *International Journal of Disaster Risk Reduction, 13*, 96–108. doi:10.1016/j.ijdrr.2015.05.002

Spielman, S. E., Tuccillo, J., Folch, D. C., Schweikert, A., Davies, R., Wood, N., & Tate, E. (2020). Evaluating social vulnerability indicators: Criteria and their application to the Social Vulnerability Index. *Natural Hazards, 100*(1), 417–436. doi:10.1007/s11069-019-03820-z

Srinivas, H. (2023, May 20). *Environmental Management and Community Participation: Enhancing Local Programmes*. GDRC Research Output: Concept Note Series E-169. Kobe, Japan: Global Development Research Center. https://www.gdrc.org/uem/iso14001/ems-comparti.html

Statistics South Africa. (2015). *Population Living in Poverty Statistics South Africa*. doi:10.15493/BEA.DATA.10000093

Statistics South Africa. (2019). *Five facts about Poverty in South Africa*. https://www.statssa.gov.za/?p=12075

Statistics South Africa. (2020). *Poverty in Perspective Statistics South Africa* https://www.parliament.gov.za

Statistics South Africa. (2021). *SA tourism industry struggles amidst COVID-19 pandemic*. https://www.statssa.gov.za/?p=14281

Statistics South Africa. (2022). *Mid-year population estimates 2022*. Statistical Release P0302. Pretoria. https://www.statssa.gov.za/publications/P0302/P03022022.pdf

Sucheran, R. (2021b). Preliminary Economic Impacts of the COVID-19 Pandemic on the Hotel Sector in South Africa. *African Journal of Hospitality, Tourism and Leisure, 10*(1), 115–130. doi:10.46222/ajhtl.19770720-90

Suindramedhi, S. (2015). Disaster Management: Lesson Learned from Japan. *GMSARN International Journal, 9*(3), 83–92.

Šuligoj, M. (2022). Work in hotels under conditions related to the COVID-19 disease: Views of Slovenian employees. *Tourism (Zagreb), 70*(2), 285–297. doi:10.37741/t.70.2.9

Süme, V., Yüksek, T., & Kaya, A. (2021). Rize Taşlıdere Havzasında heyelan kontrolü: Örnek olay incelemesi, Kireçhane ve Kırklartepe Yöreleri. *Journal of Anatolian Environmental and Animal Sciences, 6*(4), 526–531. doi:10.35229/jaes.994874

Sun, W., Bocchini, P., & Davison, B. D. (2020). Resilience metrics and measurement methods for transportation infrastructure: The state of the art. *Sustainable and Resilient Infrastructure, 5*(3), 168–199. doi:10.1080/23789689.2018.1448663

Surianto, S., Alim, S., Nindrea, R. D., & Trisnantoro, L. (2019). Regional policy for disaster risk management in developing countries within the sendai framework: A systematic review. *Open Access Macedonian Journal of Medical Sciences*, *7*(13), 2213–2219. doi:10.3889/oamjms.2019.614 PMID:31456854

Syahnur, S., & Diantimala, Y. (2021). Surviving and growing up with illegal status: The Analysis of Socio-Economic Household, Potential Conflict, the Environmental Damage, and Vulnerability of Local Community to Disaster. *Journal of Sustainable Mining*. . doi:10.46873/2300-3960.1061

Syed Arif, S. M., Pandey, V., Dhiman, S., & Sharma, P. (2021). Conceptualization of zero energy building coupled with seismically resistant structure. *International Journal of Mechanical Engineering*, *6*, 238.

Syed Zakaria, S. A., Majid, A. T., Lau, T. L., & Azimi, M. A. (2020). Exploration of Training and Education for Attaining Disaster Resilience from Sustainability Perspective Among Civil Engineering Students. In *Proceedings of AICCE'19: Transforming the Nation for a Sustainable Tomorrow 4* (pp. 361-376). Springer International Publishing.

Tabari, H. (2020). Climate change impact on flood and extreme precipitation increase with water availability. *Scientific Reports*, *10*(1), 13768. doi:10.1038/s41598-020-70816-2 PMID:32792563

Tachaudomdach, S., Upayokin, A., Kronprasert, N., & Arunotayanun, K. (2021). Quantifying road-network robustness toward flood-resilient transportation systems. *Sustainability (Basel)*, *13*(6), 3172. doi:10.3390/su13063172

Taherdoost, H. (2021). Data collection methods and tools for research; a step-by-step guide to choose data collection technique for academic and business research projects. *International Journal of Academic Research in Management*, *10*(1), 10–38.

Tajima, R. Y. O., Gore, T. O. M., & Fischer, T. B. (2014). Policy integration of environmental assessment and disaster management. *Journal of Environmental Assessment Policy and Management*, *16*(03), 1450028. doi:10.1142/S1464333214500288

Tammar, A., Abosuliman, S. S., & Rahaman, K. R. (2000). Social Capital and Disaster resilience nexus: A study of flash floods recovery in Jeddah City. *Sustainability (New Rochelle, N.Y.)*, *12*, 1–16.

Tanwar, N., & Akhai, S. (2017). Survey Analysis for Quality Control Comfort Management in Air Conditioned Classroom. *Journal of Advanced Research in Civil and Environmental Engineering*, *4*(1&2), 20–23. doi:10.24321/2393.8307.201702

Tan, Y., Liao, X., Su, H., Li, C., Xiang, J., & Dong, Z. (2017). Disaster Preparedness Among University Students in Guangzhou, China: Assessment of Status and Demand for Disaster Education. *Disaster Medicine and Public Health Preparedness*, *11*(3), 310–317. doi:10.1017/dmp.2016.124 PMID:27666525

Tao, X., Cheng, L., Zhang, R., Chan, W. K., Chao, H., & Qin, J. (2023). Towards Green Innovation in Smart Cities: Leveraging Traffic Flow Prediction with Machine Learning Algorithms for Sustainable Transportation Systems. *Sustainability (Basel), 16*(1), 251. doi:10.3390/su16010251

Tasri, E. S., Karimi, K., & Muslim, I. (2022). The effect of economic variables on natural disasters and the impact of disasters on economic variables. *Heliyon, 8*(1), e08678. doi:10.1016/j.heliyon.2021.e08678 PMID:35028461

Teegavarapu, S., & Summer, J. D. (2008). Case study method for design research. *Proceedings of IDETC/DTM ASME 2008 International Design Engineering Technical Conferences & Computers and Information in Engineering Conference.*

Teh, D., & Khan, T. (2021). Types, definition and classification of natural disasters and threat level. In *Handbook of Disaster Risk Reduction for Resilience: New Frameworks for Building Resilience to Disasters* (pp. 27–56). Springer International Publishing. doi:10.1007/978-3-030-61278-8_2

Tempelhoff, J., Van Niekerk, D., Van Eeden, E., Gouws, I., Botha, K., & Wurige, R. (2009). The December 2004-January 2005 floods in the Garden Route region of the Southern Cape, South Africa. Jàmbá. *Journal of Disaster Risk Studies, 2*(2), 93–112.

Teng, X., Teng, Y., Wu, K., & Chang, B. (2021). *Corporate Social Responsibility in Public Health During the COVID-19 Pandemic: Quarantine Hotel in China*. doi:10.3389/fpubh.2021.620930

Thompson, D. D. P. (2022). Compounding challenges for disaster resilience in small island developing states. *Disaster Prevention and Resilience,* (1), 4. doi:10.20517/dpr.2021.04

Thompson, D. (2010). *A Framework for Making Multistate Disaster Management Systems Perform: Insights from the Caribbean Emergency and Disaster Response Agency*. Now CDEMA.

Thulare, M. H., & Moyo, I. (2021). COVID-19 and street traders in the City of uMhlathuze, KwaZulu-Natal, South Africa: On responses and adaptation mechanisms. *Cogent Social Sciences, 7*(1), 2006392. doi:10.1080/23311886.2021.2006392

Tichon, J., Hall, R., Hilgers, M., Leu, M., & Agarwal, S. (2003). Education and training in virtual environments for disaster management. In *EdMedia+ Innovate Learning* (pp. 1191-1194). Association for the Advancement of Computing in Education (AACE). https://www.learntechlib.org/p/13974/

Tilbury, D., & Wortman, D. (2004). *Engaging people in sustainability*. IUCN. Available at: https://books.google.com/books?hl=en&lr=&id=FypMpjqPv0C&oi=fnd&pg=PR5&dq=people+and+environment+and++sustainability&ots=v3gPYSnqtu&sig=5DeX9edfvyAli63lCfhlUCQl-Es

Tong, J., Xiucang, L., Qingchen, C., Jiashuang, Y., & Erda, L. (2014). Highlights and understanding of climate change 2014: Impacts, adaptation, and vulnerability. *Advances in Climate Change Research, 10*(3), 157.

Ton, K. T., Gaillard, J. C., Adamson, C. E., Akgungor, C., & Ho, H. T. (2019). Expanding the capabilities of people with disabilities in disaster risk reduction. *International Journal of Disaster Risk Reduction, 34*, 11–17. doi:10.1016/j.ijdrr.2018.11.002

Torani, S., Majd, P. M., Maroufi, S. S., Dowlati, M., & Sheikhi, R. A. (2019). The importance of education on disasters and emergencies: A review article. *Journal of Education and Health Promotion*, *8*(1), 85. doi:10.4103/jehp.jehp_262_18 PMID:31143802

Trade and Investment KwaZulu-Natal. (2020). *Mining and beneficiation*. Retrieved from https://www.tikzn.co.za/open.php?cat=key_sectorsandpage=mining_and_beneficiation

Trosper, R. L. (2002). Northwest coast indigenous institutions that supported resilience and sustainability. *Ecological Economics*, *41*(2), 329–344. doi:10.1016/S0921-8009(02)00041-1

Tsangaratos, P., & Ilia, I. (2016). Comparison of a logistic regression and Naive Bayes classifier in landslide susceptibility assssments: The influence of models complexity and training dataset size. *Catena*, *145*, 164–179. doi:10.1016/j.catena.2016.06.004

Tsangaratos, P., Ilia, I., Hong, H., Chen, W., & Xu, C. (2017). Applying information theory and GIS-based quantitative methods to produce landslide susceptibility maps in Nancheng County, China. *Landslides*, *14*(3), 1091–1111. doi:10.1007/s10346-016-0769-4

Tselios, V., & Tompkins, E. L. (2019). What causes nations to recover from disasters? An inquiry into the role of wealth, income inequality, and social welfare provisioning. *International Journal of Disaster Risk Reduction*, *33*, 162–180. doi:10.1016/j.ijdrr.2018.10.003

Tsuji, Y., Satake, K., Ishibe, T., Harada, T., Nishiyama, A., & Kusumoto, S. (2014). Tsunami heights along the pacific coast of northern Honshu recorded from the 2011 Tohoku and previous great earthquakes. *Pure and Applied Geophysics*, *171*(12), 3183–3215. doi:10.1007/s00024-014-0779-x

Twigg, J. (2007). *Characteristics of a disaster resilient community*. http://www.proventionconsortium.org/themes/default/pdfs/characteristics/commu nity_characteristics_en_lowres.pdf

Twigg, J. (2015). *Disaster risk reduction*. Academic Press.

UN DESA. (2018). *Revision of World Urbanization Prospects*. United Nations.

UNDP. (2015). *Building Resilience in Zimbabwe: Towards a Resilience Strategic Framework. Zimbabwe Resilience Building Fund (ZRBF)*. UNDP Zimbabwe.

UNISDR (United Nations International Strategy for Disaster Reduction). (2015). *Sendai framework for disaster risk reduction 2015–2030*. UNISDR.

UNISDR. (2005). *Hyogo Framework for Action 2005–2015: building the resilience of nations and communities to disasters*. World Conference on Disaster Reduction (A/CONF. 206/6). Geneva: The United Nations International Strategy for Disaster Reduction.

UNISDR. (2011). *Hyogo Framework for Action 2005-2015, Building the Resilience of Nations and Communities to Disaster, Mid-term review. 2010-2011*. UNISDR.

UNISDR. (2015). *Sendai framework on disaster risk reduction 2015–2030*. UN Office for Disaster Risk Reduction.

United Nations and United Nations. (Ed.). (2008). *Achieving sustainable development and promoting development cooperation: dialogues at the Economic and Social Council*. United Nations.

United Nations Office for Disaster Risk Reduction. (2019). *Leave no one behind: A call to action for gender equality and women's empowerment in disaster risk reduction*. Retrieved from https://www.undrr.org/publication/leave-no-one-behind-call-action-gender-equality-and-womens-empowerment-disaster-risk

United Nations. (2019). *State of the world's indigenous peoples*. Retrieved from https://www.un.org/development/desa/indigenouspeoples/publications/state-of-the-worlds-indigenous-peoples.html

University of Central Florida. (n.d.). *The Disaster management cycle: 5 stages and how leadership can help prepare. Leadership and management*. https://www.ucf.edu/online/leadership-management/news/the-disaster-management-cycle/

UNWTO. (2020). *Worst Year in Tourism History with 1 Billion Fewer International Arrivals. 2021*. Available online: https://www. unwto.org/news/2020-worst-year-in-tourism-history-with-1-billion-fewer-international-arrivals

Van Assche, K., Verschraegen, G., Valentinov, V., & Gruezmacher, M. (2019). The social, the ecological, and the adaptive. Von Bertalanffy's general systems theory and the adaptive governance of social-ecological systems. *Systems Research and Behavioral Science*, *36*(3), 308–321. doi:10.1002/sres.2587

Van der Berg, A. (2015). Public-private partnerships in local disaster management: a panacea to all local disaster management ills? *Potchefstroom Electronic Law Journal / Potchefstroomse Elektroniese Regsblad*, *18*(4), 994-1033.

van der Merwe, P., Saayman, A., & Jacobs, C. (2021). Assessing the economic impact of COVID-19 on the private wildlife industry of South Africa. *Global Ecology and Conservation*, *28*, e01633. doi:10.1016/j.gecco.2021.e01633 PMID:34541259

Van Niekerk, D., Coetzee, C., & Nemakonde, L. (2020). Implementing the Sendai Framework in Africa : Progress Against the Targets (2015 – 2018). *International Journal of Disaster Risk Science*, *11*(October), 179–189. . doi:10.1007/s13753-020-00266-x

Van Niekerk, D., Nemakonde, L. D., Kruger, L., & Forbes-Genade, K. (2018). Community-based disaster risk management. Handbook of disaster research, 411-429.

Van Niekerk, D., Wentink, G. J., & Shoroma, L. B. (2018). Natural hazards governance in South Africa. Oxford Research Encyclopedia of Natural Hazard Science.

Van Niekerk, D. (2014). A critical analysis of the South African Disaster Management Act and policy framework. *Disasters*, *38*(4), 858–877. doi:10.1111/disa.12081 PMID:25196341

Van Niekerk, D. (2015). Disaster risk governance in Africa - A retrospective assessment of progress in against the Hyogo Framework for Action (2000-2012). *Disaster Prevention and Management*, *24*(3), 397–416. doi:10.1108/DPM-08-2014-0168

Van Niekerk, D., & Wisner, B. (2014). Integrating disaster risk management and development planning: Experiences from Africa. Routledge.

Varnes, D. J. (1984). Landslide hazard zonation, a review of principles and practice. IAEG Commission on Landslides. Paris: UNESCO.

Verbano, C., & Venturini, K. (2011). Development paths of risk management: Approaches, methods, and fields of application. *Journal of Risk Research*, *14*(5), 519–550. doi:10.1080/13669877.2010.541562

Vogel, L., O'Carroll, D., Edge, T., & Robinson, C. (2016). Release of Escherichia coli from Foreshore Sand and Pore Water during Intensified Wave Conditions at a Recreational Beach. *Environmental Science & Technology*, *50*(11), 5676–5684. doi:10.1021/acs.est.6b00707 PMID:27120087

Wafi, Z. K., Malek, M. F. A., Alnajjar, S. H., & Ahmad, R. B. (2015). Early warning system for Disaster management in rural areas. In *2015 International Symposium on Technology Management and Emerging Technologies (ISTMET)* (pp. 369-372). IEEE. 10.1109/ISTMET.2015.7359061

Wahyuli, S., Nurfarhaty, N., & Haeril, H. (2023). Natural Disaster Management in Local Government from the Perspective of Collaborative Governance. *Jurnal Studi Ilmu Pemerintahan*, *4*(1), 93–102.

Walsh, J. C., & High, S. (1999). Rethinking the concept of community. *Histoire sociale/Social History*. Available at: https://hssh3.journals.yorku.ca/index.php/hssh/article/view/4623

Wang, C., Yin, F., Zhao, Y., & Yin, L. (2023). Making Transportation Systems in US Cities Smarter and More Inclusive: A Synthesis of Challenges and Evaluation of Strategies. *ISPRS International Journal of Geo-Information*, *12*(2), 72. doi:10.3390/ijgi12020072

Ward, P. S., & Shively, G. E. (2017). Disaster risk, social vulnerability, and economic development. *Disasters*, *41*(2), 324–351. doi:10.1111/disa.12199 PMID:27174613

Warren, E., Sophie Hawkesworth, S., & Knai, C. (2015). Investigating the association between urban agriculture and food security, dietary diversity, and nutritional status: A systematic literature review. *Food Policy*, *53*, 54–66. doi:10.1016/j.foodpol.2015.03.004

Waugh & Tierney. (2007). Emergency Management: Principles and Practice for Local Government (2nd ed.). ICMA Press.

Wenger, C. (2017). The oak or the reed: How resilience theories are translated into disaster management policies. *Ecology and Society*, *22*(3), 18. doi:10.5751/ES-09491-220318

Wentink, G., & Niekerk, D. (2017). The capacity of personnel in disaster risk management in South African municipalities. *The Journal for Transdisciplinary Research in Southern Africa*, *13*(1), 10. doi:10.4102/td.v13i1.427

Whaley, L. (2018). The critical institutional analysis and development (CIAD) framework. *The International Journal of the Commons*, *12*(2). doi:10.18352/ijc.848

WHO. (2023). *Landslides*. Retrived from: https://www.who.int/health-topics/landslides#tab=tab_1

Williamson, D., Rasmussen, E., & Ravenswood, K. (2017). Power in the darkness: Taking a Historical and Critical Employment Relations Approach in Hospitality. *Journal of Hospitality and Tourism Management, 33*, 134–141. doi:10.1016/j.jhtm.2017.10.014

Wilson, R. S., Herziger, A., Hamilton, M., & Brooks, J. S. (2020). From incremental to transformative adaptation in individual responses to climate-exacerbated hazards. *Nature Climate Change, 10*(3), 200–208. doi:10.1038/s41558-020-0691-6

Wisner, B., Blaikie, P., & Cannon, T. (2004). *At risk: Natural Hazards, People's vulnerability and disasters*. Routledge, Elsevier.

Wisner, B., Gaillard, J. C., & Kelman, I. (2012). *Handbook of Hazards and Disaster Risk Reduction and Management*. Routledge. doi:10.4324/9780203844236

World Bank. (2019). *Project Information Document (PID)*. Report No. PIDA27189. Washington, DC: World Bank.

World Bank. (2022). *South Africa Overview*. https://www.worldbank.org/en/country/southafrica/overview

World Health Organization. (2017). *Promoting the health of refugees and migrants: Draft global action plan, 2019-2023*. Retrieved from https://www.who.int/migrants/publications/global-action-plan/en/

WTTC. (2020). *More than 197m travel and tourism jobs will be lost due to prolonged travel restrictions, according to new research from WTTC*. https://wttc.org/News-Article/More-than-197m-Travel-Tourism-jobs-will-be-lost-due-to-prolonged-travel-restrictions

Xie, Q., Wang, X., & Cong, X. (2020). How does foreign direct investment affect CO2 emissions in emerging countries? New findings from a nonlinear panel analysis. *Journal of Cleaner Production, 249*, 119422. doi:10.1016/j.jclepro.2019.119422

Xu, X., Chen, A., Xu, G., Yang, C., & Lam, W. H. (2021). Enhancing network resilience by adding redundancy to road networks. *Transportation Research Part E, Logistics and Transportation Review, 154*, 102448. doi:10.1016/j.tre.2021.102448

Yacoub, L., & El Hajjar, S. (2020). How do hotels in developing countries manage the impact of COVID-19? The case of Lebanese hotels. *International Journal of Contemporary Hospitality Management, 33*(3), 929–948. doi:10.1108/IJCHM-08-2020-0814

Yalcin, A., Reis, S., Aydinoglu, A. C., & Yomralioglu, T. (2011). A GIS-based comparative study of frequency ratio, analytical hierarchy process, bivariate statistics and logistics regression methods for landslide susceptibility mapping in Trabzon, NE Turkey. *Catena, 85*(3), 274–287. doi:10.1016/j.catena.2011.01.014

Yang, Y., & Mao, Z. (2020). Location advantages of lodging properties: A comparison between hotels and Airbnb units in an urban environment. *Annals of Tourism Research, 81*. doi:10.1016/j.annals.2020.102861

Yilmaz, A., Adamia, S., Chabukiani, A., Chkhotua, T., Erdoğan, K., Tuzcu, S., & Karabiyikoglu, M. (2000). Structural Correlation of the Southern Transcaucasus (Georgia)-Eastern Pontides (Turkey). Geological Society, London, Special Publications, 173(1), 171–182. doi:10.1144/GSL.SP.2000.173.01.08

Yılmaz, Y., Tüysüz, O., Yiğitbaş, E., Genç, Ş. C., & Şengör, A. M. C. (1997). Geology and tectonic evolution of the Pontides. *Regional and petroleum geology of the Black Sea and surrounding region, 68*, 183-226.

Yilmaz, I. (2009). Landslide susceptibility mapping using frequency ratio, logistic regression, artificial neural networks and their comparison: A case study from Kat landslides (Tokat-Turkey). *Computers & Geosciences, 35*(6), 1125–1138. doi:10.1016/j.cageo.2008.08.007

Yılmaz, Y., Güner, Y., & Şaroğlu, F. (1998). Geology of the quaternary volcanic centres of the east Anatolia. *Journal of Volcanology and Geothermal Research, 85*(1–4), 173–210. doi:10.1016/S0377-0273(98)00055-9

Yin, R. K. (2003). *Case study research: Design and methods* (3rd ed., Vol. 5). Sage.

Yordanov, V., Brovelli, M. A., Carrion, D., Barazzetti, L., Francisco, L. J. A., Comia, H. R., & Caravela, M. I. (2020). Capacity building for disaster management in Mozambique through teaching public participatory GIS and spatial data infrastructure. *The International Archives of the Photogrammetry, Remote Sensing and Spatial Information Sciences, 44*, 151–158. doi:10.5194/isprs-archives-XLIV-3-W1-2020-151-2020

Yüksek, Ö., Kankal, M., & Üçüncü, O. (2013). Assessment of big floods in the Eastern Black Sea Basin of Turkey. *Environmental Monitoring and Assessment, 185*(1), 797–814. doi:10.1007/s10661-012-2592-2 PMID:22411030

Zakour, M. J., Gillespie, D. F., Zakour, M. J. & Gillespie, D. F. (2013). Vulnerability theory. *Community Disaster Vulnerability: Theory, Research, and Practice*, 17-35.

Zaman, M. O., & Raihan, M. M. H. (2023). Community resilience to natural disasters: A systemic review of contemporary methods and theories. *Natural Hazards Review, 3*(3), 583–594. doi:10.1016/j.nhres.2023.05.003

Zhang, X., Mahadevan, S., & Goebel, K. (2019). Network reconfiguration for increasing transportation system resilience under extreme events. *Risk Analysis, 39*(9), 2054–2075. doi:10.1111/risa.13320 PMID:31039286

Zhang, Y., Wang, Y., Chen, Y., Xu, Y., Zhang, G., Lin, Q., & Luo, R. (2021). Projection of changes in flash flood occurance under climate change at tourist attractions. *Journal of Hydrology (Amsterdam), 595*, 126039. doi:10.1016/j.jhydrol.2021.126039

Zhao, Y., Wang, R., Jiang, Y., Liu, H., & Wei, Z. (2019). GIS-based logistic regression for rainfall-induced landslide susceptibility mapping under different grid sizes in Yueqing, Southeastern China. *Engineering Geology, 259*(December), 105147. doi:10.1016/j.enggeo.2019.105147

Zhou, Y., Wang, J., & Yang, H. (2019). Resilience of transportation systems: Concepts and comprehensive review. *IEEE Transactions on Intelligent Transportation Systems, 20*(12), 4262–4276. doi:10.1109/TITS.2018.2883766

Zimbabwe Vulnerability Assessment Committee. (2020). *Zimbabwe Urban Vulnerability*. ZIMVAC.

Zorn, M. (2018). Natural disasters and less developed countries. *Nature, tourism and ethnicity as drivers of (de) marginalization: Insights to marginality from perspective of sustainability and development*, 59-78.

Zou, H., He, L., Gao, F., Zhang, M., Chen, S., Wu, D., Liu, Y., He, L., Bai, H., & Ying, G. (2021). Antibiotic resistance genes in surface water and groundwater from mining-affected environments. *The Science of the Total Environment, 772*, 145516. doi:10.1016/j.scitotenv.2021.145516 PMID:33571766

Zou, Q., & Chen, S. (2019). Enhancing resilience of interdependent traffic-electric power system. *Reliability Engineering & System Safety, 191*, 106557. doi:10.1016/j.ress.2019.106557

Zscheischler, J., Martius, O., Westra, S., Bevacqua, E., Raymond, C., Horton, R. M., van den Hurk, B., AghaKouchak, A., Jézéquel, A., Mahecha, M. D., Maraun, D., Ramos, A. M., Ridder, N. N., Thiery, W., & Vignotto, E. (2020). A typology of compound weather and climate events. *Nature Reviews. Earth & Environment, 1*(7), 333–347. doi:10.1038/s43017-020-0060-z

Zschiegner, A. K. (2021). Tourism in protected and conserved areas amid the COVID-19 pandemic. *Parks*, (27), 103–118.

小池直子. (2012). *Beyond the Practicality of Time and Materialism of Life*. Available at: http://hokuga.hgu.jp/dspace/bitstream/123456789/1907/1/KOIKE.pdf

About the Contributors

Zamokuhle Mbandlwa is a Senior Lecturer at the Durban University of Technology in South Africa, specializing in Disaster and Risk Management. He has published 2 books and 50 journal articles. He has contributed significantly to the body of knowledge using different platforms such as journals, national TV channels, and Radio stations as a political commentator. He is an editorial team member of several journals. He is involved in various community projects and participates in advocacy groups.

* * *

Shalom Akhai has over 16 years of academic and research experience.

Gevher Alturk (M.Sc.), as a geomatics engineer, completed her M.Sc. study on mapping landslide susceptibility by using machine learning and statistical methods in geographic information systems environment. She currently works at the Republic of Turkey Ministry of Environment, Urbanisation and Climate Change, General Directorate of Combating Desertification and Erosion, Turkey. She continues her studies about landslide susceptibility modelling in peri-urban areas.

Syed Mohd. Arif is honorary Associate NCC Officer and faculty of Civil Engineering having around 10+ years of teaching, industry and research experience in the field of engineering and technology.

Arif Cagdas Aydinoglu (Ph.D.), as a geomatics engineer, worked about developing national geographic data models in UGent-Belgium, ITC-the Netherlands, and TUDelft-the Netherlands during his Ph.D. in Turkey. He managed projects about Turkey National GIS and worked as vice-director of GIS Research Center of Istanbul Technical University (ITU). He completed his post-doc research as a part of INSPIRE project in the European Commission Joint Research Center (JRC) in European Union IPA project and as academic coordinator of Turkey National GIS infrastructure projects. His research interests are GIS, Geographic data management, Geospatial analysis, Geospatial Data Infrastructure (SDI), and Urban GIS for planning. http://arifcagdas.com/en/.6bd0ddf6-f582-4793-86a8-892c93ef81cd

Mary Chawira is a University Lecturer in the Department of Geography and Environmental Management at the Zimbabwe Open University. She is currently undertaking PhD studies at the University of the Witwatersrand Johannesburg, South Africa. Her research interests include environmental hydrology modelling using remote sensing data, climate change and adaptation, Environmental modelling, Environmental and Social Impacts Assessment, Environmental Auditing, and GIS analytics.

Chika Eze is a senior lecturer at the Faculty of Education, Veritas University, Abuja (The Catholic University of Nigeria), Orcid ID: 0000-0002-1518-8558. She holds a PhD in Psychology from the University of KwaZulu-Natal, South Africa, M.Ed. Guidance and Counselling and a BA (Ed) Curriculum and Teaching both from University of Calabar, Nigeria. She is also a religious sister of the Society of the Holy Child Jesus, African Province. She has over 30 years' experience of teaching and learning across different educational levels including high school and tertiary institutions in different African countries such as Ghana, Kenya, Nigeria, and South Africa. At present, she lectures at Veritas University Abuja. She serves as the Deputy Director of Research in the University as well as the Coordinator of Postgraduate Studies for the Faculty of Education. Her research interest focuses on culture, gender, career development, interpersonal relationships cum personal/collective identity construction, education for transforming lives and development including post-Covid19 pandemic. She has published extensively in accredited journals.

Sunil J. Kulkarni has done his PhD in Chemical Engineering with focus on adsorption for wastewater treatment. He has done his Masters in chemical engineering from Tatyasaheb Kore Institute of Engineering and Technology, Warananagar. He has worked as an Assistant Professor in Chemical Engineering Department of Datta Meghe College of Engineering, Airoli, Navi Mumbai, India for 13 years. He has total 25 years of teaching and research experience. Dr. Sunil is involved with different academic activities across the universities in Maharashtra. His area of investigation in doctoral studies is phenol and heavy metal removal from waste water. He has also worked on kinetics and isotherms of various sorption operations. His area of research includes adsorption, clean technology and environmental biotechnology. He is on the reviewer board of many international journals and reviewed many international papers. He is also governing body member, promotional editor and executive member of many reputed international journals. Currently he is working as Associate Professor in chemical engineering and and Dean Research and Development at GIT, Lavel, India.

Francisca Kunedzimwe is a PhD holder in Environmental Conservation and Geoinformatics from Chinhoyi University of Technology, MSc in Geography and environmental Science, and BSc Geography and BA General degree. Currently working as, a Lecturer in the department of Geoinformatics at Chinhoyi University. She has skills and competencies in Environmental Conservation and GIS and Remote Sensing Projects; Education for Sustainable Development/Environmental Education & Citizen Science; Social and Environmental Standards Assessments; Biodiversity Assessments; Academic research, Teaching, Supervision and Mentorship; Safeguarding Intangible Cultural Heritage and Education 5.0 champion.

Olivia Kunguma holds great interest in science communication because she believes it is futile to develop the best scientific knowledge and not share it with the public or non-experts. She holds a PhD in Disaster Management, where she investigated the status quo and dynamics of establishing information management and communication systems in provincial disaster management centres and contributed to developing a framework for establishing and maintaining the systems. Her PhD also focused on reviewing the Disaster Management institutions at Provincial level to contribute to improving disaster governance. Dr Kunguma also holds a Master's Degree in Disaster Management, where her master's thesis was on exploring rural women's perceptions and experiences concerning water shortages in their communities and how they cope and adapt to the water shortage situation due to climate change. She also qualified for a BA and Honors degree in Integrated Marketing Communication. Dr Kunguma is a lecturer at the University of the Free State – Disaster Management Training and Education Centre for Africa (UFS-DiMTEC). She lectures in Management of Media Relations and Strategic Communication, Strategic Disaster Management, and Legal and Institutional Arrangement in Disaster Management. As a lecturer, she has graduated more than 400 postgraduate students in disaster management, supervised and graduated more than 20 Masters of Disaster Management research students, and one PhD candidate. Dr Kunguma is also a facilitator in some of the short learning programs offered by UFS-DiMTEC and an independent facilitator in Humanitarian Standards in Humanitarian Response (Sphere Project). She has published more than 20 peer-reviewed and popular articles in accredited journals. She has presented her research at local and international conferences like the World Conference on Disaster Management in Toronto, Canada. She is a registered professional member of the Disaster Management Institute of Southern Africa (PrDM08) and the South African Communications Association (SACOMM). Her research interests are analysing disaster management legislation and policies and their implementation, Strategic Disaster Management, Disaster/ Risk Communication, and institutional arrangements in disaster management. Dr Kunguma has worked on several research projects as part of my contribution to engaged scholarship, such as developing the Integrated National Drought Management Plan for South Africa, where she led the development of the Drought Information Management and Communication Plan. Dr Kunguma also worked on the review of the eThekwini Disaster Management Plan project. She was also the project leader in developing a crisis communication plan for Frances Baard District Municipality. During the Disaster Management Institute of Southern Africa conference proceedings, Olivia also implemented her 1st Art on Disasters exhibition in Durban. This project was about instilling a culture of risk avoidance through art, a resource that the communities can easily access.

Elias Mabaso is an Independent Researcher with a PhD in Disaster Risk Management with a focus on urban resilience and informality. Has published a few articles on Disaster Risk management in peer reviewed journals. More than 20 years of experience in Disaster Risk management, Response planning and resilience building. Worked for international NGOs involved in Disaster Risk management such as Save the Children UK and international organisations that include UNDP and UNOCHA. Also worked as a lecturer at the University of Zimbabwe and National University of Science and Technology teaching both under post graduate students courses that included Disaster and Development, Disaster preparedness planning and Project planning and development.

Brbra Masunga is currently working as a lecturer at Marondera University of Agricultural Sciences and Technology in faculty of Earth and Environmental Sciences within the department of Environmental Sciences and Technology. Hold a Masters in Environmental Sciences (Wageningen University) and a BSc Honors in Geography (University of Zimbabwe). Have research interests in climate change, Global environmental issues and biodiversity management.

Msekama Osia Mokhele is an emerging scholar, researcher, and lecturer in the Department of Public Management and Leadership at the Nelson Mandela University, who holds an MA in Public Management and Policy Analysis and a Magister in Governance and Political Transformation from the International University of Japan and the University of the Free State, respectively. The scholar is a joint doctoral candidate majoring in Disaster Management from the University of the Free State and Vrije University Amsterdam. His general teaching/research interests cover disaster management, public decision-making, procurement, and logistics management. His central research focus is the decentralisation of disaster management, disaster legislation and policy, disaster inclusivity and participation (which cuts across the community and marginalised groups' involvement, sectoral participation, and central-local participation). He is passionate about subjective disaster preparedness of the marginalised populations.

Genius Murwirapachena received a PhD in Economics from the Nelson Mandela University (South Africa). He is currently a Senior Lecturer in Economics at the Durban University of Technology. His research interests are in the areas of development economics, public service delivery and environmental economics. His recent publications include papers in the Canadian Journal of Development Studies, Development Southern Africa, the Environmental Economics and Policy Studies, the Water Policy, and Sustainable Water Resources Management.

Alice Ncube is the current program director, an associate professor, trainer, and researcher at the University of the Free State (UFS), Disaster Management Training and Education Center for Africa (DiMTEC, South Africa. With more ten years' experience in disaster (risk) management, her research interests are social vulnerability and climate change, international forced migration, gender issues, climate change and adaptation resilience and sustainable livelihoods of disadvantaged communities. She holds a PhD in disaster management, a masters in disaster management and postgraduate diploma in gender studies from the University of the Free State, South Africa among other qualifications form the University of Zimbabwe (B A General) , National University of Science and Technology (NUST) (Postgraduate diploma in Disaster and development Studies. She is an accredited Sphere trainer and an editorial team member of Jamba Journal of Disaster Risk Studies. Alice Ncube is the current secretary of the African Union Commission (AUC) African Science and Technology Group (AfSTAG) for Disaster Risk Reduction (DRR).

Oinuju (Uju) E. Okeke-Uzodike is a Senior Lecturer in the Department of Applied Management, Durban University of Technology, KZN South Africa. She obtained a Bachelor's degree in Chemical Engineering at Enugu State University of Science and Technology (ESUT). She later specialized in the field of Management and obtained her Masters degree from ESUT Business School-Lagos, Nigeria, and doctorate from the University of KwaZulu-Natal, South Africa. Her interest in teaching and research has grown over the years as demonstrated by her growing list of publications (including book chapters, journal articles, magazine pieces, conference proceedings). She is a member of the South African Education Research Association.

Lliosa Pahwaringira is a regional coordinator and lecturer at Zimbabwe Open University Harare regional campus. Stationed within department of Geography and Environmental Sciences. Has research interests in water, sanitation and hygiene, waste management and urban agriculture.

Mrry Ray is a third-year Ph.D. candidate at the Central University of Rajasthan Department of Education. She is studying the educational implications of Miyazawa Kenji, and her research critically examines the philosophy of Kenji's School through the interconnections of dimensions of values globally acknowledged for humanity and nature. She received an M.A. in Education from IGNOU in 2020 and a first-step Korean beginner certificate from Yonsei University in 2021. She passed the NAT Q5 level Japanese language test in 2022. She received a university grants commission-Junior Research fellowship for her Ph.D. Her research interests include Japanese educational philosophy, the Indian knowledge system, value modeling, and child education. Her publications include the book of selected Hindi poems in Notion publication. The title of the book is Mind (Mann-Parchai).

Mal Reddy is an Associate Director in the Department of Public Management & Economics at DUT. She serves on various National, Provincial and Local structures both in a professional & academic capacity. She was part of the National Disaster Management Project Team that was instrumental in the development of the National Disaster Management Education and Training Framework for South Africa. Mal served on the USAID/North-West University Steering Committee responsible for the project on the Disaster Risk Reduction Knowledge Shop, creating, sharing and exchange of information and practices in Disaster Risk Reduction. She serves on the Professional Board for Disaster Management: Disaster Management Institute of Southern Africa (DMISA). Her professional achievement was her election as the President of the Disaster Management Institute of Southern Africa (2012-2014). Mal serves on the Editorial Board of Journal of Disaster Risk Studies and Reviews for various International Journals.

Anjali Sharma is an Associate Professor at the School of Education, Central University of Rajasthan, India. She has twenty-four years of Teaching /research experience. She has published around fifty research papers in Web of Science and Scopus journals and fifteen chapters in edited National /International books. Seven Research Scholars have been awarded their degree under her supervision. She has presented around 50 papers at national and international conferences/seminars. She has completed three sponsored research projects in diverse areas. She designed an educational game and presented it at Toycathon 2021. She is a UGC nominated Master Trainer for the Community-based Participatory Research Program under

Unnat Bharat Abhiyan 2.0, Government of India. She is a UGC-nominated committee member constituted to establish guidelines for the Department of Education to implement the National Education Policy 2020. She has organized many national seminars/open-house discussions and conducted capacity-building and personality development programs, awareness programs, and workshops for homemakers in collaboration with various government autonomous agencies in India. (ORCID:).

Suleyman Sisman (M.Sc.) currently works at the Department of Geomatics Engineering, Gebze Technical University, Turkey. He, as a geomatics engineer, completed his M.Sc. study on modelling mass real estate valuation by GIS-based machine learning techniques. He continues his PhD study about machine learning techniques, spatial data analysis, sustainable land management, and Geographic Information Systems (GIS). https://www.gtu.edu.tr/tr/personel/352/48293857/display.aspx.2099dcba-644a-42b7-9ddd-a89c2c6cf09a

Rshma Sucheran is a Senior Lecturer and the Research Chair in the Department of Hospitality and Tourism, where she has been a staff member since 1997. She obtained her PhD in Geography and Environmental Science from the University of KwaZulu-Natal in 2012. She is also responsible for supervising postgraduate research in the department and has been extensively involved in designing and formulating various curriculums within the department. Professor Sucheran's primary teaching and research interests include environmental management, climate change, crises and disaster management and education and training, focusing on the global tourism and hospitality industry.

Matthys Swanepoel holds a PhD in Management Accounting. Having 15 years' solid finance and strategy experience of which 12 years at senior and executive management levels (7 years corporate and 5 entrepreneurial). Experience in management and supervision of full financial function; financial reporting and analysis; internal controls; coordination and management of budgeting and forecasting and foreign exchange as well as dealing with the South African Revenue Services (SARS). Experience includes strategic management, support services, procurement function; insurance evaluation; analyzing complex business contracts and determining risk. Additionally, performing due diligence on proposed projects; optimization of business processes; managing relations with financial institutions; as well as strategy and policy development are some of his abilities. Lecturing Management Accounting and Financial Management for 16 years at various Tertiary Institutions at different levels and is also involved in formal research. Acted as supervisor for many master and PhD students, while publishing various articles and delivered research papers at several national and international conferences.

Vivek Digambar Talnikar has completed his Ph.D. from Dr. Babasaheb Ambedkar Technological University, Lonere in 2016, Masters Degree from Harcourt Butler Technological Institute, Kanpur in 2001 and Bachelors degree in Chemical Engineering in 1996 from Pune University. He has 23 years of Teaching Experience and 2 years of Industrial experience. The research area include Reactive separations, Water Treatment and Separation Technology. He has presented several technical Papers in various National & Internal Conferences and received a Best Paper award for one of his Presentations. He has several publications for National and International Journals. He has been invited as a resource person for Guest Lectures at various Colleges and also worked as a Jury member for various technical events. He has been associated for Engineering Curriculum Desin for various Universities.

Eda Ustaoglu has a PhD in Planning and Public Policy from the School of Architecture, Planning and Environmental Policy, University College Dublin, and until 2015 she worked at the European Commission-Joint Research Centre-Ispra as a scientific officer. In the following periods, she worked as a post-doc researcher at the Department of Geomatics Engineering at Gebze Technical University and College of Social Sciences and Humanities at Koc University in Turkey. She is currently working as an Assistant Professor at the Department of Economics at Abdullah Gul University, Turkey. Over the past years, she has been involved in various projects, university studies and research activities in line with her graduate and post-graduate studies, which are directly related to the field of urban and environmental studies. Her research interests include urban economics, transport policy and planning, transport infrastructure appraisal, spatial economics and land use research.

Zelithini Zwane is currently a Lecturer at Durban University of Technology. Previously taught at Mangosuthu University of Technology and at the University of Zululand in the field of Publication Administration. My areas of expertise are in local government, human resource management and public policy.

Index

E

G

H

I

K

L

M

N

P

R

S

T

U

V

www.ingramcontent.com/pod-product-compliance
Lightning Source LLC
LaVergne TN
LVHW080601300926
849162LV00005B/1015
9798369327210